METHODS OF SURFACE ANALYSIS

METHODS AND PHENOMENA
THEIR APPLICATIONS IN SCIENCE AND TECHNOLOGY

Series Editors

S.P. Wolsky
Director of Research, P.R. Mallory and Co., Inc.,
Burlington, Mass., U.S.A.

and

A.W. Czanderna
Editor, Methods and Phenomena
P.O. Box 27209,
Denver, CO 80227, U.S.A.

METHODS OF SURFACE ANALYSIS

edited by

A.W. CZANDERNA

Editor, Methods and Phenomena
P.O. Box 27209,
Denver, CO 80227, U.S.A.

Volume 1 of
METHODS AND PHENOMENA
THEIR APPLICATIONS IN SCIENCE AND TECHNOLOGY

ELSEVIER SCIENTIFIC PUBLISHING COMPANY
Amsterdam — Oxford — New York   1975

ELSEVIER SCIENTIFIC PUBLISHING COMPANY
335 Jan van Galenstraat
P.O. Box 211, 1000 AE Amsterdam, The Netherlands

Distributors for the United States and Canada:

ELSEVIER/NORTH-HOLLAND INC.
52, Vanderbilt Avenue
New York, N.Y. 10017

First edition 1975
Second impression 1979

With 253 illustrations and 31 tables.

ISBN 0-444-41344-8 (Vol. 1)
ISBN 0-444-41640-4 (Series)

Printed in The Netherlands

CONTRIBUTORS TO VOLUME 1

J.H. Block	Fritz Haber Institut, Berlin, W. Germany
T.M. Buck	Bell Laboratories, Murray Hill, N.J., U.S.A.
A.W. Czanderna	Institute of Colloid and Surface Science, Clarkson College of Technology, Potsdam, N.Y., U.S.A.
L.E. Davis	Physical Electronics Industries, Inc., Eden Prairie, Minn., U.S.A.
A. Joshi	Physical Electronics Industries, Inc., Eden Prairie, Minn., U.S.A.
R.K. Lewis	Cameca Instruments, Elmsford, N.Y., U.S.A.
D. Lichtman	Physics Department and Laboratory for Surface Studies, University of Wisconsin-Milwaukee, Milwaukee, Wisc., U.S.A.
J.A. McHugh	General Electric Company, Knolls Atomic Power Laboratory, Schenectady, N.Y., U.S.A.
J.M. Morabito	Bell Laboratories, Allentown, Pa., U.S.A.
E.W. Müller	Department of Physics, The Pennsylvania State University, University Park, Pa., U.S.A.
P.W. Palmberg	Physical Electronics Industries, Inc., Eden Prairie, Minn., U.S.A.
M.J. Parker	E.I. du Pont de Nemours and Co. (Inc.), Monrovia, Calif., U.S.A.
W.M. Riggs	E.I. du Pont de Nemours and Co. (Inc.), Monrovia, Calif., U.S.A.
H.G. Tompkins	Bell Laboratories, Columbus, Ohio, U.S.A.
G.K. Wehner	Electrical Engineering Department, University of Minnesota, Minneapolis, Minn., U.S.A.

METHODS AND PHENOMENA

PREFACE

This volume initiates a new continuing book series *Methods and Phenomena: Their Applications in Science and Technology*. The series will be concerned specifically with detailed discussions of (1) the theory and application of new experimental methods, (2) new theoretical methods for the interpretation, evaluation or understanding of experimental problems, (3) applications broadening the utility of new or known experimental techniques or scientific phenomena, and (4) significant technical developments with unusual widespread interdisciplinary application. Each book will generally be devoted to a comprehensive discussion of a single subject of current interest. The emphasis will be on the new, the novel and the unique, with the intent of placing the reader at the frontier of activity in a given field. The subject matter of this series will encompass all scientific disciplines. The individual book editors and contributors will be leaders in their respective technical areas. Sufficient detail will be provided to give the outsider a deep insight into a new technical area. Since the discussions will also be current, they will be of value to the individual already working in that field.

Methods of Surface Analysis, Volume 1 of this series, is a first fulfillment of the intent of *Methods and Phenomena*. In this book, leaders in surface analysis discuss the latest and most important methods for attacking the age-old problem of determining the composition of surfaces. The net result is a volume of widespread interest and value.

The principal editors have established high standards for this series. Volume 1 is illustrative of the objective of this series. We firmly believe that the concept of emphasizing the method, a phenomenon, a technique, a procedure or a type of measurement, with a concurrent neglect of gadgetry, is still a refreshing idea in the world's literature in science and technology. Similar discussions on other important subjects, which are in preparation, are listed on the title page. Comments and suggestions on any aspect of this endeavor are welcome.

Finally, it is our pleasure to thank those who are helping us realize our goals. Editorial details and secretarial assistance in developing the series and volume have been handled skillfully by Mrs. Lucile Czanderna and Mrs. Margaret Dorandi. Last, and most important, our best thanks are extended to the authors whose contributions created this first volume.

Burlington, Massachusetts   S.P. Wolsky
Potsdam, New York   A.W. Czanderna
March, 1975

CONTENTS

INTRODUCTION

A.W. CZANDERNA

If the ionic species in plasmas are neglected, the universe of interest consists of three phases, solids, liquids, and gases, and none of these is infinite. It is becoming increasingly accepted that the boundary between these phases is fundamentally different and is called a *surface phase.* The possible boundaries are the solid—solid, solid—liquid, solid—gas, liquid—liquid, and liquid—gas interfaces. These interfaces are studied in surface science to develop an understanding of phenomena and to develop theories that will permit the prediction of future events. The topic of this book will deal with methods applicable to the solid—gas interface, in particular, and also to the solid—solid interface. An understanding of catalysis, corrosion, adhesion, surface diffusion, sintering, surface states, nucleation and condensation, conduction, mechanisms of failure and energy conversion devices will all be enhanced by the determination of the chemical composition of these two interfaces.

The experimental effort in studying the solid—gas interface is now very extensive; the theoretical treatment of the surface is difficult. In science, it is customary to adopt a model based on an ideal situation and compare the behavior of real systems with the ideal model. A realistic view of the boundary at a solid surface is not the obvious ideal infinite plane. On an atomic scale, the boundary includes crystal faces having composition, extent, and orientation that are fixed by the pretreatment of the solid. These pretreatments may include a variety of cleaning techniques such as outgassing, chemical reduction, flashing, ion bombardment, the deposition of thin films, cleavage, or field desorption. Residual species may accumulate at or in the boundary in trace and larger quantities, drastically altering the behavior of the boundary. Other imperfections such as an atom on the surface, a hole, an edge, a step, a crevice, a corner or a screw dislocation may also be on the surface. In an applied sense, almost everything that is interesting seems to happen on a surface or requires a surface reaction. Since this has been known for some time, why the current proliferation of interest?

For years, answers to a number of questions have been sought by surface scientists. What is the surface area? Where is the surface? How deep is the surface phase? Is the surface "real" or "clean"? What kind of a vacuum and/or thermal environment is necessary for studying various surfaces? What is the elemental composition of the surface? How much gas adsorbs onto a surface and how fast? What is the interaction between an adsorbed atom and a surface and how does this depend on the composition and structure? What are the forces that cause adsorption? What is the distribution of adsorption sites? What is the equilibrium topographical shape of a surface? Answers to the most important question, "what is the elemental composition of the surface?" could not be obtained until only several years ago!

If the discipline of surface science could have developed logically, the analysis of a surface for its chemical composition would surely have been one of the first areas of intensive research. Unfortunately, scientific progress cannot always proceed logically because the necessary tools, techniques, and methods have not been developed. For example, many reasoned guesses that a surface was clean, made in the 1960's on the basis of reproducibly obtained LEED patterns, are now known to be wrong on the basis of measurements using Auger electron spectroscopy, one of the modern methods of surface analysis described in this book. Furthermore, surfaces treated simply by outgassing and/or chemical reduction, known intuitively by international authorities to be "dirty," have been shown to be "clean"! Thus, the determination of the composition of an interface replaces opinion with facts and is a necessary part of any serious systematic experimental investigation of the solid surface.

Since 1968, the rate of growth of using methods for identifying the composition of surface has been absolutely breathtaking. Commercial instrumentation, adapted or developed for this single purpose, is now readily available. But the instruments and methods used are based on different physical principles and/or processes, which becomes a problem when the selection of a single method must be made for a specific problem. Many scientists are convinced that prerequisites to understanding experimental results include a detailed understanding of the method, the procedure, the instrument or apparatus and the nature of the sample under study. In the presentation of results in most journals, the experimental detail many readers would like is omitted. *Methods of Surface Analysis* provides a detailed description and typical application of the most widely used techniques for identifying the elemental composition of the surface.

The broad emphasis of this book is on the methods and procedures used to obtain the composition of surfaces and to identify species attached to the surface. All the methods require a vacuum environment for probing the surface and analyzing scattered or ejected particles. The application of the methods of surface analysis to obtain a composition depth profile after various stages of ion etching or sputtering is also considered. Thus, the composition at the solid—solid interface may be revealed by systematically removing atomic planes until the interface of interest is reached and then probed to determine the composition.

The first two chapters provide an overview of general phenomena. In the first chapter, the effect of ion etching on the results obtained by *any* method of surface analysis is considered. Questions such as the effect of the rate of etching, incident energy of the bombarding ion, and the properties of the solid need to be considered. The effect of the ion etch on generating an output signal of electrons, ions or neutrals and the effect of the residual gases in the vacuum environment on the results also are of interest.

The second chapter provides an overview of the important questions that should be considered when undertaking surface analysis. Of all the available methods, what are the relative strengths and weaknesses and why? How do the results from the various methods compare because of the excitation input and measured output? How does the physical principle of the different methods limit the possible results? In this chapter, as well as in the remaining chapters, other important questions about the influence of the method and the sample studied on the results are considered. For example, all samples are investigated in vacuum. But what influence does the sample geometry, type (metal, insulator, semiconductor, organic) and atomic number have on the results? Can information be gained from isotopes or about matrix effects and structure? How long does it take to collect data or to perform a depth profile? What is the signal-to-noise ratio that limits the ultimate sensitivity to identifying impurities? What is the sensitivity? What is the resolution or elemental discrimination of the method? Is the sampling depth from the surface monolayer, the top several layers or deeper? What is the depth resolution after ion etching? Is interpretation of the data simple or complicated? Can chemical and matrix effects be studied with the method? Is it likely the relevant physical processes will be understood? Is quantitative analysis a feasible hope? Answers to all these questions for every technique will not be found in Chapter 2 or in subsequent detailed chapters because they may not be known. But many of these questions are answered.

The last eight chapters provide in-depth descriptions of low energy ion scattering spectroscopy (LEIS or ISS), X-ray photoelectron spectroscopy (XPS or ESCA), Auger electron spectroscopy (AES), secondary ion mass spectroscopy (SIMS), a combined study of AES and SIMS, the atom probe field ion microscope (APFIM), field ionization mass spectrometry (FIMS) and infrared reflection-absorption spectroscopy. In this arrangement a transition is made from using input probes that do not require removal of the surface for analysis (AES, XPS, ISS) to those which require removal of the species for mass analysis (SIMS, APFIM, FIMS). The last two chapters include two significant developing methods for identifying species adsorbed onto the surface.

The editor is well aware that additional methods are emerging that may be as, or even become more, significant than those treated in this volume. The format for this book was chosen in early 1973, so it would not be unexpected that other important new methods would evolve by 1975. Professor

Lichtman's chapter presents an excellent overview about the potential of some of the currently emerging methods. Numerous investigations are under way utilizing more than one method of surface analysis on the same sample, such as ISS and SIMS, AES and SIMS, AES and XPS and even ISS, SIMS and AES. In addition, combinations of AES and LEED and of ISS and LEED, measured "simultaneously" with contact potential, thermal desorption and residual gas analysis, are providing deep insight into molecular processes at the solid surface. Systems combining quantitative information obtainable with ultramicrogravimetric methods and AES are being developed. The only limit to the application of the methods of surface analysis to science and technology seems to be qualified interested people and the financial support for their projects.

Chapter 1

THE ASPECTS OF SPUTTERING IN SURFACE ANALYSIS METHODS

G.K. WEHNER

I. Introduction

Low energy electron diffraction (LEED) requires single crystal targets which are atomically clean and well ordered up to the outermost atomic layer. By heating, such perfectly clean surfaces can be obtained with only a limited number of materials. From many other materials, it is not possible to remove oxide, carbide, nitride, or other tenacious layers by heating without melting or subliming the sample.

Farnsworth et al. [1] introduced in situ sputtering for cleaning solid surfaces in LEED studies using argon ions of less than 500 eV energy. It was found that it is necessary to remove the bombardment damage and drive out the embedded argon atoms by annealing at elevated temperature.

In composition analysis of a surface or of the "selvedge" ("near-surface region which differs from the bulk by virtue of its proximity to the surface") [2], it is usually necessary to remove remnants of cleaning acids, solvents, polishing compounds or other impurities which are acquired from the environment during handling. Sputtering has turned out to be an indispensable tool for this purpose. Beyond that, sputtering plays an increasingly important role as an in situ microsectioning technique for determining composition versus depth profiles.

Table 1 gives in matrix form a survey of compositional analysis methods based on physical phenomena. Methods for which commercial apparatus is available are underlined. *Surface* analysis methods with a depth resolution of less than ~ 100 Å are listed in the three heavily-lined quadrants $h\nu \rightarrow e$, $e \rightarrow e$, and ions $\rightarrow$ ions. Of those, only nuclear backscattering of MeV H^+- or He^+-ions is nondestructive and permits composition vs. depth analysis without substantial sputtering [3]. The ion microprobe or secondary ion mass analysis (SIMS) [4] and mass spectroscopy of sputtered neutrals [5,6] rely, of course, on sputtering. If a large analysis area is available (>0.1 cm^2) a complete mass spectrum with SIMS can be obtained with only 1/100 of a monolayer removed [7]. This "static" method then becomes nearly non-destructive. In low energy ion scattering spectroscopy (ISS) [8] sputtering of the surface is unavoidable. In fact, an initial sputtering dose is required for

TABLE 1

Survey of compositional analysis methods based on physical phenomena
(Methods for which commercial apparatus is on the market are underlined)

Excitation ⟶			
$h\nu$	e	Ions	E
Emission ↓ $h\nu$
X-ray fluorescence	Electron microprobe X-ray appearance or disappearance spectroscopy (APS, DAPS)	Surface composition analysis of neutral and ion impact Radiation (SCANIIR) Ion induced X-rays (IEX)	
e
Electron spectroscopy for chemical analysis (ESCA) or soft X-ray photoelectron spectroscopy (XPS) UV photoelectron spectroscopy (UPS) Auger electrons	Auger electron spectroscopy (AES) Ionization spectroscopy	Ion neutralization spectroscopy (INS)	
Ions
		Ion microprobe or secondary ion mass spectroscopy (SIMS) Mass spectroscopy of sputtered neutrals Ion scattering spectroscopy (ISS) Nuclear backscattering	Field ion mass spectrometry

removal of hydrogen and hydrogen-containing surface molecules which obscure the single collision events with target atoms heavier than the "probing" ion on which the interpretation of ISS spectra is based.

Whenever sputtering becomes involved in the analysis technique, a more intimate knowledge of the sputtering process and of the role which various parameters may play is often required or desirable. A wealth of sputtering data has been accumulated in the past 15 years for monoatomic metals. However, present knowledge is very meager for multi-component materials.

For instance, there has been considerable concern about how much the sputtering process may alter the composition at the sample surface. Erroneous composition analysis would occur in those methods in which the sample surface is analyzed (ESCA, ISS, AES) but not in those methods (provided equilibrium is established) which analyze the sputtered species (SIMS).

After summarizing what is presently known about the sputtering process, we will discuss the more specific role which sputtering plays in the various recently matured surface analysis methods. Wherever appropriate, only the most recent references are listed. The reader can from there, readily locate earlier, often voluminous literature on the particular subject matter.

II. The sputtering process

A. SURVEY

It has become well established that physical sputtering by noble gas ions is the result of independent binary collisions just as if the ion (or "neutralized ion", because close to a metal surface, the ion before impinging becomes neutralized by a field-emitted electron) had collided with the atoms of a gas. In this three-dimensional billiard game, the masses of the collision partners and the individual cross-sections (which depend upon the ion velocity and the electronic structure of the partners play a decisive role. Under oblique ion incidence the forward sputtered atoms can be ejected with high energy in a single collision event. Under normal ion incidence, the momentum vector for sputtering has to change in direction by more than $90°$ and this requires more than one successive collision. When the ion mass is lower than the target atom mass, the ion may be scattered backward in a single collision event. If its mass is higher, it can only be reflected backward as a result of multiple collisions. At kinetic energies in excess of ~ 100 eV, some ions begin to become embedded in the lattice. The penetration depth for a 1 keV Ar^+-ion is roughly 10 Å in Cu. The crystal structure and orientation enter the picture via the total cross-section area (shadowing effects). Whenever the projection of lattice points in a plane perpendicular to the ion beam is high, one observes a high sputtering yield but a low collection rate for ions and vice versa. A sputtered atom has to receive an energy which is sufficient to overcome its surface binding energy. Clusters have frequently been observed but the sputtered species are predominantly single atoms [9]. The heat of sublimation provides a measure for the *average* binding energy. Many atoms may be found at sites with lower numbers of nearest neighbors such as at grain boundaries or dislocations or an atom may be sitting on top of a filled plane. Such exposed atoms have a lower binding energy and are more readily sputtered than those from filled planes. A rather poorly defined sputtering energy threshold suggests that sputtering may often involve a two-step process where an atom is first lifted in one impact to a more exposed position and is subsequently sputtered in another collision event. In particular, ion bombardment-induced or -enhanced surface migration of atoms plays an important part in the sputtering of metals and alloys.

Measurements of the sputtering yield and the average velocity (which is much higher than those of evaporated atoms) of sputtered atoms, show that sputtering is a rather inefficient process. Usually not more than 5% of the incident ion energy goes into kinetic energy of sputtered atoms, and the remaining 95% appears partly as kinetic energy of backscattered ions but mostly as heat in the target. A most fortunate point for the use of sputtering as a microsectioning technique is the fact that the sputtering yields for different elements or their compounds rarely differ by more than a factor of 10. For instance, the sputtering rate for W (under Ar^+-ion bombardment) is only a factor of 2 or 3 lower than that of Al, whereas the evaporation rate for these two metals differs by more than 9 orders of magnitude (at $2000°C$).

Ion bombardment of a surface is always accompanied by emission of γ-electrons. One distinguishes between potential and kinetic γ-electron emission. The former is caused by an Auger process (radiationless transition) when the neutralization of the impinging ion provides the energy for the ejection of another electron from the solid. The γ-coefficient is of the order of 0.1 with Ar^+-ions (15.6 eV ionization energy) and is fairly independent of the kinetic energy of the ion up to 1 keV. Kinetic γ-emission becomes superimposed when the ions impinge with velocities of greater than $\sim 10^5$ m/sec and this can result in much larger γ values [10]. The γ-coefficients have to be taken into account in plasma sputtering if one wants to distinguish the bombarding ion current from the measured target current accurately. The γ-coefficient plays an important role in the charging of insulating surfaces.

Sputtering can be accomplished either with an ion beam, or with a target operated as a cathode in an abnormal d.c. glow discharge or immersed like a large negative Langmuir probe in an independently created low-gas-pressure plasma (triode sputtering). In plasma sputtering, the strong electric field in the vicinity of the target causes sputtered positive ions to be turned back towards the target and negative ions (as well as the γ-electrons) to be propelled away from the target. Sputtering of insulators is no problem with ion beams if one provides electrons (from a nearby thermionic filament) for neutralizing the accumulating positive surface charge. Bulk insulator sputtering in plasmas requires rf methods.

Many billiard game aspects of sputtering can be demonstrated when metal single crystals are sputtered because atoms are preferentially ejected in closely packed crystal directions. The resulting ejection- or deposit-patterns have been the subject of many experimental as well as theoretical studies and have contributed much to present understanding of the basic sputtering process.

For a more detailed discussion, the reader may find the survey in ref. 11

useful. A thorough discussion of theoretical aspects can be found in articles by Harrison et al. [12] and by Sigmund [13].

B. SPUTTERING YIELDS

The essential parameters which determine sputtering yields (atoms or molecules per ion) are: the kinetic energy, mass, electronic configuration, and angle of incidence of the bombarding ion and the atomic mass, electronic structure, crystal structure, orientation, binding energy of surface atoms, and the surface roughness of the target. The target temperature plays only a minor role.

Sputtering under normal incidence with Ne, Ar, Kr, or Xe ions requires a kinetic energy of at least about 4 times the heat of sublimation of the target material. For Ar, this threshold ranges from 13 eV for Al to 33 eV for W. Above the threshold energy, the yield rises first exponentially then linearly with ion energy, reaches a flat maximum, and finally decreases again with increasing ion energy. The maximum is located at lower energies for lighter ions (H^+ at ~2 keV) than for heavy ions (Ar^+—Cu at ~30 keV). This is the result of deeper penetration of the low mass ions.

Sputtering yields show characteristic periodicities with the positions of ions and target atoms in the periodic chart. Noble gas ions which behave most closely like hard spheres in the billiard game provide the highest sputtering yields. The noble metals Cu, Ag, Au for the same reason have the highest yields among vacuum-compatible metals (Zn and Cd next to Cu and Ag have still higher yields due to their lower heat of sublimation but these are undesirable elements to have in an ultra-high vacuum system). The energy transfer factor which is proportional to $mM/(m + M)^2$ plays an important role in sputtering yields. Therefore, yields for H^+- or He^+-ions impinging on heavier metal targets become very low. In Rutherford scattering with MeV H^+- or He^+-ions, the sputtering yield becomes negligibly small due to the deep penetration and low energy transfer coefficient. For the same reason, true physical sputtering with electrons is negligible unless the electrons have energies in the MeV range. A wide range of ion energies and ion species is involved in the various analysis methods. Tables 2—4 give some indication of sputtering yield values which may be encountered for polycrystalline metals. More detailed data with source references are found in various sputtering surveys [10,14,15]. Yield values often show considerable scatter which is most likely due to undefined surface roughness, preferred crystallite orientation in a nominally polycrystalline target, or the buildup of impurity layers on the target during sputtering.

At 0.13 Pa (10^{-3} torr) gas pressure, the mean free path of atoms between collisions is of the order of 4 cm. At pressures smaller than this, gas scattering of sputtered atoms back to the target becomes negligibly small. In plasma sputtering, the ion accelerating distance (sheath thickness) is usually very

TABLE 2

Sputtering yields in atoms/ion for 500 eV ion bombardment

	He	Ne	Ar	Kr	Xe
Be	0.2	0.4	0.5	0.5	0.4
C	0.07		0.12	0.13	0.17
Al	0.16	0.7	1	0.8	0.6
Si	0.13	0.5	0.5	0.5	0.4
Ti	0.07	0.4	0.5	0.5	0.4
V	0.06	0.5	0.65	0.6	0.6
Cr	0.17	1	1.2	1.4	1.5
Fe	0.15	0.9	1.1	1.1	1.0
Co	0.13	0.9	1.2	1.1	1.1
Ni	0.16	1.1	1.45	1.3	1.2
Cu	0.24	1.8	2.4	2.4	2.1
Ge	0.08	0.7	1.1	1.1	1.0
Y	0.05	0.5	0.7	0.7	0.5
Zr	0.02	0.4	0.7	0.6	0.6
Nb	0.03	0.3	0.6	0.6	0.5
Mo	0.03	0.5	0.8	0.9	0.9
Rb		0.6	1.2	1.3	1.2
Rh	0.06	0.7	1.3	1.4	1.4
Pd	0.13	1.2	2.1	2.2	2.2
Ag	0.2	1.8	3.1	3.3	3.3
Sm	0.05	0.7	0.8	1.1	1.3
Gd	0.03	0.5	0.8	1.1	1.2
Dy	0.03	0.6	0.9	1.2	1.3
Er	0.03	0.5	0.8		
Hf	0.01	0.3	0.7	0.8	
Ta	0.01	0.3	0.6	0.9	0.9
W	0.01	0.3	0.6	0.9	1.0
Re	0.01	0.4	0.9	1.3	
Os	0.01	0.4	0.9	1.3	1.3
Ir	0.01	0.4	1.0	1.4	1.6
Pt	0.03	0.6	1.4	1.8	1.9
Au	0.07	1.1	2.4	3.1	3.0
Th		0.3	0.6	1.0	1.1
U		0.5	0.9	1.3	

small and ion collisions within the sheath can be neglected. Under these conditions, the sputtering yield becomes independent of gas pressure. In ion beam sputtering where longer distances of travel are involved, the kinetic energy of the ions can be scattered over more atoms (including neutrals) and more directions. Accordingly, in ion beam sputtering, the gas pressure may already become important when it exceeds 10^{-2} Pa.

Much more stringent requirements exist with respect to the background pressure of reactive gases. With about one equivalent monolayer per second

TABLE 3

Sputtering yields in atoms/ion for 1 keV ion bombardment

	Ne	Ar	Kr	Xe
Fe	1.1	1.3	1.4	1.8
Ni	2	2.2	2.1	2.2
Cu	2.7	3.6	3.6	3.2
Mo	0.6	1.1	1.3	1.5
Ag	2.5	3.8	4.5	
Au		3.6		
Si		0.6		

TABLE 4

Sputtering yields in atoms/ion for 10 keV ion bombardment

	Ne	Ar	Kr	Xe
Cu	3.2	6.6	8	10
Ag		8.8	15	16
Au	3.7	8.4	15	20
Fe		1.0		
Mo		2.2		
Ti		2.1		

arriving at a surface at 10^{-4} Pa gas pressure, the competition between surface layer formation and sputtering becomes serious at sputtering rates of 100 Å/min which is often used in profiling. If the sample provides a high sticking coefficient to the reactive background species, the background pressure should be at least down to 10^{-6} Pa. The removal of only 1/100 of a monolayer in a static SIMS scan requires background pressures better than 10^{-10} Pa at which reactions with background gases should become negligible.

Commercially available sputter-ion guns for profiling are usually operated with Ar^+-ions at energies of 1—5 keV. Little reliable yield data are available in this energy range. If the thickness of a thin film is known, profiling provides an excellent means for calibrating sputtering rate vs. sputtering time under the actual operating conditions. As determined with AES profiling with an Ar^+-ion beam of 2 keV with 150 $\mu A/cm^2$ ion current density, the sputtering rates for stainless steels, Ta, SiO_2, Ta_2O_5, etc. lie in the range 70—130 Å/min, for Pt at 230 Å/min, and for Cu, Ag, and Au, at 300—400 Å/min. Such rates are adequate for most profiling tasks.

The question of whether it would be better to switch to a heavier noble

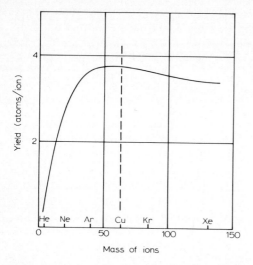

Fig. 1. Sputtering yield of Cu bombarded with 1 keV noble gas ions under normal incidence [16].

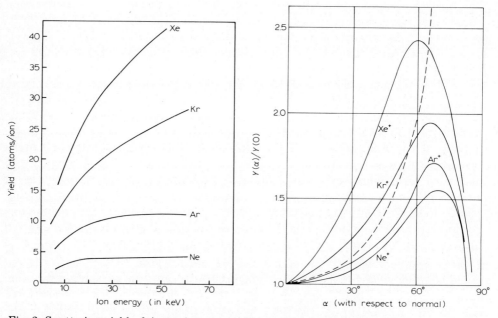

Fig. 2. Sputtering yield of Au vs. ion energy for various noble gas ions [17].

Fig. 3. Sputtering yield increase of Cu for bombardment by various noble gas ions under oblique incidence. The broken line is the 1/cosα curve which separates rate increase from rate decrease [16].

gas for achieving higher sputtering rates sometimes arises. Taken from recent measurements by Oechsner [16], performed with normal incident 1 keV He, Ne, Ar, Kr, Xe ions, Fig. 1 shows that Ar is the best choice for Cu and probably for most medium atomic weight target materials. At higher bombarding energy or higher mass materials, the sputtering yields for Xe^+ become considerably higher than for Ar^+ as shown in Fig. 2 [17].

Oechsner [16] recently reported a wealth of improved data on the influence of the angle of incidence of the ions on sputtering yields. At $60-70°$ from the normal, the yields in the energy range $0.5-2$ keV rise to maximum values which are $1.5-2.8$ times higher for high mass ions (Fig. 3) and higher for the low sputtering yield metals. With an ion beam, the ion current density decreases with the cosine of the angle α between beam and the normal to the surface, i.e. the sputtering rate-increase, if any, is much less than the yield increase. The broken curve in Fig. 3 shows the reciprocal of the $\cos \alpha$ dependence of $Y(\alpha)/Y(0)$. Only above this curve does one achieve in a beam a higher sputter removal rate. It would, of course, be erroneous to conclude that the sputtering at a target under normal ion incidence should increase with increased surface roughness. The reverse is true because sputtered atoms are preferentially ejected in a forward direction and atoms sputtered from a micro-slope are mostly ejected towards the target and not away from it. Under very oblique incidence, the sputtering yield can decrease drastically with increased sputter etch roughening of the surface due to shadowing effects. Unless the target is bombarded simultaneously with several ion guns from different directions or rotated during bombardment, it is advisable in profiling to ion-bombard the target as nearly as possible under normal incidence, even though this causes more pronounced "knock in" effects as described below.

Little yield data are available which show the influence of crystal structure and crystallite orientation. The latter subject, of course, is of much importance in etching and profiling. In the medium ion-energy range, from several 100 eV to 1 keV, the yield differences between different low index planes of a single crystal rarely exceed 50%. Even in such an extreme case as a metal in its polycrystalline or its liquid state (Sn, Al), the sputtering yield was found to differ not much more than 50%. At higher energy, channeling of ions in "open" crystal directions causes the yields to differ more substantially. In Cu at 5 keV Ar^+-ion bombardment, for instance, the sputtering yield of a (111) surface is nearly double that of a (110) surface.

As already mentioned, the target temperature seems to play no major role in the sputtering yields of metals. At low bombarding energy (<1 keV) the sputtering yields tend to decrease with increasing temperature due to the annealing of more loosely bound atoms on the surface (such as created in a previous ion impact) to positions of stronger binding. At high bombarding ion energies (>10 keV) when metals are heated to within $250°C$ of their melting point, "thermal sputtering" becomes superimposed on binary colli-

TABLE 5

Sputtering yields of oxides and their metals under 10 keV Kr^+-ion bombardment [19]

Oxide	Y (total atoms/ion) for oxide	Y (atoms/ion) for metal
Al_2O_3	1.5	3.2
MgO	1.8	8.1
MoO_3	9.6	2.8
Nb_2O_5	3.4	1.8
SiO_2	3.6	2.1
SnO_2	15.3	6.5
Ta_2O_5	2.5	1.6
TiO_2	1.6	2.1
UO_2	3.8	2.4
V_2O_5	12.7	2.3
WO_3	9.2	2.6
ZrO_2	2.8	2.3

sional sputtering so that the apparent sputtering yield increases with temperature [18].

It should be emphasized again that only few reliable sputtering yield data of alloys or compounds exist. Some recent data on oxide yields obtained by Kelly and Lam [19] for 10 keV Kr^+-ions are listed in Table 5. They found that, contrary to widespread belief, oxides have little effect as protective barriers to metals. A careful study of the sputtering yield of two phase Ag—Ni and Ag—Co alloys by Dahlgren and McClanahan [20] revealed that the low yield component, if present in an amount exceeding 10% (by volume), has the determining influence on overall yield.

With a primary ion current density J^+ [$\mu A/cm^2$], a sputtering yield Y (atoms/ion), an atomic or molecular weight A of target atoms or molecules (g/mole) and the specific weight δ [g/cm^3] of the target material, one calculates a removal rate $\dot{S}$ [Å/min] of

$$\dot{S} = 0.06 \ Y J^+ \ A/\delta$$

In profiling applications, a uniform, layer-by-layer microsectioning is usually much more desirable or important than an accurate knowledge of the sputtering rate. Uniformity is more readily obtained if the analyzed area is much smaller than the sputtered area. AES but not ESCA can meet this goal very simply by making the electron beam at least 2 orders of magnitude smaller than the sputter ion beam (which has usually mm size dimensions) and to aim with the analyzing beam towards the center of the sputtering crater. It is desirable to achieve a beam which causes a flat bottom sputtering crater. Wobbling or scanning of the ion beam can produce craters of any desired flatness and uniformity.

In AES, if the crater has a peaked shape, the alignment of the electron and the ion beam is important for achieving the maximal and reproducible sputtering rate. Deflection plates for the ion beam in combination with a Faraday collector are convenient for the alignment procedure. An occasional profile of a Ta sample which has been anodized to a known thickness is a simple way for checking that the sputtering rate has remained unchanged. A crystal oscillator film thickness sensor mounted close to the target for receiving part of the sputtered material is often a useful addition for monitoring relative sputtering rates. The sensor receives material from the whole sputtering crater but only in a certain direction of ejection. This scheme does not provide sufficiently accurate results for converting the sputtering time into a sputtered thickness scale because the crater walls are included and the angular distribution of sputtered atoms can be different for different materials.

C. SPUTTER ETCHING

Sputtering provides universal and well-reproducible etching because it is applicable to all materials. However, many details of etching are still unexplained or unpredictable even in such a supposedly simple case as that of a monoatomic metal.

The sample surface should be smooth otherwise, with a monodirectional ion beam, the etch rate will be different at differently inclined microslopes.

Sputter-etched polycrystalline metal targets show that grain boundaries or dislocations (often created by the bombardment) become delineated as furrows or pits. Differently oriented crystallites are attacked with different rates and each grain shows charateristic fine etch features (facets) which reveal the crystallite orientation.

The sputtering rate of a high-yield crystallite becomes lower as it recesses deeper below its surroundings because fewer atoms are able to clear the surface. Accordingly, the degree of surface roughness reaches a steady-state value. A single crystal or an amorphous target are etched more uniformly. The different sputtering rates of differently oriented crystallites or of grain boundaries are a disturbing fact in profiling because an interface deep below the surface appears to be broadened. The ratio of average crystallite size to film overlayer thickness plays an important role in the obtainable depth resolution at an interface.

Ion bombardment causes considerable bombardment damage and has a tendency to convert a material to the amorphous state. Well-defined ejection patterns are obtained in sputtering from metal single crystals even when they are kept at room temperature. This indicates that the surface retains, or at least resembles, its single-crystal structure under the bombardment. This fact is somewhat surprising in a material such as W which has a recrystalization temperature much above room temperature. The damage created by an ion impact must obviously anneal out so fast that the next ion which strikes there encounters a somewhat ordered surface again. This situation is different in semiconductors or insulators. When the bombardment is performed below a characteristic annealing temperature [Ge 550°K and Si 700°K for Ar$^+$-ion bombardment] [21], the surface layer becomes and stays amorphous and ejection patterns can no longer be observed. It is interesting to note that a 1 keV Ar$^+$-ion impact on Ge displaces about 50 surface atoms while only 1.7 atoms/ion are being sputtered [22]. Although this has not been demonstrated yet, one should expect the same amorphization phenomenon to occur in metals much below room temperature. Cooling of the target to liquid nitrogen temperature may provide a possible method for achieving more uniform, less orientation-dependent etch rates in polycrystalline materials.

Aside from singular features such as the often-observed cones, even an amorphous surface develops undulating features after long sputtering in the form of rilles or bumps [23] or shows the so-called "lemon-peel" pattern. The reasons for this are still poorly understood. Sputter etching of a single-crystal surface is governed by a complicated interplay between creation and movement of dislocations, surface migration of atoms, and the tendency of the surface to achieve its lowest free-energy configuration. Considerable effort has been spent in developing theories for predicting the evolution of surface topographies of impurity-free amorphous solids under ion bombardment [24—28]. The dependency of the sputtering rate on the angle of incidence of the ions plays an important role in these theories. If secondary effects such as redeposition of sputtered atoms, sputtering by sputtered atoms, bombardment-induced directional surface migration of atoms etc. are negligible, planes on a surface which are parallel to the ion beam or inclined at the angle of maximum sputtering attack would be stable and become the dominant features on an amorphous surface. Ion beam polishing has been attempted with varying degrees of success [29—32]. Oblique ion incidence, sample rotation and fairly low ion energy [33,34] seem to be beneficial for obtaining a good polishing action. Under favorable conditions, profiles with a remarkable depth resolution can be obtained. For instance, Fig. 4 shows an AES profile of a twenty-five layer (each layer 50 Å thick) single-crystal epitaxially grown periodic thin film structure (Esaki tunneling device) [35].

An extreme case of surface roughening by sputtering can arise when a minority metal species is difficult to sputter from its host metal. For instance, rather small amounts (<0.1%) of Mo, W, or Cr, etc. in Al, Au, Ag, Cu, etc. can give rise to the formation of cones which make sputtering a rather poor micro-sectioning technique. For example, Fig. 5 shows what happened to an Au surface which was seeded with a rather small flux of Cr atoms during sputtering. The Cr atoms can be furnished either as an internal impurity or deposited from another Cr source during the sputtering of a pure Au target [36]. It is obvious that such a surface exhibits a rather low sputtering yield and any interface must become washed out because atoms

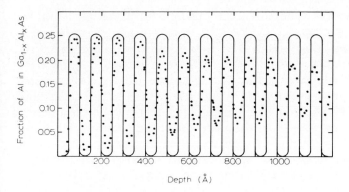

Fig. 4. Composition vs. depth profile of a superlattice structure GaAs—$Ga_{1-x}Al_xAs$ obtained with AES [35].

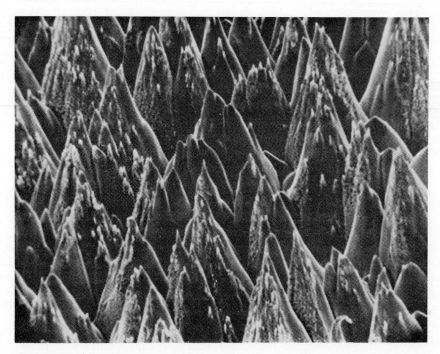

Fig. 5. SEM picture of sputtered Au target onto which a small amount of Cr was deposited during sputtering. The cone height is several microns.

References pp. 35—37

sputtered back and forth between the cones can linger on the surface for a long time. The angles of incidence of the ion beam and analyzing beam on such a surface have a large influence on the results. Surface migration of the low sputtering yield species seems to play an essential part in the formation of cones and this cone formation can be avoided if the sample is kept below a critical temperature during sputtering. It is unlikely that cone formation is a problem in profiling of thin, smooth films in which only hundreds of Å thickness are involved. But the fact that certain atomic species are more difficult to sputter from a substrate of other atoms than from their own may give erroneous composition results as discussed in the next section. Of interest is the fact that pillars with a very narrow apex angle are produced at the surface of organic materials like Teflon, blood cells, etc. as observed under ion or neutral atom bombardment in the SEM [37].

D. COMPOSITION CHANGES CAUSED BY ION BOMBARDMENT

Composition changes of a monoatomic metal surface bombarded by noble gas ions at a low impurity background pressure are confined to the implantation of the noble gas ions. Incorporation of noble gas is more pronounced at a rough surface because gas atoms may become buried underneath other sputtered and redeposited atoms. Since the sputtering yield is usually above unity and the implantation depth is so shallow, little noble gas can accumulate on the surface in profiling with ion beams in the keV range. Any embedded noble gas is readily detectable with AES or ESCA. With ISS (90° scattering) analysis, however, detection is possible only if a lighter noble gas than the embedded species is used. If noble gas peaks cause interference with AES analysis peaks, a rather infrequent occurence, it is always possible to switch to another noble gas for bombardment.

Any metal surface, except possibly Au, becomes very reactive and acquires a high sticking coefficient for impurity atoms after cleaning by ion bombardment. Therefore, the impurity background pressure requirements become very stringent for keeping the surface clean after bombardment.

A much more complicated situation arises in multi-component materials. For instance, if such a sample is cleaved in UHV in the analysis apparatus, the topmost surface composition will most likely be different from that of the bulk, even without any ion bombardment. If the samples cleave intergranularly, the new surface consists of the original grain boundaries which have a composition which is often different from the bulk. But even if the sample cleaves intragranularly, different thermodynamic driving forces arise for different constituents because the surface tries to achieve its lowest surface-free energy.

An answer to the question of what the ion bombardment does to the surface composition of a multi-component sample is of much wider interest. This is particularly true in those methods in which the sample surface is analyzed (as in AES, ISS, ESCA) instead of the sputtered material (SIMS). In the absence of significant bulk thermal diffusion, it is obvious that the material which is sputter-removed from a multi-component target must have the same composition as the bulk target, once equilibrium is established. However, the high-sputtering yield elements must become depleted at the target surface.

It should take very little sputtering (only some tens of monolayers) of a compound or a one-phase alloy to establish the new sputter-equilibrated surface layer which then recedes with an unaltered composition different from the bulk material. In multiphase materials in which the components exist as separate crystallites, the time for reaching equilibrium can be much longer because topographical changes at the surface become important. For example, high-yield crystallites become recessed until their sputtering rate is reduced to the point that the constituents leave in the same ratio as they exist in the bulk.

Any composition change caused by different sputtering rates of different species in a one-phase alloy or in a compound is most pronounced in the outermost layers from where the vast majority of sputtered atoms are ejected. Thus ISS, which analyzes only the topmost layer, should be most seriously concerned with this problem. In AES and ESCA, the information is gathered from a layer which is between 4 and 20 Å thick depending on the energy of the ejected electrons. For many elements, the latter two methods offer the possiblity of comparing analysis peaks which originate at different average depths. The composition changes at the sample surface are of no concern in bulk analysis with SIMS or in glow discharge mass spectrometry after equilibrium conditions are established.

Experimental data show that the bulk-sputtering yield values of the components are not useful for predicting the composition of the altered suface. Aside from the ion energy and the masses of the collision partners, the phase diagram of the material, the "surface phase diagram" [38], the directionality of bonds and the sample temperature (which determine surface migration) become additional important factors. At this stage, the situation becomes so complicated that one has to depend largely on experimental evidence.

An important difference exists between bombardment in a plasma or with an ion beam. A much lower electric field exists at the target surface in ion-beam sputtering than in plasma sputtering where the ion sheath adjacent to the sample surface creates a strong electric field. Atoms sputtered as positive ions are more likely to be pulled back to the negative target surface in a plasma than in ion-beam sputtering while negative ions are accelerated away from the sample. This should lead to enrichment of the sample surface with electropositive atoms.

Changes in composition can also arise at the surface of an insulator film from charging and subsequent induced motion of impurities. Some interesting studies on this were recently performed by McCaughan and Kushner

[39]. They bombarded a 5000 Å thick SiO_2 film on Si (which was covered with 10^{14} Na atoms/cm^2) with Ar^+-ions of 500 eV and found that the positive charge delivered to the SiO_2—Na surface had caused many of the Na atoms to move as Na^+ to the interface between SiO_2 and Si. Hughes et al. [40] studied SiO_2 on Si films (3000 Å) in which Na ions were implanted (10^{15}/cm^2). The profiles obtained with the ion microprobe were radically different for O^+ and O^- ion bombardment. The O^+ bombardment caused the major part of the Na to move to the SiO_2—Si interface. Such ion migrations in insulators can, of course, also be induced by the X-ray photons in ESCA or even more so by the bombarding electrons in AES [39].

Few experimental studies of differential sputtering and of the composition changes at an ion-bombarded target surface have been made so far. This subject deserves much more attention in view of the final goal: namely, to advance surface analysis from semiquantitative to truly quantitative techniques. Using an optical emission spectroscopy technique, Anderson [41] was the first to demonstrate the large amount of sputtering which may be required for reaching equilibrium. When he sputtered an Ag—Cu eutectic consisting of crystallites of Cu and of Ag, it took about one micron thickness of material removal before the high-yield Ag crystallites were sufficiently recessed for the Ag/Cu ejected atom ratio to become the same as the bulk composition ratio.

Henrich and Fan [42] studied a mixture of MgO and Au particles (<50 Å particle size) under 500 eV Ar^+-ion bombardment with AES. They found that the 69 eV Au Auger peak decreased by a factor of 20 before a constant surface composition was reached. The time for reaching this equilibrium was of the same order as required for sputtering roughly 50 Å thickness.

Palmberg [43] cleaved a MgO crystal in the AES apparatus under UHV conditions and determined with AES the Mg/O ratio before and after sputtering. He found hardly any composition change after the 1 keV Ar^+-ion bombardment.

Valuable information on differential sputtering can be obtained by analyzing not only the ion-bombarded sample but also the sputter-deposited coating, as it has been collected on a nearby substrate after equilibrium is established. This can be accomplished without opening the system with a V-shaped arrangement of target and substrate by slightly rotating the sample holder. The vacuum or background pressure requirements are, of course, more stringent when analyzing a deposit and the analysis has to be performed in situ immediately after sample analysis. If the ratio of Auger peak heights of different constituents is identical at sample and deposit, it can be safely concluded that preferential sputtering is no problem.

Using this method and AES, we observed negligible differential sputtering in the case of TiC samples with different Ti/C ratios. After equilibrium was reached, the ratio of the two Auger amplitudes (Ti at 422 eV, and C at 272 eV) of two typical samples, was 0.92 and 1.12 respectively. The correspond-

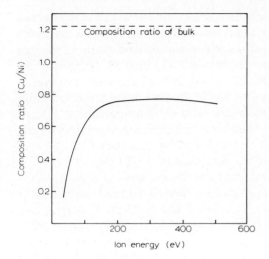

Fig. 6. Composition of 55% Cu—45% Ni target surface after Ar⁺-ion bombardment of various energies [44].

ing deposits gave ratios of 0.92 and 1.14, respectively.

Composition changes at the surface of a Constantan [55% Cu—45% Ni] sample have been studied as a function of the bombarding ion energy (plasma sputtering) by comparing the Cu 920 eV and Ni 716 eV Auger peaks [44]. It was found, as shown in Fig. 6, that the Cu/Ni ratio at the sample surface changed rapidly with >200 eV Ar⁺-ion bombardment from 1.2 to 0.75, indicating that the sputtering yield of Cu from the Constantan must be about 1.6 times higher than that of Ni. At the lower ion energies, the surface becomes much more depleted in Cu and at an ion energy of 35 eV, the altered surface layer contains only 14% Cu atoms instead of 55%, i.e. the sputtering yield ratio of Cu/Ni has risen to more than 7. The conclusion from this experiment is that low ion energies cause more problems with differential sputtering than higher energies.

Shimizu et al. [45] extended the AES studies to various ratios of Cu/Ni and found that their results (at 500 eV Ar⁺-ion bombardment) can best be fitted if one assumes that the Cu sputters with a yield about 1.9 times larger than that of Ni.

Cu and Ni are side by side in the periodic table and therefore provide a rather specialized case. One would normally not expect the sputtering ratio to remain constant for various mixture ratios because a quite different situation arises in sputtering when an atom is mostly surrounded by its own or by another species with quite different mass or different electronic shell structure.

Using AES, Färber and Braun [46] have recently studied the one-phase Ag—Au alloy using samples which covered the whole composition spectrum

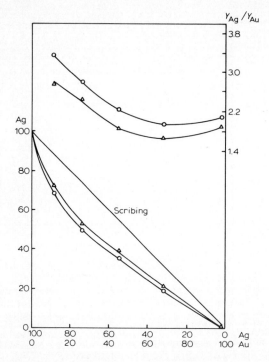

Fig. 7. Composition and sputtering yield ratio at the surface of Ag–Au alloys after 0.5 keV (○) and 2 keV (△) Ar⁺-ion bombardment [46].

from pure Ag to pure Au. They found that in situ cleaving and in situ scribing with a steel or diamond tool gave identical Ag/Au peak ratios over the whole composition range. One can safely assume that the "near surface" composition in this particular case is identical with the true bulk composition of the alloy. Figure 7 shows the Ag/Au ratio change after ion bombardment (0.5 and 2 keV) and the corresponding change in the sputtering yield ratio. This yield ratio was found to be about 3 at the Ag-rich side and ∼1.8 at the Au-rich alloy. Of interest is the minimum at about 60% Au which indicates that Ag is sputtered from the alloy at a higher rate than Au when the Ag atoms are either much in the minority (right side) or even more so much in the majority (left side). In the minority case, increased Ag sputtering probably arises from reflections of Ag atoms from the many surrounding Au atoms. In the majority case, the few Au atoms have little chance to become reflected from the many surrounding lighter Ag atoms and this increases the Ag/Au sputtering ratio.

Scribing is unfortunately not a valid method for obtaining a near surface composition which is the same as the bulk composition in multi-phase alloys because material from crystallites of one component may be more easily smeared over the other than vice versa.

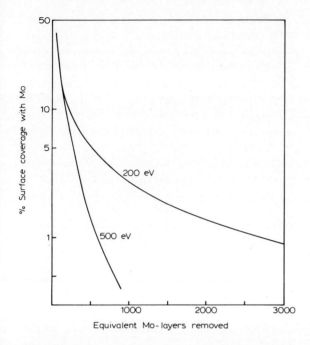

Fig. 8. Sputter removal of 12 monolayers of Mo from Al with 200 and 500 eV Ar$^+$-ions [47].

Some interesting aspects of differential sputtering were uncovered when we used AES to study the sputter removal of thin Mo films from various substrates [47]. It was found that the sputter removal of a Mo layer from a W substrate proceeds as if the Mo were sputtered from a Mo substrate. However, this is not the case for Mo removal from Au or Al, especially at low-bombarding ion energy. For instance, the Auger spectrum from a 12-atom layer-thick Mo film on W showed only Mo and the W was obviously completely covered with a Mo film thicker than the escape depth of W Auger electrons. Deposition of 12 Mo layers on Al under the same conditions gave a Mo Auger signal of only 10% of its full value, indicating Mo agglomeration into islands. Under Ar$^+$-ion bombardment, the Mo coverage first increases due to spreading of the multilayered Mo islands. As shown in Fig. 8, the remarkable fact is that it becomes very difficult to remove the last traces of Mo from the Al and after removal of the equivalent of 3000 layers of Mo, the Mo Auger peak is still 1% of its bulk value. Accordingly, the yield for sputtering Mo atoms from Al neighbors must be three orders of magnitude lower than from Mo neighbors. The 1% value is already reached after removal of 600 equivalent monolayers at 500 eV bombarding ion energy. Again, a higher bombarding ion energy apparently causes less difficulty with differential sputtering.

References pp. 35—37

Surfaces which contain such "difficult to sputter" species, as mentioned above, tend to become covered with cones. The most likely explanation for this phenomenon is that such species undergo surface migration until they find quiet places (like leaves in a courtyard). Slopes on steps or protrusions receive a lower bombarding flux density and are less agitated. Apparently, these "difficult to sputter" species accumulate there, are mixed together with sputter-deposited substrate atoms and this mixture has a low sputtering yield. Support for this picture is obtained from the fact that cooling of the sample to impede or stop surface migration of atoms suppresses the cone formation. Cones develop abundantly at scratches (slopes), and the scanning Auger microprobe shows that the cone-causing species are concentrated at the cones.

Such extreme cases as that of Mo on Al are fortunately rare and in analysis they become noticeable only at very low coverage. Low target temperature and high bombarding ion energy reduce this differential sputtering phenomenon.

In many practical cases of depth profiling, it is often less important to know absolute quantitative amounts of constituents but rather to see how the constituent distribution has changed after heat treatment, or corrosion, or what elements have accumulated at interfaces between different layers, etc. In such tasks, some differential sputtering is rarely of much concern.

Experimental evidence obtained with AES seems to indicate that some compounds like TiC and MgO show very little differential sputtering effects. Other compounds like TiO_2, ZrO_2, Fe_2O_3, CuO, etc. become enriched with the metal component under ion bombardment at their surface [48]. Composition changes under ion bombardment are, in general, more pronounced on rough (such as pressed powder) surfaces. The reason for this is that many atoms in depressions will be sputtered back and forth many times before being ejected and very small differences in sticking coefficients of different constituents can cause large composition changes at the surface. Turos et al. [49] using MeV He^+ backscattering spectrometry reported recently that 2 keV Ar^+-ion bombardment causes preferential Si sputtering from a SiO_2 film.

Aside from differential sputtering, surface composition changes at a sample surface can be caused by differential knock-in effects. We observed this effect recently using AES with a sample of Al containing a small amount of Ag which had diffused in from the surface. As shown in region I in Fig. 9, the Ag Auger peak height vs. depth was recorded while sputtering with a 500 eV Ar^+-ion beam. Switching (region II) to 2 keV Ar^+-ion bombardment and continuing to record the Ag profile shows a rapid drop of the Ag signal which then recovers partly after some sputtering. An obvious explanation is that the 2 keV bombardment preferentially knocks the Ag atoms deeper inside, i.e. deeper than the analysis depth. Eventually, the sputter etching catches up with the Ag-enriched region but the Ag signal stays below its 500

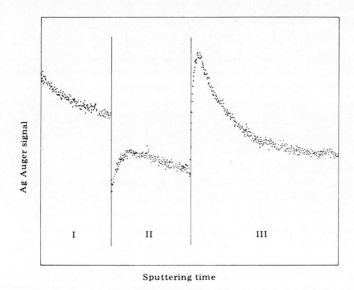

Fig. 9. Knock-in of Ag in Al matrix.

eV value due to the ongoing knocking-in of the Ag. Switching back to 500 eV (region III) causes the Ag signal to rise above its previous 500 eV value because the surface advances during sputtering into the region which was previously enriched with Ag by the 2 keV bombardment. After sputtering through this region, the Ag signal goes back to its level corresponding to 500 eV bombardment. This knock-in effect substantiates the billiard game concepts in sputtering. Without doubt, such differential knock-in of atoms can play an important role at interfaces in profiling. The use of sputtering in combination with a surface analysis technique for determining ion implantation profiles such as in Si, for example, has seldom been very successful because the knocked-in implanted atoms cause longer tails than present in the original profile. A profile of a K film on a Si wafer, for instance, shows a very long tail of K which seems to get pushed deeper and deeper into the Si substrate. Such spurious tails show up most readily in methods with very high sensitivity such as that obtainable in SIMS. Conditions for this knock-in process are more favorable if the matrix atoms have a lower mass than the minority atom but this is, of course, not a necessary condition. Schulz et al. [50], for instance, studied B implanted at 20 keV in Si and found that the depth at which the B signal decreased to 10^{-3} of its maximum value shifted from 1600 Å, to 1800 Å, to 2200 Å, and finally to 2700 Å when in the SIMS analysis the Ar^+-ion beam energy was increased from 5 keV to 10 keV, to 30 keV, and 50 keV, respectively.

It is interesting to note that differential sputtering effects decrease with higher ion energy while the opposite is the case in differential knock-in.

References pp. 35—37

Staib's [51] unsuccessful attempt to observe the 10 Å periodicity of K in a depth profile of a mica crystal is most likely the result of such a knock-in process.

E. THE RATIO OF SPUTTERED IONS/NEUTRALS

The ratio of sputtered ions/neutrals is not of much concern if sputtering is used for surface cleaning or profiling in combination with ESCA, AES, or ISS. In SIMS, however, where only ions are detected with the mass spectrometer, this ratio is of vital importance. It has been known for a long time that clean metals emit very few sputtered ions when bombarded by noble gas ions. Mueller's field ion emission studies indicated that an ion close to a clean metal surface cannot survive because it pulls an electron from the surface by field emission and becomes neutralized. The other extreme is an ionic compound like an alkali halide where the sputtered species consist mostly of positive and negative ions and are sputtered as such. Over the years, a more refined picture has emerged with the extremes referred to as "kinetic" (clean metal) and "chemical" ionization. The ion yield of systems in which kinetic ionization dominates is usually much lower than in chemical ionization. This is best illustrated with some data collected with 3 keV Ar^+-ions by Benninghoven et al. [52] as shown in Table 6. In the oxidized state (chemical ionization), the yield for Al, V, and Cr is close to unity which is typically three or four orders of magnitude larger than the yield for the corresponding clean metals. It becomes immediately obvious that the spread in ion yield poses a serious problem for quantitative analysis. It is, for instance, disturbing to see in SIMS profiling of a Ta_2O_3 film on Ta that the Ta signal decreases to a very small value when entering the pure Ta. The complexity of the situation is further illustrated by Benninghoven's results of the ion yield from an oxidized W surface vs. bombarding ion energy which

TABLE 6

Yields of metal ions from cleaned and oxidized metal surfaces [52]

Metal	$\gamma(M^+)$ cleaned surface	$\gamma(M^+)$ oxidized surface
Mg	0.0085	0.16
Al	0.02	2.0
V	0.0013	1.2
Cr	0.005	1.2
Fe	0.001	0.38
Ni	0.0003	0.02
Cu	0.0013	0.0045

showed an exponential increase (exponent between 2 and 3) of the ion yield when the primary Ar^+-ion energy was raised from 150 eV to 3 keV [53].

A pronounced difference exists with respect to the positive or negative ion yield. The presence of a reactive species such as oxygen, either contained in the solid, or introduced into the system as a gas or as the primary bombarding ion, enhances positive ion yields [54,55]. Negative ions, however, are enhanced by the presence of Cs [55] and thereby it is possible to produce large negative ion yields even with pure Au.

Various ideas about the basic concepts of kinetic and chemical ion emission have been described by Castaing and Hennequin [56], Evans [4], Anderson [55], Schroeer et al. [57], Joyes [58], and Blaise [59]. The kinetic process requires lattice particles to be emitted in the excited metastable state from which they can be auto-ionized in the de-exitation process by an Auger transition after being separated from the surface. In chemical ionization, the species are ejected as ions because the neutralization probability is reduced when reactive species such as oxygen reduce the number of available free conduction band electrons. On the other hand, Cs on the surface either enhances the negative ion emission by lowering the electronic work function and thereby increasing the electron availability near the surface, or the polarization of the topmost layer (Cs^+—metal atom$^-$) increases the probability for direct negative metal atom emission.

Although many attempts have been made, the complexity of the phenomenon is such that an all-encompassing theory which would predict the positive and negative ion yields of various species is not yet in sight.

III. Specific particle bombardment aspects

A. IN ISS

In ISS, a surface is bombarded with a beam of noble gas ions and the energies of the ions scattered under a fixed scattering angle are analyzed. The masses of the collision partners determine the energy of the scattered ions in billiard ball fashion. Details of the performance of a commercially available instrument are described by Honig and Harrison [60]. This instrument works with a 90° scattering angle, 0.5—3 keV ions, 10 $\mu A/cm^2$ ion current density and an ion beam diameter of the order of 1 mm at the target. In a complete spectrum scan, about 0.1—1 monolayer is sputtered from the sample when He ions are used. The intensity of the scattered beam is proportional to the differential scattering cross-section which is a function of the atomic numbers of the colliding species, the ion energy, and the scattering angle, $1 - P_n$ where P_n is the neutralization probability, which is a function of scattered ion velocity, and a geometric factor which describes the masking of one atom species by another. The neutralization term $1 - P_n$ [$\approx 10^{-3}$] is

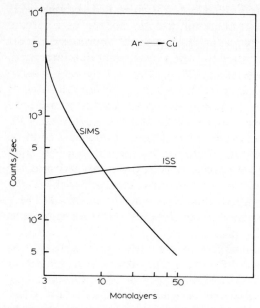

Fig. 10. Copper signal in SIMS and ISS as a function of sputter removal of surface oxides [61].

poorly understood. Fortunately, it varies far less for different species as does the neutralization of sputtered ions. This is demonstrated in Fig. 10 which shows how the contamination removal by sputtering decreases the SIMS Cu signal by orders of magnitude but has little influence on the ISS signal [61]. It remains to be shown that the neutralization process is independent of the matrix. The differential scattering cross-section for oxygen or carbon is about a factor of 10 lower than for heavy mass atoms.

Beyond doubt, among all the methods listed in Table 1, ISS has the highest sensitivity to the topmost surface layer. Aside from the difficulties discussed in section II, D, the topmost layer of a sample must certainly acquire a different composition in UHV than it had in air or its environment for which the surface composition is sought. Aside from this, a small amount of sputtering is always necessary in ISS before the peaks appear because adsorbed gases such as hydrocarbons shield the lattice scattering sites. In composition depth profiling, the sputter etching soon breaks up the surface to such an extent that sampling the topmost layer provides no particular advantage.

Interpretation of ISS spectra is often complicated and the analysis sensitivity is decreased by a number of secondary bombardment related effects. Sequential scattering of a primary ion from two neighboring lattice atoms produces shoulders on the high energy side of a peak. In fact, it was demonstrated that a well-annealed Ni (110) surface initially shows only multiple scattering [62] when bombarded with 1 keV Ar^+-ions at $45°$ incidence. It takes a dose of at least 10^{15} ions/cm² to break up the surface or to create sufficient defects for the single binary peaks to appear. It becomes difficult to separate adjacent masses heavier than Si when He projectiles are used. Higher mass resolution for atoms heavier than Si is obtained by switching to Ne. With no provision for mass separation in the primary beam, switching of gases can lead to contamination with the previously used gas due to memory effects in the vacuum system and can often cause interfering peaks. Tails appear on the low energy side of the peaks when higher energy He is used. They are probably caused by penetration of the ion below the topmost layer which produces energy losses. Doubly-charged ions in the beam arrive at the sample with double the energy. Such ions can acquire an electron from the surface and be scattered as singly-charged ions. This results in peaks at twice the normal energy. A broad peak due to sputtered lattice particles appears at the low energy end. The detectability of low mass impurities is then severely limited when Ne or heavier ions are used for analysis. Recent studies of a Ni single crystal on which S was absorbed [63] have demonstrated that a rather serious matrix effect in ISS can arise from the shadowing of one species by another. In particular, this is the case at small scattering angles ($<30°$) but, in fact, this phenomenon can be used advantageously for obtaining structural information by ISS [64].

The ratios of Al/O or Si/O analyzed with 1.5 keV He^+-ions, are found to correspond well to the expected values for Al_2O_3 or SiO_2 if the scattering cross-sections are taken into account. In other cases, however, wide discrepancies from the expected values are found such as at the Se face of a (0001) CdSe crystal where the ratio of the atoms Cd/Se changed from 7 at 1.1 keV to 1 at 2.6 keV [60]. ISS analysis of a PbO powder never showed an O peak because the O atoms always seem to remain well shielded by the Pb even during sputtering [60]. The high sensitivity of ISS to the topmost layer was demonstrated with the different peak ratios from opposite faces of polar crystals such as: CdSe [60], CdS [65], and ZnTe [66].

Honig and Harrison [60] point out that present commercially available equipment could be considerably improved by incorporating mass analysis of the primary beam, limiting the analysis area to the bottom of the sputtering crater (i.e. excluding in profiling the crater walls) and making the scattering angle variable. The recent incorporation of a mass analyzer (3M), thereby adding the capability for SIMS analysis, is bound to help much in the interpretation of results from both methods. In general, peaks in ISS spectra are much less well-defined than in the other surface analysis methods and they ride usually on a rather noisy background. This phenomenon is probably intrinsic to ISS and is not likely to be overcome because the billiard game is played not strictly with single atoms but with an assembly of atoms in which the influence of the neighbors of the struck atom is never completely negligible [67,68].

B. IN SIMS

The energy of the incident ions (either noble or reactive gas) ranges from 1 to 20 keV [4]. In the SIMS method, the sample area from which the sputtered atoms or molecules are mass analyzed can be made as small as 1 micron in diameter in commercially available ion microprobes at impinging ion current densities of the order of 10^{-2} A/cm^2. At the other extreme, Benninghoven's static SIMS surface analyzer [7] operates with only 10^{-9} A/cm^2 but with a large ion beam or analysis area of 0.1 cm^2.

SIMS achieves much higher sensitivities than all other surface analysis techniques. It is possible to reach detection limits of $<10^{-6}$ of a monolayer or $<10^{-14}$ g for many elements with the static SIMS method provided that an analysis area larger than 0.1 cm^2 is available. Such high sensitivities provide an excellent opportunity to study ion implantation profiles, but one often observes long tails which are caused by inherent and unavoidable knock-in effects induced by the sputtering process.

Morabito and Lewis [69] drew attention to the fact that for obtaining a desired detection sensitivity, a minimum sample volume needs to be consumed and mass analyzed. For instance, in order to detect 10 p.p.m. of Al (the ratio ions/neutrals is about 10^{-3}) with a precision of 3% from a 100 μ diameter Ar ion bombarded area, a thickness of at least 130 Å must be sputter-removed for analysis. The same sensitivity can be achieved with a 2 mm diameter analysis area by sputtering only $\frac{1}{3}$ Å in thickness.

With respect to the sampling depth in SIMS, it is known that the sputtered atoms come essentially from the top one or two atom layers at bombarding ion energies up to 1 keV. The average analysis depth is estimated to be about 10 Å at 3 keV, although the author was unable to locate concrete experimental data on this point. By studying the time dependence of the intensity of various ion species Benninghoven [7] showed that species which are located in the outermost surface layer only decrease strictly exponentially with time. Species, however, which may become freshly exposed as sputtering proceeds follow other nonexponential time dependencies.

As pointed out above, SIMS analyzes the sputtered material and not the sample surface as the other methods. However, differential sputtering can cause erroneous results before equilibrium conditions are established. It is possible that the angular ejection distribution of different atomic species is different in multicomponent materials, particularly compounds and this can lead to other analysis errors.

As discussed in detail above, an overriding and disturbing consideration in quantitative SIMS analysis is the widely varying ion/neutral atom ratio. From a pure Au target, nearly all sputtered atoms are neutral, but even minute traces of Na etc. give rise to a large Na etc. ion signal. Interpretation of SIMS data is further complicated by the fact that the spectra contain many molecular and multiply charged ions. Colby and Evans [70], for instance, list the following possibilities: (a) multiply charged matrix ions (X^{2+}, X^{3+} etc.), (b) matrix—self polymer ions (X_2^+, X_2^+, XY^+, X_2Y^+ etc.), (c) primary ion—matrix ions (XO^+, X_2O^+ etc.), (d) hydride ions (XH^+, XH_2^+ etc.), (e) noble gas—matrix ions such as XAr^+. A skillful researcher can, of course, retrieve from these species a great deal of interesting chemically oriented information such as related to catalysis, in situ oxidation, or adsorption phenomena [7]. A sometimes important advantage of SIMS is the fact that H is detectable.

Another complication unique to the SIMS technique results from the ejection energies of the sputtered ions. Sputtered atoms are ejected with a wide spread of energy and the average energies are one or two orders of magnitude higher than thermal evaporation energies. The energy spread influences transmission and peak shape in single focusing sector instruments or quadrupole mass filters. Atomic ions usually have higher kinetic energies than molecular ions and this fact has been used for "simplifying" the spectra by letting only the ions of selected energy enter through an energy window into the mass analyzer [71,74].

One obtains better depth resolution in profiling at low bombarding ion energy. Various approaches are used for obtaining the analysis from a flat-bottomed crater while excluding the crater walls: (a) defocusing and aperturing the primary ion beams such that nearly vertical-sided crater walls are obtained with a flat-bottomed crater; (b) accepting only ions which come from the bottom of the crater by mechanical apertures in the focal plane of the analyzed beam; and (c) rastering the primary beam and electronic gating of the detection system such that ions are only collected from the central crater region.

The goal for depth profiling is to be able to raise the sputtering rate to above 50 Å/min. Triode ion gun sources [72,74] have been developed for commercial instruments which combine telefocusing with high ion current density and desired (trapezoidal) beam profiles.

An important advance for improving the signal-to-noise ratio was made independently by Schubert and Tracy [73] and Maul et al. [74]. They recognized that when energetic atoms or ions scattered or sputtered from the target enter the quadrupole analyzer, ternary ions can be released from surfaces such as the quadrupole rods. Such ions can reach the detector and contribute significantly to the background signal. This effect becomes increasingly severe with higher bombarding ion energy. The solution for this is to incorporate a deflection ion energy filter in front of the quadrupole. This filter prevents neutrals or fast ions from entering through the aperture into the quadrupole chamber.

Charging of bulk insulator targets in SIMS should hardly pose serious problems because positive ion bombardment and ejection of γ-electrons can only shift the surface potential to more positive values. Positive charging, however, can readily be counteracted by attraction of electrons from a nearby thermionic filament, or the sample can be flooded with electrons from an electron gun which allows both the energy and the number of flooding electrons to be controlled.

C. IN ESCA AND AES

In both ESCA and AES, the bombarding X-ray photons or electrons are strictly intended only to perform the analysis part. Surface cleaning or microsectioning is accomplished with ion bombardment from a separate ion or plasma gun. This separation of functions provides an important practical advantage in depth profiling insofar as the sputter microsectioning can be interrupted at any time for studying composition details such as at an interface. Furthermore, it is possible to gain additional valuable information on ion bombardment-induced surface composition changes such as caused by differential sputtering or knock-in effects by analyzing the surface after it had been bombarded with high or with low bombarding ion energies. A possible new artifact, however, can arise when the analyzing beam and the ion beam strike at different angles with respect to the target surface. In usual AES arrangements, the ion beam strikes the target surface at nearly normal incidence while the analyzing electron beam impinges 60° away from the surface normal. Consequently, analysis signals can be detected from microslopes of a rough surface which receive a very low ion flux density and are barely being sputtered at all.

X-rays are difficult to focus with sufficient intensity onto small target areas. This has prevented ESCA from much being used for depth profiling because rather large diameter ion beams (>1 cm) with uniform current density then become a necessity. Only recently has the analysis area in a commercially available ESCA instrument been reduced to 1 mm diameter.

Only in AES with its high sensitivity and signal-to-noise ratio is it possible to obtain a complete spectrum in 0.1 sec and display it in real time on an oscilloscope. Furthermore, simultaneous use of Auger analysis and ion sputter etching provides the capability for automatic composition vs. sputtering time (depth) plotting. With ESCA one has to obtain depth profiles by sequential sputter etching and analyzing because the X-ray source cannot tolerate a 5×10^{-5} torr Ar pressure. Electrostatic charging can occur in the case of insulator samples in both AES and ESCA. In ESCA, the sample surface is mostly charged positively by the ejected electrons but electrons impinging on the sample from other surfaces, such as an X-ray window, can enter into the picture. Although high resolution ESCA is very sensitive to even minor charging, or to potential gradients arising at the analysis surface, it is usually possible to solve this problem by electron flooding as described under SIMS, or by creating favorable conditions for surface photoconduction to a metal film surrounding the analysis area [75,76]. In AES, the charging can cause serious difficulties because a surface with a secondary electron emission coefficient below unity is driven towards a negative potential which is more difficult to counteract. Positive ion bombardment is of little help because the electron current density is usually orders of magnitude larger than the ion current density. Even with lowering of the bombarding electron energy and with oblique electron incidence (both increase the secondary electron emission) it is often not possible to overcome charging in materials such as polymer target samples. Charging is not a major problem as long as the whole Auger spectrum is shifted by a fixed determinable amount of energy. However, analysis becomes impossible when the charging becomes unstable and erratic.

Composition or bonding changes arising from the photon or electron bombardment are of concern in both ESCA and AES. Photon- or electron-induced desorption of species has been a subject of thorough studies [77]. Well known, of course, is the build-up of carbon compounds by polymerization under electron beam or photon bombardment if the background pressure of hydrocarbons is not sufficiently low. Another rather dramatic phenomenon concerns the drilling of holes by electron beams in alkali halides [78—80]. It was, for instance, observed that at $280°$C KCl or NaCl loses one molecule per 300 eV electron. Dissociation or break-up of molecules under electron beam bombardment does occur even at surfaces of rather stable insulators such as SiO_2 or Al_2O_3 [80]. We were quite surprised to notice, for instance, that the C in carbonates (such as in a pearl or in limestone) is not detectable with AES but clearly visible with ESCA.

With respect to such artifacts, ESCA is a much more gentle method than AES. This is not because photons cause less radiation damage than electrons of comparable energy, but because the power delivered to the analysis volume in ESCA is many orders of magnitude lower than in AES. In fact, as the electron beam in AES is more and more concentrated to higher flux density and higher beam energy, electron beam heating and possible changes of diffusion profiles become a problem. Differences detected in the apparent composition between ion beam "on" or "off" and oscilloscope observation of the Auger spectrum after the analyzing electron beam has been shifted to a new, previously non-electron bombarded, spot on the target can often give valuable information on artifacts caused by the electron beam.

IV. Outlook

The main drawback of SIMS as discussed above and in many previous surveys is the widely varying material and matrix dependent ratio of ions to neutrals. Many attempts have been made for eliminating this problem. Using oxygen ions for the sputter bombardment enhances "chemical ionization" and causes the ion yields of all species to rise to higher but much less widely spaced values. The use of such a reactive species as oxygen, however, is often undesirable for many other reasons. Not only are the sputtering yields low, but analysis of the sample for oxygen becomes impossible.

An obvious solution to the problem would be to detect and analyze the sputtered neutrals instead of ions because the sputtering yields of different materials are much closer to each other than the ion yields. High sensitivity mass analyzers, however, require *ions* and, therefore, it is necessary to convert the sputtered atoms or molecules, after they are ejected, into ions. This is difficult to achieve because sputtered species are ejected from the target with such high velocity that they traverse any ionization volume, such as a plasma or an electron beam, so fast that they have little chance to become ionized.

Two approaches have been used for accomplishing substantial ionization of sputtered atoms. In one method, first published by Coburn and Kay [81], called "glow discharge mass spectroscopy", a glow discharge plasma at a pressure of ~ 1 to 10 Pa serves for sputter etching, for slowing down the sputtered species by gas collisions (the average mean-free path at 10 Pa is 0.4 mm) and for ionizing the sputtered atom species by Penning ionization. In the other method, first published by Oechsner and Gerhard [82], the ionization efficiency in a low gas pressure rf-excited plasma (10^{-2} to 10^{-1} Pa) is increased greatly by operating the discharge in the cyclotron resonance mode by superimposing a uniform magnetic field created by a pair of Helmholz coils (10^{-3} Wb/m^2 at 26 mHz) [83]. The low gas pressure approach has the advantage that backscattering of sputtered atoms to the target is avoided and differential pumping of the quadrupole becomes unnecessary.

With both methods it is possible to decrease substantially the wide span of sensitivity to different elements and to eliminate the disturbing SIMS matrix effects. In both methods, however, the target is immersed directly in a plasma for ion bombardment. It is difficult to achieve uniform sputter etching over the entire sample area unless the flat specimen is mounted flush with the target holder. The lack of spatial resolution, the limitation on geometry of the sample, etc. make it unlikely that these methods will find wide use in practical surface analysis or profiling tasks.

The matrix shown in Table 1 is complete and it is unlikely that major new surface analysis methods will emerge in the future. For practical analysis tasks, five methods dominate the field: ESCA, AES, SIMS, ISS, and nuclear backscattering. The latter stands separate from the others insofar as it is nondestructive and requires no ion bombardment for cleaning and not even for depth profiling. In SIMS, sputtering is essential; in ISS, it is unavoidable. In ESCA and AES sputtering is added for surface cleaning, in situ microsectioning and compositional depth profiling. Each method has its shortcomings and advantages. A general tendency, therefore, is to make more than one technique available for being able to analyze the same specimen area with different analysis schemes. Comparisions are, in fact, often essential for uncovering the shortcomings of one particular method.

Sputtering plays an important role in nearly all techniques. The most serious remaining problems in sputtering, which need a better understanding, concern the ratio of ions/neutrals (which is of vital importance in SIMS) and of differential sputtering or knock-in effects which cause composition changes at the target surface.

The goal of every composition analysis method is to advance from qualitative to semi-quantitative to quantitative determination of the constituents. This, together with more automated data acquisition, will most likely be the subject of major thrust in the near future. As our knowledge in surface physics advances, it will become more and more important to increase spatial resolution for analyzing areas of less than micron dimensions and to gain information on chemical shifts or subtle binding energy differences. AES in the first case and ESCA in the latter are likely to emerge as the most important analytical methods for practical problems.

Acknowledgments

Professor Jack Judy helped much in making the manuscript more palatable to American readers. My sputtering knowledge was acquired during many years of consistent support from the Office of Naval Research. The more recent familiarity with surface analysis methods, in particular with Auger electron spectroscopy, I owe mostly to NSF and NSF—RANN research and equipment support. Stimulating discussions and arguments with my post-doctoral (Fulbright exchange) research fellows, Dr. G. Betz and Dr. P. Braun, have helped much to clarify some points and to alert me to some often-overlooked but important Austrian publications.

References

1 H.E. Farnsworth, R.E. Schlier, T.H. George and R.M. Burger, J. Appl. Phys., 29 (1958) 1150, 1195.
2 H.D. Hagstrum, Science, 178 (1972) 275.
3 W.D. Mackintosh, in P.F. Kane and G.B. Larrabee (Eds.), Characterization of Solid Surfaces, Plenum Press, New York, 1974.
4 C.A. Evans, Jr., Anal. Chem., 44 (13) (1972) 67A.
5 J.W. Coburn, E. Taglauer and E. Kay, J. Appl. Phys., 45 (1974) 1779.
6 H. Oechsner and W. Gerhard, Surface Sci., 44 (1974) 480.
7 A. Benninghoven, Surface Sci., 35 (1973) 427.
8 E.P.Th. Suurmeijer and A.L. Boers, Surface Sci., 43 (1974) 309.
9 R.F.K. Herzog, W.P. Poschenrieder and F.K. Satkiewics, Radiat. Eff., 18 (1973) 199.
10 G. Carter and J.S. Colligan, Ion Bombardment of Solids, Heinemann Educational Books, London, 1968.

36

11 G.K. Wehner and G.S. Anderson, in L.I. Maissel and R. Glang (Eds.), Handbook of Thin Film Technology, McGraw-Hill, New York, 1970, Chap. 3.
12 D.E. Harrison, W.L. Moore and H.T. Holcombe, Radiat. Eff., 17 (1973) 167.
13 P. Sigmund, Phys. Rev., 184 (1969) 383.
14 R. Behrisch, Ergebnisse der exakten Naturwissenschaften, Bd. XXXV, Springer, Berlin, 1964, p. 297.
15 L.I. Maissel, in L.I. Maissel and R. Glang (Eds.), Handbook of Thin Film Technology, McGraw-Hill, New York, 1970, p. 4—1.
16 H. Oechsner, Z. Phys., 261 (1973) 37.
17 P.K. Rol, D. Onderlinden and J. Kistemaker, Trans. Third Int. Vac. Congr., Vol. 1, Pergamon, Oxford, 1966, p. 75.
18 R.S. Nelson, Phil. Mag., 11 (1965) 291.
19 R. Kelly and N.Q. Lam, Radiat. Eff., 19 (1973) 39.
20 S.D. Dahlgren and E.D. McClanahan, J. Appl. Phys., 43 (1972) 1514.
21 G.S. Anderson and G.K. Wehner, Surface Sci., 2 (1964) 367.
22 R.L. Jacobson and G.K. Wehner, J. Appl. Phys., 36 (1965) 2674.
23 I.H. Wilson, Radiat. Eff., 18 (1973) 95.
24 W.J. Witcomb, J. Mater. Sci., 9 (1974) 1227.
25 G. Carter, J.S. Colligon and M.J. Nobes, J. Mater. Sci., 6 (1971) 115.
26 D.J. Barber, F.C. Frank, M. Moss, J.W. Steeds and I.S.T. Tsang, J. Mater. Sci., 8 (1973) 1030.
27 A.R. Bayly, J. Mater. Sci., 7 (1972) 404.
28 G. Carter, J.S. Colligon and M.J. Nobes, J. Mater. Sci., 8 (1973) 1473.
29 E.G. Spencer and P.H. Schmidt, J. Vac. Sci. Technol., 8 (1971) 52.
30 J.B. Schroeder, S. Bashkin and J.F. Nester, Appl. Opt., 5 (1966) 1031.
31 I.S.T. Tsang and D.J. Barber, J. Mater. Sci., 7 (1972) 687.
32 P.D. Davidse, J. Electrochem. Soc., 116 (1969) 100.
33 H. Yasuda and K. Nagai, Jap. J. Appl. Phys., 11 (1972) 1713.
34 C.R. Guliano, Appl. Phys. Lett., 21 (1972) 39.
35 L. Esaki, Science, 183 (1974) 1149.
36 G.K. Wehner and D.J. Hajicek, J. Appl. Phys., 42 (1973) 1145.
37 M.J. Fulker, L. Holland and R.E. Hurley, Scanning Electron Microsc. Proc. IIT Research Institute, Chicago, Ill., April, 1973, Part III.
38 S.H. Overbury, P.A. Bertrand and G.A. Somerjai, to be published in Chem. Rev.
39 D.V. McCaughan and R.A. Kushner, in P.F. Kane and G.B. Larrabee, (Eds.), Characterization of Solid Surfaces, Plenum Press, New York, 1974, Chap. 22, p. 627.
40 H. Hughes, R.D. Baxter and B. Phillips, IEEE Trans. Nucl. Sci., NS19 (1972) 256.
41 G.S. Anderson, J. Appl. Phys., 40 (1969) 2884.
42 V.E. Henrich and J.C.C. Fan, Surface Sci., 42 (1974) 139.
43 P. Palmberg, Physical Electronics Ind., Eden Prairie, Minn., unpublished results.
44 M. Tarng and G.K. Wehner, J. Appl. Phys., 42 (1971) 2449.
45 H. Shimizu, M. Ono and K. Nakayama, Surface Sci., 36 (1973) 817.
46 W. Färber and P. Braun, Vak. Tech., 23 (1974) 239.
47 M. Tarng and G.K. Wehner, J. Appl. Phys., 43 (1972) 2268.
48 L.I. Yin, S. Ghose and I. Adler, Appl. Spectrosc., 26 (1972) 355.
49 A. Turos, W.F. van der Weg, D. Sigurd and J.W. Mayer, J. Appl. Phys., 45 (1974) 2777.
50 F. Schulz, K. Wittmaack and J. Maul, Radiat. Eff., 18 (1973) 211.
51 T. Staib, Radiat. Eff., 18 (1973) 217.
52 A. Benninghoven and A. Mueller, Phys. Lett., 40A (1972) 169.
53 A. Benninghoven, C. Plog and N. Treitz, Int. Conf. Ion—Surface Interactions, Garching, 1972.

54 A. Benninghoven, Z. Naturforsch, 22A (1967) 841.
55 C.A. Anderson, Anal. Chem., 45 (1973) 1421.
56 R. Castaing and J.F. Hennequin, in A. Quale (Ed.), Advances in Mass Spectrometry, Vol. V, Institute of Petroleum, London, 1972, p. 419.
57 J.M. Schroeer, T.N. Rhodin and R.C. Bradley, Surface Sci., 34 (1973) 571.
58 P. Joyes, in R. Behrisch et al. (Eds.), Ion Surface Interaction, Sputtering and Related Phenomena, Gordon-Breach, London, 1973, p. 139.
59 G. Blaise, in R. Behrisch et al. (Eds.), Ion Surface Interaction, Sputtering and Related Phenomena, Gordon-Breach, London, 1973, p. 147.
60 R.E. Honig and W.L. Harrison, Thin Solid Films, 19 (1973) 43.
61 W. Grundner, W. Heiland and E. Taglauer, Appl. Phys., 4 (1974) 243.
62 W. Heiland and E. Taglauer, Radiat. Eff., 19 (1973) 1.
63 E. Taglauer and W. Heiland, Appl. Phys. Lett., 24 (1974) 437.
64 H.H. Brongersma and P.M. Mul, Surface Sci., 35 (1973) 393.
65 W.H. Strehlow and D.P. Smith, Appl. Phys. Lett., 13 (1968) 34.
66 R.F. Goff and D.P. Smith, J. Vac. Sci. Technol., 7 (1970) 72.
67 D.S. Karpuzov and V.E. Yurasova, Phys. Status Solidi, 47 (1971) 41.
68 E.S. Mashkova and V.A. Molchanov, Radiat. Eff., 16 (1972) 143.
69 M. Morabito and R.K. Lewis, Anal. Chem., 45 (1973) 869.
70 N. Colby and C.A. Evans, Appl. Spectrosc., 27 (1973) 274.
71 C. Feldman and F. Satkiewicz, J. Electrochem. Soc., 120 (1973) 1111.
72 K. Wittmaack, Nucl. Instrum. Methods, 118 (1974) 99.
73 R. Schubert and J.C. Tracy, Rev. Sci. Instrum., 44 (1973) 487.
74 J. Maul, F. Schulz and K. Wittmaack, Advances in Mass Spectrometry, Vol. 6, Applied Science Publishers, England, 1973, p. 493.
75 M.F. Ebel and H. Ebel, J. Electron Spectrosc. Related Phenom., 3 (1974) 169.
76 M.F. Ebel, Vak. Tech., 23 (1974) 33.
77 T.E. Madey and J.T. Yates, Jr., J. Vac. Sci. Technol., 8 (1971) 525.
78 H. Steffen, R. Niedermeyer and H. Mayer, Thin Films, 1 (1968) 223.
79 P.D. Townsend and J.C. Kelly, Phys. Lett., 26A (1968) 138.
80 C.C. Chang, in P.F. Kane and G.B. Larrabee (Eds.), Characterization of Solid Surfaces, Plenum Press New York, 1974, p. 549.
81 J.W. Coburn and E. Kay, Appl. Phys. Lett., 18 (1971) 435.
82 H. Oechsner and W. Gerhard, Phys. Lett., 40A (1972) 211.
83 H. Oechsner, Plasma Phys., 16 (1974) 835.

Chapter 2

A COMPARISON OF THE METHODS OF SURFACE ANALYSIS AND THEIR APPLICATIONS

D. LICHTMAN

I. Introduction

Several years ago while discussing the various techniques of surface analysis, I found it quite possible to review virtually all existing methods in a reasonable period of time. The last few years, however, have seen a tremendous explosion of interest in surface studies accompanied by a very large increase in the number of people working in the field. This expansion has led to a simultaneous increasing development of new methods of surface analysis. Just recently, a report was prepared [1] in which all the known techniques for surface measurements were tabulated (summer of 1973). The list tabulated 56 separate techniques. In a subsequent conversation with the author, he reported that the day after the report was printed there appeared in the literature a description of an entirely new technique, so that his list was already obsolete. It therefore seems quite clear that by now no one person can hope to seriously review in any detail all of the techniques of surface analysis within a reasonable effort.

Since the purpose of this chapter is to compare various techniques, and the status of surface science being what it is, I will attempt to cover a reasonable fraction of the existing methods trying to choose those to include the most popular and widely used as well as many of the new techniques, which may have considerable future potential. This chapter will primarily consider the relative merits of the various techniques and therefore I will not describe either the complete fundamental basis of each technique or the results obtained by those using them but will provide references to basic papers and review articles for that purpose. In this discussion, I will describe the techniques in just enough detail to enable the reader to follow the discussion. I will concentrate on those aspects of the methods which will hopefully enable the reader to judge their suitability for various applications. Thus, I will be concerned with such aspects as the relative complexity of the experimental system, the economics of the system, the kinds of information available and not available, the kinds of material suitable for analysis and the probable future potential of each method, at least as I see it now.

40

The primary purpose in surface analysis is to identify the surface elemental composition (qualitatively), to determine the amount and nature of species adsorbed and to elucidate the properties of the surface atoms or adatoms. A surface is generally considered as that part of the bulk from roughly 1 to 10 atomic layers from the surface monolayer. The usual procedure when outlining methods of surface analysis is to prepare a table or figure of some kind, indicating the various kinds of probes to the sample under study and various forms in which information is obtained from the sample. I have chosen the diagram as shown in Fig. 1. This figure indicates the fundamental basis of surface analysis techniques. As can be seen, we consider eight basic input probes which give rise to one or more of four types of particles that leave the surface carrying information about it to a suitable detector. The input probes can be particle beams of electrons, ions, photons or neutrals or non-particle probes such as thermal, electric fields, magnetic fields or sonic surface waves. All of the input probes (with the exception of magnetic fields) give rise to emitted particle beams, i.e. electrons, ions, photons, or neutrals. The various surface analysis techniques can therefore be classified according to the type of input probe and the type of emitted particle (e.g. electrons in, ions out; thermal in, neutrals out, etc.). In analyzing the emitted particles, one can consider four possible types of information; identification of the particle, spatial distribution, energy distribution and number. Any or all of these forms of information are then used to develop a better understanding of the surface under study. (Of course the nature of the probe beam is also important in evaluating the resulting data.) Through the remainder of the chapter, we will consider all the various probes of the surface and

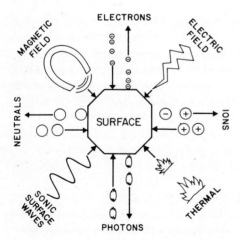

Fig. 1. Pictorial representation of surface analysis techniques. Various combinations of probes in and particles out determine the various surface analysis techniques.

the methods for analyzing the particles which leave the surface and provide the information about it.

II. Classification of the methods for surface analysis by the incident particles used to produce an output of detectable particles

A. THERMAL INPUT WITH NEUTRALS OUT

I will begin with a discussion of the technique popularly known as flash filament, which utilizes basically thermal input. I choose this technique first because I believe it is the simplest method; it is very often the technique I would recommend to people starting in surface analysis. As a technique, it has been used for many years and several excellent papers are available describing the details [2—6]. The method is based on the following simple premise. Gases will adsorb onto a substrate with certain binding energies. If the substrate temperature is sufficiently low, the binding energy will retain the adsorbed gas on the surface. At room temperature, this situation will prevail for chemisorbed species while at low temperatures, such as that of liquid nitrogen, physisorbed species can also be retained on the surface. If the substrate temperature is then increased, the adsorbed components will desorb at a rate determined by their binding energy, a frequency factor, the order of the desorption process and the heating rate of the substrate. The analysis then involves the detection of the components leaving the surface. Except for the very special case of thermally induced ion desorption [7], the adsorbed species desorb as neutral gases and must therefore be detected by neutral beam detection techniques. If one wishes merely to detect the quantity of desorbing material, one generally uses a pressure indicator such as an ion gauge. If more information, such as chemical identification, is desired, then the common detection device is a mass spectrometer. Thus, one can perform a surface analysis with nothing more than a substrate which can be heated, a source of gas for adsorption and an ion gauge for detection. This system is extremely simple, compact and inexpensive. If one wishes to enter the field of surface analysis, a system like this can generally be set-up within a day, assuming that a vacuum system is available. It is an excellent way to start learning the details and problems of surface analysis with a minimum expenditure of time and money. That is not to say that this technique is trivial and useful only for preliminary learning purposes. By utilizing suitable sensitive mass spectrometric analysis and properly controlled substrate heating and gas introduction, significant data can be obtained. The extended literature describing the utilization of flash filament indicates its significant value. The term flash filament used to describe this general technique arises from the fact that many researchers use wires or ribbons (filaments) and cause desorption by rapid heating of these substrates (flashing). Obviously,

42

the experiment can be performed with large extended surfaces and with a large range of heating rates. However, the term flash filament is still used to describe the basic technique in which thermal input produces desorbed neutral particles as output. The flash filament method is not only used as a starting technique, but also as a secondary technique combined with other methods, since it can be added readily to most other surface analysis systems. Thus, in recent years, many experimenters using other surface analysis methods include flash filament capability in their systems.

Under suitable conditions, thermal energy input can give rise to electron emission (normally referred to as thermionic emission) and ion emission (normally referred to as surface ionization). Although both of these processes have been investigated in great detail, they have had very little application in surface analysis and will not be discussed.

B. ELECTRONS IN

Of the four basic particle probe beams (electrons, ions, neutrals and photons), electron probes have been used longer and more extensively than any of the others. This is because electron beams of controlled energy and density can be easily generated. Those interested in using particle probes will often consider an electron probe first, since they are the simplest, are readily available or can be constructed easily and are quite inexpensive.

Electron probing of the surface can cause the emission of all four types of emitted particles. These processes are summarized in Fig. 2 in which the popular designations for some of the techniques are used. Of the four types

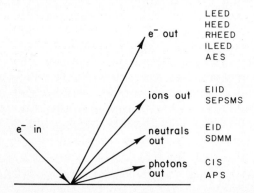

Fig. 2. Sketch summarizing the general group of techniques involving electron probes in. LEED, low energy electron diffraction; HEED, high energy electron diffraction; RHEED, reflected high energy electron diffraction; ILEED, inelastic low energy electron diffraction; AES, Auger electron spectroscopy; EIID, electron impact ion desorption; SEPSMS, electron probe surface mass spectrometer; EID, electron induced desorption; SDMM, surface desorption molecule microscope; CIS, characteristic isochromat spectroscopy; APS, appearance potential spectroscopy.

of emitted particles, the analysis of emitted electrons is most commonly used because of the relative ease of detecting electrons. We will consider now some of the more common methods utilizing electron probes.

1. Electrons out

If one probes a surface with a beam of electrons and analyzes the electrons leaving the surface, there are three parameters available for determination. One can measure the number of emitted electrons, their energy or their spatial distribution. In some cases, only one of these parameters is measured, while in others, two or all three are determined. If one detects specifically the reflected primaries (repelling all electrons which have lost energy by suitable repelling potentials) and considers the spatial distribution of these reflected primaries, one has the technique designated low energy electron diffraction (LEED) [8—12]. In this technique, one is attempting to determine the existence of periodic structure on the surface which will give rise to diffraction peaks in space. LEED is a well-known technique and has been used for almost 50 years. To utilize electrons with a wavelength comparable with the expected periodicity of the surface, one is restricted to electron energies in the range of 20—200 eV. This energy range, considered quite low for electrons, is the basis for the designation LEED. The experimental arrangement can be seen in Fig. 3. To perform experiments in which one merely determines the spatial distribution of the scattered primaries obviously does not involve a complex facility. The major consideration is that the trajectory of low energy electrons can be strongly affected by fairly small magnetic fields. Since their trajectories are of paramount importance, it is essential to maintain a region of minimum magnetic field in the vicinity of the experiment. Thus, to perform a LEED experiment and obtain a spot pattern is not particularly difficult. The proper analysis of the spot pattern is another matter and on the basis of very considerable literature, it is clearly not simple.

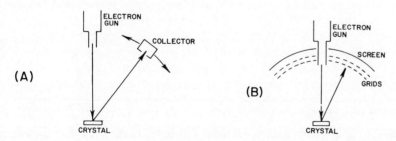

Fig. 3. Simple schematic showing typical systems for LEED studies as a typical surface analysis facility. (A) System for measuring numerical values of scattered electrons. (B) System for displaying diffraction pattern.

References pp. 70—73

In recent years, it has been determined that it is quite difficult to obtain meaningful data from the spatial distribution only. Thus, it is also important to measure quite precisely the quantity of electrons arriving at each spot. This requires the use of a movable Faraday cup which adds considerable complexity to the system. However, the measurement not only of the spatial distribution but also of the number of the reflected primaries at each spot as a function of the primary electron energy shows promise of providing significant information about the spatial periodicity of the surface structure under study. Thus, simple LEED experiments are not difficult to carry out but do not provide much unambiguous information. In the more complex LEED experiments, the determination of all three electron parameters (spatial, energy and quantity) is considerably more difficult but provides much significant data.

LEED provides information on the spatial positioning of the substrate atoms but it does not provide chemical identification. One must also remember that the probing electron beam not only scatters to produce the LEED pattern, but will also interact with the adsorbed species causing changes of these components which must be taken into account.

Although most people interested in electron diffraction use low energy probing electrons, experiments can be done using high energy electrons and are called high energy electron diffraction (HEED) [13,14]. In this technique, one is generally working with very thin films, using high energy electrons to penetrate the film and observing the transmission electron diffraction pattern. This technique clearly provides information more related to describing the bulk rather than the surface and, therefore, is not as surface sensitive. It is not utilized very much in specific surface analysis research.

One can utilize high energy electrons and make them surface sensitive by probing the surface at a glancing angle. The scattered electrons are then generally reflected from the surface layers and appear on the same side as the probe beam. This technique is designated as reflection high energy electron diffraction (RHEED) and a number of researchers are utilizing this technique for surface analysis [15,16].

The discussion so far has been related to analyzing the reflected electrons which have lost essentially no energy in the scattering process. Thus, many people refer to LEED experiments as ELEED, meaning elastic low energy electron diffraction. It is well known that electrons interacting with solids can also undergo inelastic collisions. It is also possible for electrons to undergo some inelastic collisions and some elastic collisions such that they are ultimately scattered back from the sample. One can then do a spatial analysis of these electrons by permitting only electrons which have lost certain discrete amounts of energy to reach the detector system. If one does a spatial analysis of scattered electrons which have lost specific amounts of energy, one has the process designated ILEED [17], inelastic low energy electron diffraction. This technique can provide additional information about the substrate but is clearly much more difficult. One must utilize carefully controlled energy selection and one will clearly have to deal with a much smaller signal. The technique has seen very limited use, at least partly, because of the considerable complexity of the experimental system.

The various techniques described above basically provide information only about the periodicity of the substrate structure. If the structure has no periodicity (over distances of at least 10—20 atom spacings), the technique will provide no useful information from a surface analysis point of view. For periodic structures, the technique can provide considerable information. The significance of the data obtained is somewhat related to the willingness to do careful quantitative measurements of numerical values in addition to the spatial distribution.

In the techniques described above, one is concerned primarily with elastically scattered electrons. Many experiments have also pursued the study of the inelastically scattered as well as the secondary electrons. In this general approach, one is concerned with the energy analysis of the emitted electrons. One of the most popular of the recently developed techniques is Auger electron spectroscopy (AES) [18—23]. The technique can be described simply with reference to Fig. 4. An incoming electron with sufficient energy can remove a core electron from an atom. The excited state atom then will relax back towards the ground state by several possible paths. One of these paths is for an electron from a higher energy level to drop into the core state. The energy liberated can then be emitted as a photon giving rise to a conventional characteristic X-ray emission. It is also possible for the excess energy to be transmitted to another of the electrons in the atom which then has

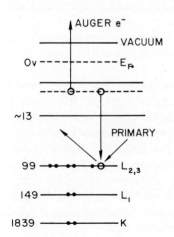

Fig. 4. Energy level diagram showing mechanism for Auger electron emission.

References pp. 70—73

sufficient energy to leave the atom and the material. Its kinetic energy is then uniquely determined by the energy difference indicated in the figure. This electron is known as an Auger electron. Energy analysis of these electrons is known as AES and can result in chemical identification in exactly the same manner as analysis of emitted characteristic X-rays provides this information. The technique requires the generation of a beam of electrons of energy sufficient to produce core level excitation of the atoms being studied (this generally means energies in the range of 1000—3000 eV). This kind of beam generation is very simple and straightforward. The second part of the problem is energy analysis of the Auger electrons. Devices for energy analysis of electrons are well-known and, again, fairly simple and straightforward. The difficulty in the technique is to measure a small number of Auger electrons from the large general secondary electron emission spectrum. Therefore, the Auger electrons produce a very small bump on a plot of the intensity of secondary electrons versus energy. The breakthrough in using this technique came with realizing that the way to make these small increments stand out is to differentiate the signal. The differentiation is obtained by relatively simple electronic means and the ready availability of lock-in amplifiers made this a very practical process. Thus, devices can be constructed or purchased commercially which enable one to detect Auger signals from substrates and from adsorbed species of perhaps 1% of a monolayer.

The use of AES therefore provides information about the chemical identity of the components being probed. To obtain a maximum number of Auger electrons, one generally uses a primary beam energy of about 2500 eV. This energy beam will penetrate the substrate to a significant depth. However, Auger electrons produced inside the substrate will have a high probability of undergoing ineleastic collisions on the way out and therefore will not appear in the desired Auger peak. Thus, in practice, Auger electrons can be detected from a maximum depth of roughly 10—30 Å. This detection depth will obviously be a function of the material being studied but the range for almost all materials is in this three to ten atom layer depth region. The contributions of the various layers decrease exponentially so the signal is very strongly surface monolayer dependent. If one were to analyze the X-rays being emitted, one would have contributions from a considerable depth since the mean free path of the X-rays is much greater than the penetration depth of the primary beam. Thus, Auger analysis has become a most popular technique for surface analysis since it is so surface dependent.

If one wishes merely to obtain peaks for simple atomic chemical identification, then the system is quite straightforward. If, on the other hand, one wishes to obtain additional information such as quantitative interpretation and ability to specify the chemical state of the detected atom, one must, as expected improve the operation of the system. Quantitative data, that is, the ability to relate an Auger peak height or integrated peak shape to a specific quantity of material, requires very careful calibration as well as very careful and controlled operation of the analyzer. If one wishes to determine whether the detected signal is from the atom itself or from an atom which is part of a more complex molecule, one requires high resolution. For example, the energy levels of oxygen will be somewhat different for the atomic state, the molecular state and oxygen in various molecules such as carbon monoxide, silicon dioxide, etc. These small variations will give rise to Auger electrons of slightly different energies and peaks of somewhat different shape. Well resolved peaks can be analyzed, in conjunction with suitable calibration information, to provide information about the specific molecular structure of the surface under study. Clearly, this kind of information is of very great value in understanding surfaces and the reactions of adsorbed species with surfaces. One can see that Auger Electron Spectroscopy (AES) has great potential in surface analysis and will undoubtedly continue to develop for many years to come. Simple chemical identification can be accomplished with a fairly simple device. If one wishes to obtain more detailed information, a more carefully constructed system is required. In general, one can say that AES basically provides chemical identification of the surface under study, and with considerable effort, information about the quantity of material and possibly the molecular identification.

Electrons will cause a large variety of transitions to occur in the bombarded atoms of the substrate as well as in the adsorbed species. An energy analysis of emitted electrons may then provide information about the interaction between the incident electrons and the substrate and its electronic states. Since electron probes are so easy to construct, it is tempting to utilize them and look for all possible processes that can be analyzed. AES is an example and since it is at present the most popular it will suffice to indicate the general approach. Some of the other specific techniques are described in the literature [24—26].

It is perhaps worth mentioning a specific combined technique. The grids used in the standard LEED system can be utilized as an energy analyzer. Thus, one can combine a standard LEED system with some additional electronics and have the ability to do AES as well as LEED on the same sample, thus obtaining both chemical identification and information about the periodic structure with the combined system. This type of system sounds exceedingly useful and, superficially, one would expect that it might be one of the most popular systems used. Unfortunately, using the LEED grid structure for an energy analyzer structure leads to a small signal-to-noise ratio, requiring relatively long analysis times in order to obtain Auger data. The other kinds of energy analyzers, such as cylindrical mirror type, enable one to scan 100—1000 times faster. This is sufficiently significant so that most people have decided to utilize the high speed analyzers. However, if one has the time, the combined system can be used to provide both LEED and Auger in a system that is not too complex.

48

2. Ions out

Electrons can undergo inelastic collisions with atoms or molecules raising them to various excited and/or ionized states. Since the mass ratio of electrons to atoms is very small, kinetic energy exchanges are minimal. Electronic transitions, however, have a fairly high cross-section and involve energies in the range of zero to about fifty electron volts. Therefore, bombarding atoms or molecules with electrons of energy of perhaps 75—100 eV will cause virtually all possible electronic transitions. If the atoms or molecules are adsorbed on a substrate surface, bombarding electrons will still interact with these adsorbed components, again causing the various electronic transitions to occur. If the adsorbed component is raised to an excited or ionized state, it will still remain bound to the surface. However, if an adsorbed molecule is raised to a dissociative state, then the fragments will generally have sufficient excess kinetic energy to be able to leave the surface. If the process is one of dissociative ionization, then the fragment may leave the surface as an ion. A diagram is shown in Fig. 5 of some of the energy states of the gas-phase hydrogen molecule which would be similar to those of an adsorbed molecular species. Some of the dissociative states can be seen which would give rise to the desorbing components. One therefore has a process very similar to that of conventional gas-phase excitation. In this situation, one obtains ion fragments from adsorbed species as opposed to

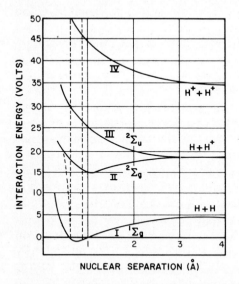

Fig. 5. A number of potential energy curves for the hydrogen molecule. The broken lines indicate the region of Franck-Condon transitions. Curves III and IV indicate states which lead to fragments. Although these curves represent gas phase hydrogen, adsorbed molecules have similar energy level structures.

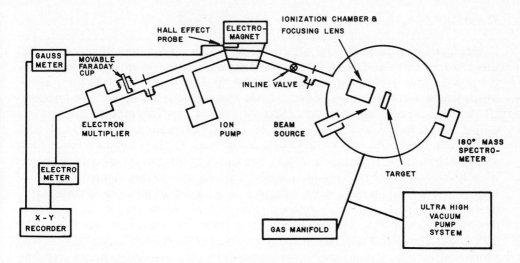

Fig. 6. General schematic of typical electron induced ion desorption system as a typical surface analysis system.

both ion fragments and parent molecule ions from gas-phase interactions. One can then either measure the total number of desorbing ions or one can direct the ions through a mass spectrometer for determination of both type and quantity. The technique therefore involves probing the surface with an electron beam of energy in the range of 100 eV and mass analyzing the desorbing ion components. The technique is referred to as electron induced ion desorption (EIID) or sometimes electron probe surface mass spectrometry (EPSMS). Once a mass spectrometer is available, setting up the probing portion is fairly simple. A schematic showing the basic components of a typical system is shown in Fig. 6. To utilize this technique to its maximum efficiency, one should use a very low density probe beam; that is, current densities of 1 $\mu A/cm^2$ or less. Since the typical desorbing probability is approximately 10^{-5} ions per electron, one must utilize a sensitive mass spectrometer employing an electron multiplier detector.

This technique evolved from attempts to desorb components from a substrate material. Since an atom is bound into a solid with a total binding energy in the order of 15—25 eV, it was considered energetically feasible to remove substrate atoms by bombardment with perhaps 100 eV electrons. Apparently an incoming electron cannot break all the bonds holding an atom into a solid by electronic transitions alone. If one were to use sufficiently high energy electrons so that the atom could be dislodged by kinetic energy exchange, one would require electron energies of at least hundreds of kilovolts. These electrons would dislodge atoms fairly deep in the bulk of the material and they would not therefore be able to leave the substrate. Thus, low energy electrons cause desorption of adsorbed species only. The ion signal one obtains is from the outermost adsorbed monolayer. Thus, data are only obtained from adsorbed components; this is one of the few techniques that is sensitive only to the adsorbed monolayer. The signals obtained are therefore quite simple in form and one has a good opportunity to follow the behavior of adsorbed components as parameters are varied. There is considerable literature describing many detailed experiments [27—32].

There are two major disadvantages to this technique. One is that not all adsorbed species give rise to electron induced ion desorption. It is not yet clear why some adsorbed components give rise to large signals (e.g. some states of adsorbed CO and oxygen) while some yield no electron induced desorbing ions (such as adsorbed nitrogen). A second disadvantage is that no direct information is obtained about the substrate. However, the technique is basically straightforward and can provide quite useful information about adsorption and desorption processes.

An interesting extension of EIID involves using a scanning probe beam [33]. Here, one scans the surface with a focused beam utilizing a television type raster. The desorbing ions are sent through the mass spectrometer, the particular signal desired is detected, and its amplitude is used to intensity modulate an oscilloscope run synchronously with the scanning beam. Thus, one obtains a picture indicating the spatial distribution of the various desorbing components. The technique provides the ability to "see" the surface monolayer at least in rough spatial resolution. As promising as it seems, little progress has been made with this technique. The principal difficulty is that the ion signal obtained is small and one must use considerable care in the structure of the analyzing region and maximize the electronic detection to obtain reasonable output. Some effort is being put into this approach and one should be able to obtain some quite useful information about spatial

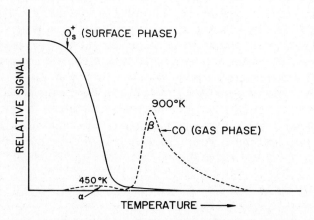

Fig. 7. O_s^+ ion signal from adsorbed CO and CO^+ gas phase signal as a function of surface temperature between $300°$K and $1300°$K at $45°$K/sec heating rate. The curves show how typical data appear using two techniques simultaneously (i.e. EIID and flash filament).

distribution of some adsorbed species. Also, the technique would lend itself to direct observation of surface migration of adsorbed components which yield electron induced ion desorbing signals.

Another simultaneous technique combined with EIID is the flash filament. If one has two mass spectrometers or a single one with suitable switching capability, both gas-phase and surface-phase data can be obtained simultaneously as shown in Fig. 7 with data obtained for CO adsorbed on molybdenum [34]. One can see that the electron induced surface O^+ signal comes primarily from the weakly bound α state of adsorbed CO. Combined EIID and flash filament can give considerable insight into some of the surface processes occurring and the possible types of components in the adsorbed monolayer.

3. Neutrals out

Electrons which raise adsorbed molecular species to dissociative states should certainly give rise to desorbing neutral fragments as well as desorbing ions. One can therefore study the desorbing neutrals in terms of their identity and number again to provide information about the adsorbed monolayer. Very little progress has been made with this technique to date. The basic problem is the following. Ions desorbing from a surface can be directed with appropriate electric fields to the detector and high collection efficiency is possible. Desorbing neutrals, on the other hand, will leave the surface in all directions. Only a fraction of them will pass through an auxiliary ionization chamber where a fraction of these can then be ionized. In a typical system, the ratio of collection efficiency for desorbing neutrals to desorbing ions will be as low as 10^{-2} to 10^{-4}. Since the desorbing ion signal is already small, the desorbing neutral signal is often too small to be detected. In addition, the secondary ionizing beam will also ionize the residual gas components whose identity is quite often the same as the adsorbed species. Therefore, one must now be able to detect the small desorbing component above a background component at the same m/e value. The technique of analyzing electron induced·neutral desorption (EID) has the promise of providing very considerable information about the surface monolayer. The problem of detecting these extremely small signals in a system whose pressure must be maintained in the ultra high vacuum region, presents quite a formidable task. The advent of pulsed techniques and lock-in amplifiers should enable some progress to be made in this area.

Although it does not fit exactly in this section, one should mention another technique which utilizes emitted neutral particles. This approach is called by its developer "the molecule microscope" [35]. In this system, neutral atoms or molecules leave the surface, generally due to thermal evaporation. A small pinhole permits components from a small region to pass to the detector and be registered. By scanning the pinhole over the surface, spatial variations in the rate of evaporation of molecules from surfaces can be plotted. Other methods of causing the desorption of neutrals are being considered. This is a new development and may provide another method for determining spatial variations of surface components.

4. Photons out

The problem of detecting photons is related to two major difficulties. First, desorbing photons, being neutral particles, present the same problems as desorbing neutral atoms. That is, they leave the surface in all directions and only a very small fraction can generally be intercepted and detected. The second problem is related to the noise level. If one generates photons in the visible or near ultraviolet region, one would require a very carefully designed system to prevent scattered light from filaments and other sources from producing a large background noise level. One can reduce the problem by generating high frequency photons in the X-ray region. This approach leads, however, to the use of high energy probing electrons which penetrate the substrate so deeply that the X-ray data are not too surface sensitive. In addition, the bombarding electron beam generates a considerable back-ground level of photon noise due to *Bremstrahlung*. Nevertheless, various devices such as the electron microprobe [36], characteristic isochromat spectroscopy [36] and appearance potential spectroscopy [37] have been developed. These systems are not simple and provide information from a region considerably thicker than the surface monolayer. Although the gener-ation of the electron probe beam is simple, the detection and analysis of the emitted photons is quite difficult. This general approach will undoubtedly continue to receive limited development as far as surface analysis is con-cerned because it is generally not suitable for surface monolayer restricted analysis.

C. IONS IN

Developing a particular technique is certainly related to the expected technical characteristics and probable useful output data, but the level of difficulty in performing the experiment is also of considerable consequence. Thus, although probing with ion beams has considerable potential in surface analysis work, the greater difficulty of producing a suitable controlled beam of ions compared with electrons is at least partially responsible for the considerably fewer experiments utilizing ions beams at this time. As in the case of electron probes, ion beams can cause emission of all four types of particles (Fig. 8). As in the case of electron beams where most of the analyti-cal systems detect emitted electrons, ion beam probes give rise primarily to systems utilizing the detection of emitted ions. A major difference in using ion probes compared with electron probes is that considerable kinetic energy

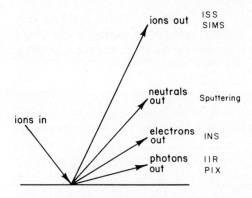

Fig. 8. Sketch summarizing the general group of techniques involving ion probes in.

exchange can occur because the ion particles have masses comparable with those of the substrate. This effect may cause considerable sputtering and therefore produce considerable change of the surface under study. Extra care must therefore be taken when using ion probes so that the surface being studied is not destroyed before information about it has been obtained.

1. Ions out

As in the case of electron analysis, one can determine the number, energy and spatial distribution of the emitted ions. In the case of emitted ions, however, there is the additional factor of mass analysis which was obviously of no concern in the case of emitted electrons. Thus, a probing ion beam can give rise to reflected primary ions, as well as secondary positive and negative ions from the substrate material. Since probing of a surface with an ion beam gives rise to both positive and negative ions of many possible values of mass-to-charge ratio as well as many reflected primaries, measurement of the total number of secondary ions alone provides virtually no useful information. Thus, in all techniques in this category, measurement of number is always combined with measurement of at least one of the other characteristics. Although some information can be obtained by spatial analysis of the emitted ions, this approach has received very little attention. The great difficulty in doing this experiment coupled with the limited information that it might provide as far as surface analysis is concerned, has prevented any serious effort in this direction. Thus, the efforts in ion probing have been to determine the number combined with either energy analysis and/or mass analysis.

The dominant technique involving energy analysis is called ion scattering spectrometry (ISS) [38—40]. In this technique, it is assumed that the bombarding ion undergoes a simple binary collision with a substrate atom and loses energy according to simple kinetic energy exchange. On that basis, one

54

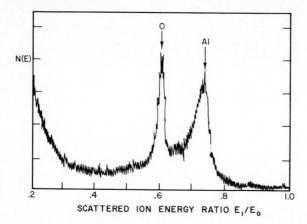

Fig. 9. Spectrum produced by energy analysis of scattered primary ions. In this case, the primary beam was 1500 eV He ions bombarding an aluminum oxide substrate. (Adapted from data provided by 3M Company.) This plot shows the form of results of the relatively new technique, ISS.

can calculate the energy of the scattered primary ion as a function of the scattering atom and the angle of scattering. For a fixed angle of scattering (often taken as $90°$), the ratio of the energy of the reflected primary to its initial energy is a simple function of the ratio of the mass of the scattering atom to the mass of the bombarding ion. The experiment therefore involves probing the surface with a primary beam in the energy range of several keV and energy analyzing the reflected primary beam. One then obtains a spectrum such as that shown in Fig. 9. It is most interesting to note that one obtains peaks very close to the value predicted on the basis of the simple binary collision assumption. If the primary beam were to penetrate the substrate under study, the possibility of undergoing a simple binary collision inside the bulk material and then scattering back out without losing any additional energy is very small indeed. Therefore, this technique is very surface sensitive and one would expect the signal obtained to be related almost completely to the surface monolayer with perhaps an extremely limited contribution from the second or third atom layers. The significant advantage of this technique is the simplicity of the overall system and the high degree of sensitivity to the surface monolayer. The difficulty is related to the basic problem of generating a controlled ion beam while maintaining the degree of vacuum around the sample generally required for many surface experiments. As one can see from the figure, the resolution of the system obtainable with reasonable signal-to-noise ratio is rather limited so that, at present, the technique is only suited to simple surface compositions. The resolution of the present day practical instrument does not enable one to ascertain the form of the substrate atom which does the scattering. That is, scattering oxygen atoms cannot be specified as to whether they are adsorbed

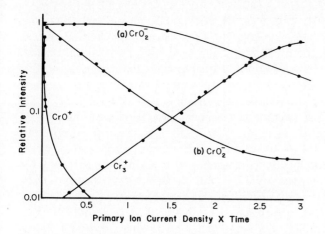

Fig. 10. Change of secondary ion intensities during sputtering of different oxide layers on chromium. Curve (b) corresponds to an oxygen dose of about 100 L while the other curves, including (a), correspond to an oxygen dose of 1200 L. (Adapted from ref. 45.) These curves show depth profile analysis, one of the exciting new forms of information obtained with surface analysis instruments.

oxygen, part of a surface oxide, part of an adsorbed CO, etc. As a final problem, we have the general situation that bombarding kilovolt ions cause considerable physical damage to the substrate. Thus, to do a careful surface analysis experiment, one should use the lowest density probe beam possible which obviously requires very sensitive detection capability. Although we have indicated a number of limitations, the technique does provide chemical identification of the surface components with a rather simple experimental arrangement.

In spite of the fact that ion probing can be fairly destructive to the surface, one can also utilize the process of sputtering to some advantage. Thus, one can attempt the surface analysis experiment utilizing the lowest possible probe beam density. After the spectrum is taken, one can increase the ion beam density and intentionally remove one or more atom layers by sputtering. One can then reduce the probe beam density to the analysis level and obtain a relative compositional spectrum of the new surface. Repeating this process will ultimately give rise to a depth profile such as is shown in Fig. 10. Calibrated experiments indicate that this approach yields surprisingly good depth spatial resolution. Thus, one has the advantage of surface composition information and depth profile analysis utilizing a single probe gun and a fairly simple analyzer.

If, instead of energy analyzing the reflected primaries, one utilizes mass analysis of the sputtered ions, one has the technique generally referred to as secondary ion mass spectrometry (SIMS) [41—45]. In this technique, the probe gun portion is the same as that used in ISS and one merely replaces the energy analyzer with a suitable mass spectrometer. In this experiment, one can determine the number and type of both positive and negative ions emitted from the surface. Some significant differences from ISS are apparent. First, the number of different kinds of emitted ions is obviously much greater. Secondly, the resolution of the analyzer is significantly greater. The large number of secondary ion peaks leads to a much more complex spectrum, but on the other hand, provides the possibility of determining more information about the detailed surface chemistry. This aspect of the experiment has been pursued in some detail and is suitably described in Chapters 5 and 6. As compared with ISS, this technique requires a somewhat more complex detection system but more information is obtained about the surface composition. As in the case of ISS, the technique is quite surface sensitive (see Chap. 5) since relatively few ions involved in the sputtering process inside the substrate can penetrate through the substrate material and arrive at the analyzer. Obviously, this system is also well suited to depth profile analysis. In both techniques, one might further expand the system by combining it with flash filament and/or EID. One might also extend the system to utilizing a focused ion beam which is scanned over the substrate providing surface spatial variation information. Of course, this added feature involves considerably increased complexity to the system.

One can see some similarities of available information from these techniques and AES. There are obviously advantages and disadvantages in each approach as indicated in the individual discussions. At the moment, AES is more widely used than ISS or SIMS but a significant part of the reason is the much greater familiarity of most researchers with electron beam generation and detection. As the technology of controlled ion beam generation improves, I expect that ion probing techniques will become more widespread. I believe this will be true especially for SIMS since the large number of secondary ions produced can possibly provide considerable molecular information. In addition, the availability of an ion probe means that depth profile and sputter cleaning are automatically available. One can also add that hydrogen detection is not possible with AES but is available to SIMS.

2. Neutrals out

Ion bombardment of surfaces will also lead to considerable desorption of neutral atoms and molecular fragments. One can obviously analyze these components in the same manner as is done with ions in ISS and SIMS for the same kinds of information. However, in this case, there is a considerable increase in the difficulty of the experiment in much the same manner as the difficulty encountered in analyzing the neutrals desorbed by electron impact. The signal-to-noise ratio available with ion desorption is already small. If one must now attempt to ionize a small fraction of the desorbing neutrals and then analyze those components, one will clearly have a very difficult problem indeed. Thus, surface analysis techniques involving ions in and neutrals out are receiving very limited attention. Most of the work in this area is related to studying the basic sputtering process and is not directly related to surface analysis in the context considered in this discussion.

3. Electrons out

Although ion probes cause electron emission from surfaces, this particular technique has not been pursued very much. Basically, this is because it is easier to produce secondary electrons by primary electron probes. However, one should mention the significant work done by Hagstrum in the technique he calls ion neutralization spectroscopy [46,47]. Here, the surface is probed with low energy ions. When the ion approaches the surface, it is neutralized and the neutralization energy can be transmitted to an electron at the surface giving it sufficient energy to be emitted. Analysis of the energy of these emitted electrons can provide information about the electronic nature of the surface under study. The experimental system is not simple and the technique has been exploited only in a limited number of cases. The early work of Hagstrum was quite significant but it now appears that many of the new electron in, electron out spectroscopies, as well as the photon in, electron out spectroscopies to be discussed later may well provide similar information. I therefore do not expect that this particular technique, that is, ions in, electrons out, will be pursued very actively in the future.

4. Photons out

The main effort in the development of the technique involving ions in and photons out is related to the detection and analysis of X-rays [48]. Of course, there has been considerable effort in the use of X-ray analysis in conjunction with devices such as the ion microprobe (IMXA). In this system, compositional information is obtained by X-ray analysis, rather than mass analysis, of secondary ions. If one utilizes X-ray analysis, then the depth of material contributing to the signal is much greater than the case of mass analysis of secondary ions. Therefore, the technique of ion microprobe X-ray analysis is not nearly as surface sensitive as other approaches. In addition, X-ray analysis capability is generally more complex and costly than secondary ion mass analysis. Furthermore, unless one has a considerable resolution, the information obtained is related to atomic components and does not provide molecular information as SIMS does. Thus, the IMXA system is quite complex, costly and somewhat limited in providing information about surfaces; this approach has very limited appeal.

Some recent efforts have been directed towards using proton-induced X-ray analysis (PIX) [49,50]. This particular approach has several advantages over the general ion probe X-ray system. First, protons produce the least amount of damage to the material under study compared to other ion bom-

58

bardment particles. Secondly, proton scattering and inelastic collision cross-sections are probably known better than most reactions due to the considerable effort in the nuclear physics area. Thus, one has the possibility of utilizing known quantitative data to help greatly the analysis of obtained results. The major disadvantage of this technique is that the proton penetration is considerable and the X-rays are obtained from a rather significant volume of the material. Thus, the technique is quite useful for bulk analysis but has somewhat limited value for providing specific surface information. Since there is considerable capability in some areas of physics research to generate and control proton beams, I expect that there will be some effort towards further developments of techniques such as PIX but I suspect that it will be related more to bulk rather than surface analysis. Efforts at utilizing low energy proton beams, which would undergo minimum penetration and therefore be more surface sensitive, would undoubtedly improve the value of this technique for the general surface studies area. One must consider that the generation of proton beams and X-ray analysis dictate a rather extensive, fairly costly system. It is most probable that those who would pursue this approach would have proton beam capability already available in their laboratories.

D. PHOTONS IN

The possible use of photons as the probing beam for surface studies is certainly intriguing. Two major characteristics of photons make them well suited for this kind of experiment. First, they have negligible momentum compared with the components being studied and therefore produce minimum disturbance of the surface. Secondly, they are neutral and therefore greatly reduce the problems of maintaining optimum electric and/or magnetic fields in the vicinity of the target. Their use would also greatly reduce the concerns of charging of the substrate under study. Obviously, there must

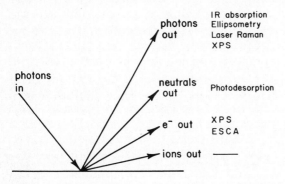

Fig. 11. Sketch summarizing the general group of techniques involving photon probes in.

be some limitations or photon probing would be the dominant technique in surface analysis. The difficulties arise because of the problem of producing sufficiently intense photon beams in the wavelength range of interest and the cross-section for reactions is generally small, therefore producing minimal signals when one attempts to probe only the surface monolayer or two. However, recent developments in photon production (lasers, etc.) and sensitive detection techniques have led to increased efforts to developing photon probing methods for surface analysis. Some of the approaches being pursued are indicated in Fig. 11. The problems are quite different for different wavelength regions and will be considered in the appropriate subsections.

1. Photons out

As with all the probing beam techniques, major developments occur in detecting the same kind of particle coming out as was used in the probe beam itself. Thus, considerable effort is related to probing the surface with photons and analyzing the reflected photon signal. Experiments using photon probes can generally be divided into the four wavelength regions embraced by infrared, visible, ultraviolet and X-rays. Photons in each of these ranges provide information on quite different aspects of the surface. Infrared photons can interact with the surface monolayer as they are able to excite vibrational states. The technique simply involves irradiating the surface with photons of varying wavelength, detecting the reflected signal and looking for absorption bands. The problems in developing this technique are described in the literature and have recently been solved rather well [51—55] so that a number of programs utilizing this approach are in progress. Information about the vibrational states enables one to perhaps obtain definite information about the molecular structure of the adsorbed monolayer.

The technique of using infrared absorption for obtaining information about molecular structure is a well established one and is in rather wide use. The difficulty in applying it to studying controlled surface monolayers results from the minimum signals obtained from the low cross-section of the small number of atoms available in a monolayer. One further problem with this technique is that it can work only with adsorbed components which are infrared active. Recent progress in this area has now begun to provide data using this technique for certain adsorbed components on relatively small, well prepared samples. I would expect that there will be further development in this area since it is one of the very few that can provide specific information about the detailed molecular structure of adsorbed components. Infrared monochromators are straightforward but moderately expensive. Thus, the overall complexity and cost of this kind of system will undoubtedly prevent its widespread utilization.

In the region of visible light, there are basically two techniques that are used in surface analysis research. These are Raman scattering and ellipsometry. In ellipsometry, the technique is simply to shine well controlled light onto the surface and analyze the reflected light in terms of changes in polarization, phase, etc. One then attempts to learn something about the surface from its apparent index of refraction which has affected the reflected light. The basic difficulty with this technique is that one or two monolayers of material will have an exceedingly small effect and therefore cause minute changes in phase and polarization directions in the light beam. There has been some effort to utilize this approach to obtain information about surfaces [56,57], but it is obviously quite difficult to obtain useful data and it seems somewhat unlikely that there will be much extended effort in this direction.

Raman scattering is a well-known technique in which light interacts with matter involving various possible transitions. The scattered photon will have lost or gained a quantum of energy related to the elementary transitions involved. Analysis of the change in frequency of the emitted photons then enables one to determine the magnitude of the elementary transitions and therefore gain information about the material being studied. It is obviously a very powerful tool which did not see very rapid development because of the lack of the required high density monochromatic light source. With the advent of the laser, considerable interest developed in this technique and it is now becoming a rather widely used research tool [58—62]. Although the basic instrumentation is becoming reasonably straightforward, it is still moderately expensive and this has prevented an extensive proliferation. Most early work using laser Raman scattering has been related to bulk-type analysis. Efforts are now being extended to use this technique for surface analysis. Again, the problem is related to the very small number of scattering atoms in a monolayer so that the signal-to-noise ratio will be extremely small. If the technique can be adapted to surface monolayer analysis, it can have a considerable impact on the ability to learn a great deal about surfaces. Efforts are now under way by a number of researchers to adapt this approach to surface studies and I expect there will be at least some success in the future. The problem of signal-to-noise appears rather formidable and it is somewhat difficult to say at present whether the efforts will yield only minimal results or if the technique can possibly develop very significantly.

In the region of ultraviolet (UV) radiation there are a considerable number of experiments involving the measurement of scattered UV photons relating primarily to absorption bands. The problem here is that UV will generally penetrate considerably into the material being studied and the information is therefore much more related to bulk properties rather than to the surface. Experiments related to the near surface region can be carried out but at the moment, the probability of using ultraviolet photons in, photons out to study the surface monolayer seems rather unlikely. In addition, although infrared and visible photon beams are easily generated and controlled, ultraviolet photon beams are orders of magnitude more difficult to generate and control. This particular problem will be discussed further in the next several sections.

2. Electrons out

The process of photons in, electrons out is obviously the well-known photoelectric effect. This process has been studied for many decades and has involved the use of photons primarily in the high frequency end of the visible range and the ultraviolet range. The technique is quite straightforward since one determines the number and energy of emitted electrons as a function of the wavelength of the incident radiation. In recent years, with the advent of sophisticated high resolution electron energy analyzers, this basic technique has been more widely used. To obtain still more information about the material under study, and specifically to obtain chemical identification, the use of higher energy photons, in the X-ray range, has been incorporated. The technique then is known by the general name of X-ray photoelectron spectroscopy (XPS) [63—65]. This technique is used to determine the electronic structure of solid surfaces as well as chemically indentify surface components. When used in this latter manner, it is generally known by the name electron spectroscopy for chemical analysis (ESCA) [66—70]. In this form, the technique is quite similar to high resolution AES except that X-ray photons are used as the probe instead of electrons. The availability of high resolution electron energy analyzers has led to quite a resurgence in studying the basic photoelectric effect since considerable information about the surface can now be obtained. The use of photon probes instead of electrons has the advantages and disadvantages discussed in the beginning of this section. It seems clear from recent reports that considerable effort will go into photons in, electrons out techniques for surface analysis in the coming years.

3. Neutrals out

Considerable effort has been expended in attempting to determine whether true photodesorption occurs, that is, whether an incoming photon can cause the desorption of an adsorbed atom or molecule by a quantum process [71—75]. It appears that, as of today, true photodesorption has been verified [76]. It seems rather clear that the process occurs only in the wavelength range generally below 3000 Å and, therefore, one must only consider using ultraviolet photon probe beams. The long time required to make useful measurements in this region is related to the great difficulty in generating, controlling and using radiation in the wavelength range below 3000 Å and even more preferably below 2000 Å. The great difficulty in working with radiation in the extreme UV range has prevented extensive research in this general area. Some increased activity has developed in recent years due to the evolvement of electron storage rings as synchrotron radiation sources. These facilities provide excellent probe beams over the entire extreme UV region, but since only a few such facilities exist in the world, there is limited access for scientists wishing to pursue these experiments. A second problem seems to be that most experiments were attempted on pure metal surfaces. It now appears that photodesorption occurs basically on semiconductor surfaces and with extremely small or zero cross-section values on metals [76]. This new insight should now enable workers to pursue the photodesorption process more thoroughly. Knowledge about the photodesorption process has both theoretical as well as practical value. However, the considerable difficulty in generating the photon beam, the limited number of substrates useful in this process, and the somewhat restricted amount of information one might gain about the surface will probably result in a rather limited overall effort in this approach. The practical needs for information about this process will probably motivate at least some continued research effort to obtain proper quantitative results. The technique is clearly just entering its quantitative stage and perhaps some unexpected interesting results could motivate considerably more interest in pursuing this approach.

4. Ions out

At the moment, there appears to be no physical mechanism which could lead to the process of photons in, ions out in any of the frequency ranges from the infrared to X-ray regions. It was at first thought that photons would interact with adsorbed molecules in a manner similar to that of probing electrons. Since probing electrons produce desorbing ions, it was expected that one might observe the same effect with comparable energy photons. This has been attempted but desorbing ions have never been seen. Since we now understand that photodesorption occurs by absorption of the photon in the substrate and subsequent desorption by a secondary process, one no longer expects to see photo-induced ion desorption. Thus, I do not expect this particular combination to develop.

E. NEUTRALS IN

The last of the particle beams to be considered is neutral atoms or molecules. Techniques involving neutral probing beams have had the least amount of widespread use of the various probe beam approaches. This is because of the relatively great difficulty in producing and controlling neutral particle beams and of the limited amount of information available from the specific experiment. This should not imply that neutral beam probe experiments are not useful and have not contributed significant information, but the difficulty of the experiment and limited data obtained has limited the work to a relatively small number of surface scientists. Clearly, this being the only type of particle beam probe which does not lend itself to energy control is a major restriction. Therefore, it is virtually impossible to do experiments as a function of bombarding particle energy, a fact which severely limits the technique.

1. Neutrals out

Again, the dominant approach involves the analysis of similar particles out as is used for the probe beam. Thus, efforts in this area are related to the technique neutrals in, neutrals out. One basically analyzes the scattered neutral particle beam by determining, primarily, the number and spatial distribution essentially as a function of the scattering surface [77—80].

Certain specific experiments can be performed using this technique, although it is obviously much more difficult to produce and detect neutral particle beams as compared with, for example, electrons in, ions out. On the other hand, the neutral particles are generally of very low energy (i.e. thermal) and therefore cause minimal damage, if any, to the surface being studied. One area of effort has been to use this technique to study surface chemical reactions by making the probe beam one of the chemical components of the reaction to be studied.

I expect that the considerable difficulty in operation of this technique and the limited area of application will restrict its use for surface analysis work. Clearly, there are specific experiments for which this technique can be used to provide useful information about surfaces and surface reactions. I would therefore expect a continuing effort on the part of a limited number of surface scientists utilizing this mode of surface analysis.

2. Ions out

Since the molecular beams generally used involve energies in the thermal range, insufficient energy is available for the production of desorbing ions. In principle, one could develop this technique by utilizing high energy neutral probes. This approach would be quite complex and it is difficult to say what advantage it might have over using comparable energy ion probes. Thus, the probability of development of this approach is close to zero.

3. Electrons out

The same comments apply here as in the case of ions out.

4. Photons out

It does seem at least physically possible to use moderate energy neutral beam probes with the emission of photons due to a suitable inelastic collision at the surface. Again, the experimental difficulties would be quite severe and it seems quite unlikely that any significant effort would be expended in this direction. Of course, if one utilizes quite high energy probe particles, considerable desorbing photons may be produced but again, it is difficult to see the advantage over using probing ion beams; the generation of high energy neutral beams is certainly more complex.

III. Electric and magnetic fields in

We now leave the area of particle beam probing and consider a number of other techniques which are generally quite specific in experimental detail. In the case of electric and magnetic fields in, we are considering the application of essentially d.c. fields or, at most, slowly varying fields as opposed to electromagnetic radiation. Since the techniques involved here are quite varied, it will be easier to consider the details of the methods as each approach is considered.

A. ELECTRIC AND MAGNETIC FIELDS OUT

I will continue to use the same format for discussion although it will be clear that these techniques are different in many ways from the particle in, particle out approach. First, in terms of applied magnetic fields, the simplest experiment in principle is to measure magnetic saturation generally as a function of adsorbed species. That is, one might consider the situation of a very thin film of a magnetic material such as nickel. If one then adsorbs a gas such as hydrogen, it may interact with the electrons of the surface nickel atoms thereby changing the overall magnetization and one may be able to detect this effect. This approach has been attempted and is described in the literature [81,82]. The technique involving the application of a magnetic field and the measurement of changes in substrate magnetization is certainly relatively simple and straightforward. The difficulties in this technique are that one must use magnetic substrate material and the substrate must be extremely thin so that changes in the surface monolayer will have a measurable effect on the overall magnetization. For any realistic substrate film, the changes will be very small and therefore signal-to-noise problems will be significant. These problems have restricted the application of this technique to a limited number of experiments. For those specialized substrate—gas interfaces that are amenable to this approach and especially for those scientists who are concerned with the basic mechanisms of magnetism, the technique will undoubtedly continue to attract an occasional enthusiast. This will undoubtedly occur since, although the number of experiments that are applicable is limited, the considerable simplicity of the technique is quite attractive.

A more widespread range of experiments is available when one applies an electric field in the region of the surface under study. In this case, we are considering the surface under study as one side of a diode structure. For example, one can have a capacitor arrangement and measure changes in capacitance as one adsorbs or desorbs gaseous species from the surface. Again, one can see that the experiment is extremely simple in principle. However, it is clear that to obtain measurable changes in capacitance due to an adsorbed monolayer, one must prepare a very special capacitor and be able to measure extremely small changes. One further restriction is the limited amount of information that can be obtained using this technique. A more useful modification of this kind of measurement involves measurements of changes in work function by contact potential differences. This approach again basically involves a diode structure in which one generally determines changes in contact potential difference as specific gases are adsorbed or desorbed from the surface of interest. As one can visualize, the experimental facilities are relatively simple and the technique is certainly applicable to a fairly wide variety of surface materials and adsorbed gas components. Thus, a fair amount of effort has been expended in the development and application of this technique [83—87]. Perhaps the major limitation which prevented the widespread utilization of this approach, is that one merely obtains an averaged change in contact potential difference. The difficulty in interpreting the change observed as it relates to the specific adsorbate—surface interaction is quite significant. Since, in many cases, gases adsorb into several different states, each of which may produce different and possibly electrically opposite changes, the net observed change can often be interpreted in several ways. However, since the experimental requirements are relatively simple, I would expect there will be continued effort to utilize this approach. It may well be that its use will be expanded mostly as an auxiliary technique to be used with other surface analysis methods in much the same way as flash filament is used by many scientists in conjunction with other techniques.

In the general area of applied fields, one should also mention the techniques of NMR and EPR. In this case, one basically applies a d.c. magnetic field and analyzes for absorption of a transmitted rf signal. The absorbed signal frequency would then correspond to the frequency of magnetic resonance induced by the applied magnetic field on the electrons or nucleons of the atom being studied. To obtain a reasonable signal-to-noise ratio, an interaction with a considerable number of atoms is required so the techniques have been used primarily with bulk analysis. In recent years, there has been some effort to apply both EPR [88—92] and NMR [93—95] to surface analysis. In addition to a fairly complex and costly experimental system (primarily the magnet and rf equipment required), the problem of obtaining sufficient signal-to-noise for an adsorbed monolayer is significant. Since there has been some success in applying these techniques to surface studies, one must assume that there will continue to be some effort expended in this direction. The reason for the willingness to pursue a rather costly complex system is that the results obtained can be used to identify uniquely the electronic configuration of the adsorbed species which is of considerable value. As can be seen from the literature, even the sample preparation is a formidable task. I would expect that a small number of researchers will continue to pursue this technique since in certain specific systems, which are amenable to this type of analysis, the information available is quite useful.

B. ELECTRONS OUT

Perhaps one of the most precise surface analysis approaches is related to the techniques known as field emission microscopy (FEM) and field ion microscopy (FIM). The basic mechanism in these techniques involves the application of sufficiently high electric fields to a surface such that electrons can tunnel through the work function barrier at the surface. Since the tunneling probability is dependent on the exact state of the surface where the electron passes, the number of tunneling electrons will vary from point to point as the surface details vary so that one can obtain considerable information about spatial variations with resolutions that approach the dimensions of the individual surface atoms. In addition to spatial variations which provide "pictures" of the surface under study, measurement of the absolute number of tunneling electrons can provide quantitative information on the detailed potential configuration at the surface. The technique is well established and considerable information exists in the literature on both FEM [96—101] and FIM [102—107] (and Chaps. 8 and 9).

In FEM, electric fields are applied causing electron tunneling out of the surface. The number of these electrons is measured and/or they are accelerated to a fluorescent screen to provide a picture of the surface variations in work function. In FIM, the field is reversed and one causes tunneling of electrons from a gas atom into the surface, thereby producing a positive ion. This positive ion is then accelerated to a fluorescent screen to again provide pictures of variations in surface potential. The principle problem with these techniques is that one requires electric fields of millions of volts per centimeter to produce detectable results. It is clearly impractical to attempt this type of experiment with a conventional flat surface. Thus, to obtain these fields, one must use extremely fine points; that is, samples which have been tapered to point radii in the neighborhood of several hundreds of angstrom units. Thus, the techniques are restricted to those materials which can be fabricated in this form and which can withstand electric fields of the magnitude required without being torn apart. For these reasons, most of the experiments performed to date have been done with the refractory metals. The requirement to produce the special sample form does provide, however, the most controlled clean surface available for study. Thus in experiments utilizing FEM and FIM, one has the maximum information about the detailed atomic structure of the surface being studied. With fields applied, one is also able to maintain the cleanest possible surface (but see Chap. 8). At its full level of capability, FIM enables one to observe positions of individual atoms.

In one modification of this technique called the atom probe [108,109], one is able to image the surface, choose a particular atom of interest, apply a pulsed excess electric field sufficient to remove the atom and then, by time-of-flight mass spectrometry, identify the mass of the atom in question. As research has continued to reveal more of the basic mechanisms of these techniques, the ability to utilize other types of material as sample points has been considerably extended (Chap. 9). Careful experiments have been able to provide not only detailed information about the substrate surface but also to monitor adsorption of gases, surface migration, etc.

Although these techniques have obvious considerable value in surface analysis, they have not experienced very wide application. The reasons for this limited use are manifold. For one, the types of substrate that can be studied *is* limited. Secondly, one must be concerned with extrapolation of information from very special, extremely fine point samples to the possible behavior of practical, large, relatively flat structures. Further, the tips used are delicate and experiments must be performed in carefully controlled UHV environments so that samples being studied are not rapidly destroyed. This means that although the systems are fairly simple, one must use the best UHV methods. In the case of FIM, the ability to obtain optimum results requires the use of liquid helium or liquid hydrogen temperatures in addition to the best ultrahigh vacuum techniques. Thus, the systems do not require complex or expensive components but must be put together and utilized with considerable care. The need to use liquid helium is also not trivial and requires the basic capabilities for handling that material. The items discussed above are responsible for the limited use of these techniques. However, the ability to produce the cleanest most controlled surface for study and the ability to "see" details to a resolution in the order of angstrom units, makes these surface analysis approaches unique. There will undoubtedly be a continuing effort to extend the use of FIM and FEM to more and more systems and the results obtained will certainly add useful information to the overall area of atomistic surface science.

In the area of fields in, electrons out, one should perhaps include a relatively new technique that most probably belongs in this category. The technique is called inelastic electron tunneling spectroscopy (IETS) [110—112]. It involves using a metal—oxide—metal tunneling junction sandwich structure. In a typical sample, one might use aluminum metal, aluminum oxide and lead as the sandwich. If an adsorbate is present on the aluminum oxide surface and one applies a bias potential across the two metal electrodes, electrons tunneling through the sandwich can undergo inelastic collisions with the adsorbate and raise it to allowed excited vibrational states. If one monitors the current versus voltage, one can detect steps where these inelastic collisions occur and this information can enable one to determine the molecular structure of the adsorbate in a manner similar to that used in infrared absorption spectroscopy. It is a most interesting new technique which obviously has considerable potential. It does not lend itself to immediate use by all workers since one must learn to fabricate appropriate junction sandwiches with oxide layers of the order of 25 Å thick. One must be able to introduce the adsorbate of interest onto the oxide layer and cover it with the appropriate metal film. In addition, the measurements must be made with one metal in the superconducting state so that liquid helium temperatures must be available. These experimental requirements indicate that the technique will not become widely used but will certainly be adapted by a number of people for the study of certain systems.

In general, it seems clear that the techniques of fields in usually involve study of limited specialized systems but do provide considerable information about those systems.

In this general category of surface analysis techniques, one can also mention the measurement of the change in electrical conductivity of a thin film as gases are adsorbed or desorbed. The premise here is that adsorbed gases will interact with conduction electrons at the surface, thus changing the number of available carriers and producing a change in measured resistivity or conductivity of the film. To have a significant effect, it is obvious that the film must be extremely thin, i.e. certainly no more than several hundred angstrom units thick and preferably less. One can see that this technique is obviously quite simple and involves only one major difficulty; that is, producing a controlled, reasonably well understood film that approaches thicknesses preferably below 100—200 Å. This difficulty is by no means a minor matter and has resulted in rather limited exploitation of this approach, even though it is one of the simplest in terms of components, cost, etc. It has many of the same basic problems related to the attempt to measure magnitization saturation changes due to adsorbed monolayers as described previously. It is again a technique which lends itself more readily to certain combinations and has been attempted with some success on a number of systems [113,114]. Should the ability to produce well-controlled, uniform, very thin films greatly improve, the further application of this approach may well increase. The other problem involved in this kind of measurement is that of making good electrical contact with the film so that conductivity measurements are reliable, since the changes produced are generally very small. Considerable work in the field of microelectronics is producing a good deal of progress and this part of the problem will undoubtedly be solved with time.

IV. Surface waves in

Very recently, a whole new approach to surface studies has been developed utilizing controlled surface acoustic waves to cause desorption. It has long been known in the high vacuum field that accidental striking of the walls of a chamber generally results in observed pressure bursts in the system. Now, specific controlled experiments are being carried out in which acoustic waves are generated and transmitted along the surface of the sample [115]. The first experiments used piezoelectric materials in which electrical oscillations are converted to surface Raleigh waves. It is found that this process causes desorption of adsorbed components.

A. NEUTRALS OUT

To date, this technique has led to the observation of neutral component desorption from the surface. The first experiments were performed on single crystal quartz and the preliminary results have been quite encouraging. The basic experimental system is relatively simple with the major difficulty involving the ability to fabricate the appropriate interdigital transducer elements on the substrate to be studied. A considerable advantage to this technique is that it is applicable to many insulating materials which have not been studied very well since most other techniques do not lend themselves readily to the study of insulators. Since one can also put thin films of metals or semiconductors onto the substrate and still transmit the surface acoustic wave, measurements are possible on all kinds of surfaces. Desorption observations can be made without bulk heating or bombardment of the surface with external particles. It appears at present that the major advantage and potential to this approach lies in its application to many kinds of surfaces which have had little or no previous study (most insulating materials and many semiconductors and metals which have been difficult to study by other techniques). Since the technique is new, it is difficult to foresee all of its possible applications but one would certainly expect some continuous level of effort utilizing this approach. The reasonable difficulty in fabricating the transducer will undoubtedly prevent its rapid widespread utilization. It should also be mentioned that proper use of the technique involves the application of rather low power levels in the surface acoustic wave to prevent other factors from becoming dominant, such as bulk thermal heating, severe mechanical disruption of the sample, etc. This limitation leads to the desorption of very small quantities of gas, therefore requiring ultrahigh vacuum capabilities so that small pressure differentials can be observed. To date, only neutrals have been observed desorbing and at the moment, I do not foresee using this technique to study desorption of the other particle beams.

V. Conclusions

I have attempted to survey and compare a variety of surface analysis techniques which hopefully represent many of those in present use, but certainly not all that have ever or will ever be used. There has been minimal discussion of experimental details since this information is available in the references or the following chapters. The principal effort has been to compare advantages, disadvantages and areas of application among the various techniques, admittedly from a very specific point of view. I would certainly expect that other surface scientists would be more enthusiastic about some techniques which I perhaps did not discuss very thoroughly. It has been my observation that there are almost as many attitudes about surface analysis research as there are workers in the field. Thus, this discussion is one man's view. Techniques which have not been discussed at all or perhaps in a very limited manner, do not necessarily indicate a negative implication but represent the fact that it is generally unreasonable in finite time and space to be 100% complete. My general feeling, which is strongly supported by the tremendous mushrooming in numbers of techniques and numbers of people in surface analysis, is that this field is just beginning to open up in much the same manner as solid state physics did some two to three decades ago. I would expect, then, that this field will continue to grow in breadth and depth and will contribute to at least as many, and possibly more areas, as has solid state physics. A most encouraging sign is that good young (and old) theoreticians are beginning to enter the field. The interaction between the experimentalist and the theoretician will now, I believe, greatly speed up the development of our understanding of surface phenomena.

References

1 H.F. Dylla, Comparison of Techniques for Surface Measurements, M.I.T. Res. Lab. Electron. Quart. Progr. Rep. 110, July 15, 1973.
2 P.A. Redhead, Vacuum, 12 (1962) 203.
3 A. Schramm, Vide, 103 (1963) 55.
4 G. Ehrlich, Advan. Catal. Relat. Subj., 14 (1963) 271.
5 J.P. Contour and R. Prud'homme, Bull. Soc. Chim. Fr., 8 (1969) 2693.
6 L.A. Pétermann, in S.G. Davison (Ed.), Progress in Surface Science, Vol. 3, Part 1, Pergamon, Oxford, 1972.
7 D. Lichtman, J. Vac. Sci. Technol., 2 (1965) 91.
8 P.J. Estrup and E.G. McRae, Surface Sci., 25 (1971) 1.
9 G.A. Somorjai and H.H. Farrell, Advan. Chem. Phys. 20 (1971) 215.
10 E.N. Sickafus and H.P. Bonzel, Progr. Surface Membrane Sci., 4 (1971) 115.
11 J.W. May, Advan. Catal., 21 (1970) 151.
12 W.D. Robertson, J. Vac. Sci. Technol., 8 (1971) 403.
13 R.D. Heidenreich, Fundamentals of Transmission Electron Microscopy, Wiley, New York, 1964.

14 P.B. Hirsch, A. Howie, R.B. Nicholson, D.W. Pashley and M.J. Whelan, Electron Microscopy of Thin Crystals, Butterworth, Washington, 1965.
15 E. Bauer, Techniques of Metals Research , Vol. II, Part 2, Wiley, New York, 1969, Chap. 15.
16 P.E. Hojlund Nielsen, Surface Sci., 35 (1973) 194.
17 E.J. Sheibner and L.N. Tharp, Surface Sci., 8 (1967) 247.
18 L.A. Harris, J. Appl. Phys., 39 (1968) 1419.
19 C.C. Chang, Surface Sci., 25 (1971) 53.
20 N.J. Taylor, Bull. Acad. Phys. Soc., 16 (1971) 1353.
21 P.W. Palmberg, J. Vac. Sci. Technol., 7 (1970) 76.
22 R.E. Weber, Solid State Technol., 13 (1970) 49.
23 G.L. Connell and Y.P. Gupta, Mater. Res. Stand., 11 (1971) 8.
24 J.J. Melles, L.E. Davis and L.L. Levenson, J. Vac. Sci. Technol., 10 (1973) 140.
25 W.M. Mularie and W.T. Peria, Surface Sci., 26 (1971) 125.
26 H.E. Bishop and J.C. Riviere, Appl. Phys. Lett., 16 (1970) 21.
27 D. Lichtman and R.B. McQuistan, Progr. Nucl. Energy Ser. 9, 4 (2) (1965) 95.
28 D. Menzel, Angew. Chem. Int. Ed. Engl., 9 (1970) 255.
29 T.E. Madey and J.T. Yates, Jr., J. Vac. Sci. Technol., 8 (1971) 525.
30 J.H. Leck and B.P. Stimpson, J. Vac. Sci. Technol., 9 (1972) 293.
31 P.A. Redhead, Appl. Phys. Lett., 4 (1964) 166.
32 D.R. Sandstrom, J.H. Leck and E.E. Donaldson, J. Chem. Phys., 48 (1968) 5683.
33 G.D. Rork and R.E. Consoliver, Surface Sci., 10 (1968) 291.
34 D. Lichtman, R.B. McQuistan and T.R. Kirst, Surface Sci., 5 (1966) 120.
35 J.C. Weaver and J.G. King, Proc. Nat. Acad. Sci. U.S., 70 (1973) 2781.
36 R. Castaing, Advan. Electron. Electron Phys., 13 (1960) 317.
37 R.L. Park and J.E. Houston, J. Vac. Sci. Technol., 10 (1973) 176.
38 W.H. Strehlow and D.P. Smith, Appl. Phys. Lett., 13 (1968) 34.
39 D.P. Smith, Surface Sci., 25 (1971) 171.
40 W. Heiland and E. Taglauer, J. Vac. Sci. Technol., 9 (1972) 620.
41 R.E. Honig, J. Appl. Phys., 29 (1958) 549.
42 R. Castaing, R. Jouffrey and G. Slodzian, Compt. Rend., 251 (1960) 1010.
43 R.V. Rybalko, V.Ya. Kolot and Ya.M. Fogel, Sov. Phys. Solid State, 10 (1969) 2518.
44 A. Benninghoven, Surface Sci., 28 (1971) 541.
45 A. Benninghoven, Surface Sci., 35 (1973) 427.
46 H.D. Hagstrum, Phys. Rev., 150 (1966) 495.
47 H.D. Hagstrum and G.E. Becker, Phys. Rev. B., 4 (1971) 4187.
48 C.W. White, D.L. Simms and N.H. Talk, Science, 177 (1972) 481.
49 R.G. Musket and W. Bauer, Appl. Phys. Lett., 20 (1972) 411.
50 W.F. Van DerWeg, W.H. Kool and H.E. Roosendaal, Surface Sci., 35 (1973) 413.
51 A.M. Bradshaw and O. Vierle, Ber. Bunsenges. Phys. Chem., 74 (1970) 630.
52 H.G. Tompkins and R.E. Greenler, Surface Sci., 28 (1971) 194.
53 M.A. Chesters and J. Pritchard, Surface Sci., 28 (1971) 460.
54 J.T. Yates, Jr. and D.A. King, Surface Sci., 30 (1972) 601.
55 J.T. Yates, Jr., R.G. Greenler, I. Ratajczykowa and D.A. King, Surface Sci., 36 (1973) 739.
56 R.R. Stromberg and F.L. McCrashing, in G. Goldfinger (Ed.), Clean Surfaces: Their Preparation and Characterization for Interfacial Studies, Dekker, New York, 1970.
57 F. Abeles, J. Vac. Sci. Technol., 9 (1972) 169.
58 E.M. Flanigen, H. Khatami and H.A. Szymanski, Advan. Chem. Ser., 101 (1971) 201.
59 C.L. Angell, J. Phys. Chem., 77 (1973) 222.
60 P.J. Hendra, J.R. Horder and E.J. Loader, J. Chem. Soc. A, (1971) 1766.
61 R.O. Kagel, J. Phys. Chem., 74 (1970) 4518.

62 E.J. Loader, J. Catal., 22 (1971) 41.
63 See ref. 37.
64 K. Seigbahn, D. Hammond, H. Fellner-Feldegg and E.F. Barnett, Science, 176 (1972) 245.
65 C.S. Fadley and D.A. Shirley, J. Res. Nat. Bur. Stand. U.S., A, 74 (1970) 543.
66 D.L. Ames, J.P. Maier, F. Watt and D.W. Turner, Discuss. Faraday Soc., 53 (1972) 277.
67 C.R. Brundle and M.W. Roberts, Surface Sci., 38 (1973) 234.
68 W.N. Delgass, T.R. Hughes and C.S. Fadley, Catal. Rev., 4 (1970) 179.
69 S. Hagstrom, Vak. Tech., 22 (1973) 183.
70 C.J. Todd, Vacuum, 23 (1973) 195.
71 J. Peavey and D. Lichtman, Surface Sci., 27 (1971) 649.
72 R.O. Adams and E.E. Donaldson, J. Chem. Phys., 42 (1965) 770.
73 P. Genequand, Surface Sci., 25 (1971) 643.
74 H. Galron, Vacuum, 22 (1972) 229.
75 J.W. McAllister and J.M. White, J. Chem. Phys., 58 (1972) 1496.
76 G.W. Fabel, S.M. Cox and D. Lichtman, Surface Sci., 40 (1973) 571.
77 J.N. Smith, Jr. and H. Saltsburg, H., H. Saltsburg et al. (Eds.), Fundamentals of Gas—Surface Interactions, Academic Press, New York, 1967, pp. 370—391.
78 J.N. Smith, Jr. and H. Saltsburg, J. Chem. Phys., 40 (1964) 3585.
79 W.H. Weinberg and R.P. Merrill, J. Chem. Phys., 56 (1972) 2881.
80 J.P. Toennies, Vak. Tech., 22 (1973) 185.
81 I.E. Den Beston, P.G. Fox and P.W. Selwood, J. Phys. Chem., 66 (1962) 450.
82 H.J. Bauer, D. Blechschmidt and M. Von Hellermann, Surface Sci., 30 (1972) 701.
83 G.J.H. Dorgelo and W.M.H. Sachtler, Naturwissenschaften, 46 (1959) 576.
84 C.L. Balestra and H.C. Gatos, Surface Sci., 28 (1971) 563.
85 J.E. Boggio, J. Chem. Phys., 57 (1972) 4738.
86 T.A. Delchar, Surface Sci., 27 (1971) 11.
87 D.L. Fehrs and R.E. Stickney, Surface Sci., 17 (1969) 298.
88 R.J. Kokes, in R.B. Anderson (Ed.), Experimental Methods in Catalytic Research, Academic Press, New York, 1968.
89 R.J. Kokes, J. Phys. Chem., 66 (1962) 99.
90 D.J. Miller and D. Haneman, Phys. Rev. B, 3 (9) (1971) 2918.
91 R.B. Clarkson and A. Cirillo, J. Vac. Sci. Technol., 9 (1972) 1073.
92 J. Lunsford, Catal. Rev., 8 (1973) 135.
93 D.E. O'Reilly, H.P. Leftin and W.K. Hall, J. Chem. Phys., 29 (1958) 970.
94 J.J. Fripiat, C. Van der Meersche, R. Touillaux and A. Jelli, J. Phys. Chem., 74 (1970) 382.
95 D. Freude, D. Müller and H. Schmiedel, Surface Sci., 25 (1971) 289.
96 E.W. Müller, Ergeb. Exakt. Naturw., 27 (1953) 290.
97 R.H. Good, Jr. and E.W. Müller, Handbuch Physik, Vol. 21, Springer-Verlag, Berlin, Heidelberg, New York, 1956, p. 176.
98 R. Gomer, Field Emission and Field Ionization, Harvard Univ. Press, Cambridge, Mass., 1961.
99 R. Vanselow, Appl. Phys., 2 (1973) 229.
100 B.E. Nieuwenhuys, D.Th. Meijer and W.M. Sachtler, Surface Sci., 40 (1973) 125.
101 R. Vanselow and R.D. Helwig, Appl. Phys., 1 (1973) 223.
102 E.W. Müller, Advan. Electron. Electron Phys., 13 (1960) 89.
103 J.J. Hren and S. Ranganathan (Eds.), Field Ion Microscopy, Plenum Press, New York, 1968.
104 E.W. Müller and T.T. Tsong, Field Ion Microscopy, Elsevier, New York, 1969.
105 G. Ayrault and G. Ehrlich, J. Chem. Phys., 59 (1973) 3417.

106 C.C. Schubert, C.L. Page and B. Ralph, Electrochim. Acta 18 (1973) 33.

107 W.R. Graham, F. Hutchinson and D.A. Reed, J. Appl. Phys., 44 (1973) 5155.

108 E.W. Müller, Naturwissenschaften, 57 (1970) 222.

109 S.S. Brenner and J.T. McKinney, Rev. Sci. Instrum., 43 (1972) 1264.

110 J. Lambe and R.C. Jaklevic, Phys. Rev., 165 (1968) 821.

111 B.F. Lewis, M. Mosesman and W.H. Weinberg, Surface Sci., 41 (1974) 142.

112 B.F. Lewis, W.M. Bowser, J.L. Horn, Jr., T. Luu and W.H. Weinberg, J. Vac. Sci. Technol., 11 (1974) 262.

113 K. Kawasaki, T. Sugita and S. Ebisawa, J. Chem. Phys., 44 (1966) 2313.

114 P. Wissmann, Thin Solid Films, 13 (1972) 189.

115 C. Krischer and D. Lichtman, Phys. Lett. A, 44 (1973) 99.

Chapter 3

LOW-ENERGY ION SCATTERING SPECTROMETRY

T.M. BUCK

I. Introduction

A. GENERAL REMARKS

The basic measurements of low-energy ion scattering for surface analysis are quite simple. As depicted in Fig. 1, a monoenergetic, well-collimated beam of ions with energy of $\sim 0.1-10$ keV strikes the target surface, and the energy distribution of ions scattered off at some particular angle is measured. The energy spectrum obtained (see, for example, Fig. 2) provides information on the mass, or chemical identity, and the number of surface atoms through the energy position and magnitude, respectively, of peaks in the spectrum. In the case of single crystal targets, surface structure information may be derived from the positions and relative sizes of different peaks as influenced by the angles of incidence and reflection.

These various kinds of information are obtained with varying degrees of accuracy and effort. Mass identification is quite straightforward; peaks for higher masses occur at higher energy in a predictable manner. Quantitative

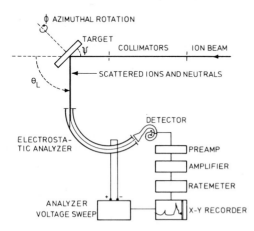

Fig. 1. Schematic diagram of ion scattering experiment.

References pp. 100—102

determination of the number of surface atoms is not so simple since the scattered ion yield depends on scattering cross-sections and neutralization efficiencies, neither of which are very well known at present. Neutralization of the scattered particles is important because electrostatic energy analysis is used almost exclusively in low-energy work. However, quantitative analysis of foreign atoms at a surface can be achieved by calibration of the system against standards. Structure information, for example about atom locations and defects is derived from "shadowing" effects and multiple scattering, i.e. sequential single scattering from two or more atoms, and it involves accurate orientation of single crystals and varying amounts of computational effort.

An outstanding feature of low-energy ion scattering is its fine degree of surface selectivity: the detected ions have scattered from only the first layer or two of atoms at the surface. This behavior seems to result from a combination of two factors. First, large scattering cross-sections lead to attenuation or depletion of the ion beam as it penetrates the solid, and second, more complete neutralization occurs as particles penetrate more deeply. Both of these effects reduce the yield of ions scattered from inside the solid, thereby emphasizing the peaks due to scattering from surface atoms.

B. HISTORICAL

The scattering of low-energy ions from solid surfaces has been studied from several viewpoints over a period of almost 20 years. For example, Brunnee [1] in 1957 reported that alkali ions incident at 0.4—4 keV on Mo surfaces and scattered over large solid angles had maximum energies predicted for single, binary, elastic collisions, and also lower energies believed to result from multiple collisions. Panin [2] performed experiments on Mo and Be targets with beams of various ions, e.g. H^+, He^+, N^+, O^+, Ar^+, at incident energies of 7.5—80 keV, using a more restricted analyzer aperture. He, also, found that many features of the observed energy spectra could be accounted for by assuming that incident ions interacted with target atoms in the solid in the same way as in elastic scattering from free atoms, and that the laws of classical mechanics were applicable in describing peaks in the spectra. Peaks for a given combination of ion and target atom occurred at fixed values of E/E_0, independent of the incident energy. (E_0 = incident energy, E = scattered energy.) He also noted that surface contamination had a strong influence on the spectra. Mashkova and Molchanov [3] studied the angular distribution of Ar^+ (E_0 = 30 keV) scattered from Cu, and obtained reasonable agreement with a theoretical calculation based on a screened Coulomb interaction potential. Fluit et al. [4], by measuring the velocities of Ar^+ and Ar scattered from Cu, obtained a similar angular distribution and confirmed that the energy distribution was peaked at the value expected for single, elastic, binary collisions. Datz and Snoek [5] used high resolution electrostatic analysis to obtain detailed energy spectra of Ar and Cu ions of charge up to

+5 emitted from a single crystal of Cu. With incident energies of 40—80 keV, they used a beam aligned along a $\langle 110 \rangle$ axis to minimize multiple collisions and obtain sharper peaks. The early work on scattering phenomena, per se, has been reviewed [6—8].

The first demonstration of low-energy ion scattering as a surface analytical technique, which is the subject of this chapter, was provided by the work of Smith [9] who used beams of He^+, Ne^+, and Ar^+ at energies of 0.5—3.0 keV on targets of polycrystalline Mo and Ni. He obtained sharp peaks corresponding to scattering from substrate surface atoms, and also from adsorbed species, oxygen and carbon. Furthermore, he inferred structure information from the relative heights of the C and O peaks for CO adsorbed on nickel. Subsequently, Strehlow and Smith [10] distinguished the Cd and S faces of a CdS single crystal by relative peak heights. Thus, low-energy scattering was shown to have possibilities for both chemical composition and surface structure analysis, which have occupied most workers in the field since then. Smith emphasized the importance of using noble gas ions, He, Ne, and Ar rather than reactive gas ions such as H, N, and O to obtain distinct peaks. The reactive gas ions gave broad continuous spectra and this difference in behavior was attributed to a neutralization effect. It was assumed that the low energy "tails" represented ions which had penetrated into the solid and lost energy in several collisions before scattering back to the detector, and that such ions were less likely to be neutralized in the case of reactive gas atoms.

A more recent example of Smith's [11] results is shown in Fig. 2. There are two spectra, one for He scattered from a thin (~ 50 Å) oxide film on a polycrystalline Al foil, and the other taken on the bare Al surface exposed after heavy ion bombardment. The clear resolution of the Al and oxygen peaks above the background contrasts markedly with spectra which would be obtained at higher energy, say 100 keV or 1 MeV. The Al peak heights

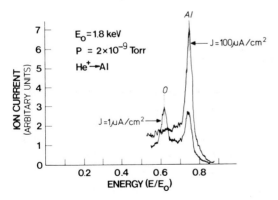

Fig. 2. Typical low-energy ion scattering energy distributions. $^4He^+$ scattered from 50 Å film of Al_2O_3 and pure Al. After Smith [11].

correspond to the atomic densities of Al in Al_2O_3 and Al metal. However, such good quantitative agreement is not generally obtained, for reasons to be discussed later.

C. COMPARISON WITH ION SCATTERING AT HIGHER ENERGIES

Concurrent with the work at low energies, there has been vigorous development of ion scattering at higher energies, up to 3 MeV, for analysis of surfaces and surface layers. One of the early investigations was that of Rubin [12] using a magnetic spectrometer. Later development was stimulated by the advent of solid-state (silicon surface barrier) detectors, and by interest in ion channeling in single crystals [13]. No attempt will be made in this chapter at a comprehensive comparison, which has been made in other review papers [14,15] but a few paragraphs to provide some perspective on the capabilities of low- versus higher-energy (0.1—2 MeV) techniques may be useful.

Low-energy scattering is a true surface technique. The characteristic sharp peaks of low-energy spectra represent scattering from depths of only a few Angstroms, which is a very useful attribute in some applications. The higher energy versions of ion scattering probe much deeper. The spectra obtained, especially with the light ions H^+ and He^+ which are normally used, are generally very broad, exhibiting substantial scattering yields from very low energies, representing deep penetration into the solid, up to a maximum energy corresponding to scattering from surface atoms. At still higher scattered energies, there may be peaks in the spectrum due to foreign surface atoms heavier than the atoms of the substrate. The greater sampling depth makes high-energy scattering more useful for thin film and surface layer analysis than for true surface analysis. Layers of a few thousand Angstroms' thickness can be analyzed with depth resolution of ~200 Å by 2 MeV ^{4}He scattering without the progressive layer removal which is required in low-energy analysis.

Quantitative interpretation of scattering yields is simpler at high energies since the Rutherford scattering law with the simple Coulomb interaction potential applies, whereas electron screening of the nuclei must be taken into account at low energies.

A notable difference in technique between high- and low-energy scattering has to do with energy analysis. At high energies, the silicon surface barrier detector and multi-channel pulse height analyzer system accumulates counts at all energies of interest above some minimum (~20 keV), and registers both ions and neutrals. On the other hand, the electrostatic analyzer normally used in low-energy work is usually a single channel instrument, selecting the particles in only one small ΔE at a given E. This results in rather inefficient utilization of the scattered yield but does not necessarily lead to longer analysis times since heavy beam currents can be used without interference from pulse "pile-up" problems, which limit the permissible beam current in MeV scattering. Furthermore, as mentioned previously, the electrostatic analyzer collects the ions but not the neutrals although the neutrals constitute the majority of the scattered particles at low energy. This discrimination in favor of the ions, however, while decreasing the effective yield still further, contributes to the unusual surface sensitivity of low-energy scattering as pointed out above, and the loss in effective scattering yield is offset by larger scattering cross-sections in comparison with high-energy scattering.

Low-energy scattering offers a considerable saving in equipment and personnel requirements. It is complementary to other surface techniques and can be used for routine composition analysis or for fundamental structure analysis on a wide variety of surfaces.

Examples of surface analysis and other low-energy ion scattering phenomena reviewed in this chapter are illustrated mainly by ion energy spectra, since electrostatic energy analysis has been used almost exclusively. However, a later section will present some neutral and neutral plus ion spectra obtained recently by a time-of-flight technique.

II. Experimental equipment

A. GENERAL REQUIREMENTS

The experimental facilities needed vary with the type of work contemplated. The bare essentials are an ion beam, a vacuum chamber, a target, and an energy analyzing system. This section discusses features which are desirable and fairly typical, although not necessarily essential for all applications. For example, a target orientation capability is much more important for surface structure studies on single crystal targets than for chemical composition analysis on polycrystalline samples. A system which has most of the usual features is shown schematically in Fig. 3 [16].

A monoenergetic ion beam at a current of 10—200 nA is commonly used. The beam diameter is usually 1—2 mm FWHM. An energy spectrum of scattered ions is obtained by stepping or continuously scanning the voltage on an electrostatic analyzer (ESA) and measuring counts in a detector at the output of the ESA. The beam current should be steady, within a few percent, so that each point in the spectrum corresponds to the same integrated flux, or dose, of ions per cm^2. Larger fluctuations can be tolerated if a current integrator and scaler are used to normalize the dose. In either case, secondary electron emission may cause errors since electrons leaving the target produce a current of the same sign as positive ions striking the target. Secondary emission can be suppressed by a positive potential (50—100 V) applied to the target, or by a negative potential on a surrounding shield. Alternatively,

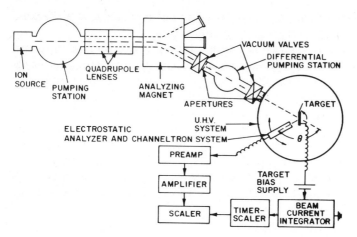

Fig. 3. Schematic of apparatus for a low-energy ion scattering system. After Ball et al. [16].

the observed target current can be corrected to ion current by comparison with the current to a Faraday cup in place of the target. In many cases, one may wish only to compare two or more peaks in the same spectrum. If so, steady current or accurate current integration will suffice to normalize ion dose throughout the spectrum. However, when different spectra, perhaps taken on different targets on different days, are to be compared, a knowledge of the absolute ion flux to the target is needed and this requires correction for secondary electron emission if adequate suppression has not been used.

Energies of 0.5—2 keV are sufficient for what is usually called low-energy scattering, although energies above and below this range are useful for some purposes. Since the useful information is derived from an energy spectrum, all factors which can degrade the energy resolution should be considered in the design or purchase of a system. Energy spread in the incident beam can originate in the ion source and accelerating potential power supply. Angular spread is important also since the energy of a scattered particle depends on the scattering angle, i.e. the angle its trajectory makes with the direction of the incident particle. Angular spread depends on the size and separation of collimating slits and on properties of electrostatic or magnetic lenses in the ion optical system through which the beam is transported from source to target. Energy resolution of 1—3% is usually feasible and adequate for most purposes.

The beam should contain only one ion species, e.g. ^{4}He or ^{20}Ne, in a single charge state. Foreign ions will cause additional peaks either by their different mass or in the case of different charge states, their energy after acceleration. Unwanted ions and neutrals can be removed from the beam by a switching magnet as in Fig. 3, or by a Wien velocity filter [17] in which opposing magnetic and electric fields deflect ions of unwanted velocities (and mass or energy) out of the in-line beam. The latter method is preferable if the accelerator serves only one scattering chamber.

Stray magnetic fields near the beam should be minimized by shielding and by avoiding magnetic materials in beam pipes and valves. Insulating materials which can be charged by the beam should also be excluded. Low-energy ions are very easily deflected from their intended paths by stray fields.

Beam line apertures, target center, and ESA aperture should be aligned carefully, by a telescope or laser beam. It is desirable to have a means of monitoring the shape and position of the ion beam at one or more points between the ion source and the target. Control and efficient transport of low-energy ion beams can be very difficult if target current is the only diagnostic measurement. Fluorescent viewers or sectored metal apertures are helpful in beam control.

B. ION SOURCE

Ions are produced in an ion source usually by electron bombardment of a gas at a pressure of $\sim 5 \times 10^{-6}$ to 10^{-3} torr. The noble gases He, Ne, or Ar are most commonly used in low-energy ion scattering. The source is held at the positive accelerating potential and ions are drawn from it through a small aperture by a negatively biassed electrode. A beam is formed by a lens system.

Source parameters important in surface analysis include: (a) the energy spread, which should be no more than a few volts, (b) the ion current obtainable at the source, a few μA at least, and (c) an angular divergence of $\pm 1°$. Accurate control of the gas leak into the source, and convenient replacement of the filament which supplies the electrons are also important.

Several types of ion sources have been used successfully. Smith [9] used a duoplasmatron source [18]; Ball et al. [16] an oscillating electron discharge source (OED) [19]. Grundner et al. [20] (GHT) have described in detail the testing and optimum performance of an electron impact source. Brongersma and Mul [21] used the Colutron source [22] and Suurmeijer et al. [23,24] the Nier-type source [25]. All of these except the Nier source are rated for high currents. The duoplasmatron and OED sources are reported to have energy spreads of ~ 10 eV [19] while the Colutron and GHT sources have considerably smaller spreads of $\leqslant 1$ eV. Detailed descriptions and comparisons of ion sources for many applications can be found in refs. 18, 19, 26 and 27.

C. VACUUM SYSTEM AND SCATTERING CHAMBER

Good vacuum conditions are required for low-energy scattering and must be better than the minimal needs for work at higher energies. Pressures of $\sim 1 \times 10^{-9}$ torr or lower should be maintained in the scattering chamber during an experiment since the technique is so surface sensitive that adsorbed layers of background gases can seriously reduce the scattered ion yield from the solid surface to be analyzed. For composition analysis on "practical" or technological samples, these background gases should be minimized by adequate pumping and preliminary baking of the chamber walls. For surface structure studies, one should also be able to clean the target surface in situ, maintain an ordered surface by annealing, and admit controlled amounts of gases for adsorption studies.

As illustrated in Fig. 3, vacuum pumping is required for the scattering chamber itself, and in one or more other sections of the system. The UHV chamber of Fig. 3 employs two sorption pumps for roughing, a mercury diffusion pump during bake-out followed by Ti sublimation pumping, and sputter-ion pumping to hold low pressures, $\leqslant 1 \times 10^{-9}$, once achieved. Oil diffusion pumps are used at the source and differential pumping sections. The greatest demands on pumping speed usually arise from (a) the need for fast pump-down after sample changes, (b) gas evolution during target heating, and (c) the noble gas introduced by the ion beam itself.

Other systems developed by various experimenters [20,21,23] are similar in general layout although differing in details of vacuum components. A commercial system [28] which is used in a number of industrial laboratories for analysis of practical surfaces [29,30] omits the differential pumping between source and target. The entire system is pumped to $\sim 10^{-9}$ torr after which it is back-filled to 10^{-5} torr with the noble gas required for the beam, while removal of active gases is continued, by Ti sublimation and liquid N_2 cryopumping.

D. ELECTROSTATIC ANALYZER AND ION DETECTOR

The energy distribution of the scattered ions is usually obtained by means of an electrostatic analyzer (ESA) which acts as energy/charge filter. As the voltage between the analyzer plates is stepped or continuously scanned through the region of interest, scattered ions of corresponding E_1/q are transmitted through the analyzer and counted by a detector at the output aperture. For a curved plate analyzer the relationship between energy and analyzer voltage is $E_1/q = (V_a R)/(2d)$ where E and q are the kinetic energy and charge of ions transmitted, V_a is the potential difference between the plates, R the mean radius of the plates and d their separation. The outer and inner plates are held at $+V_a/2$ and $-V_a/2$, respectively, so that the central trajectory through the analyzer is at ground potential. With an analyzer of 10 cm radius and 0.5 cm plate separation, 100 eV singly charged scattered ions would be passed when $V_a = 10$ V. The analyzer would also pass 200 eV doubly charged ions and any foreign ions having $E/q = 100$ V. However, these ambiguities are usually not a serious problem; the common manifestations are small extra peaks due to doubly charged ions of the beam species and humps at low E/q due to various ions desorbed from the target surface by the beam.

Energy resolution, $\Delta E/E$ of the analyzer depends primarily on the ratio of the sum of the aperture widths to the analyzer radius, and on the selected energy itself. It is usually in the range of $(0.01-0.03)E_1/q$ with the lower limit imposed by solid angle (counting rate) considerations. In the most common type of ESA, the relationship between resolution and E_1 causes a distortion of the spectrum, i.e. the energy window of the analyzer is proportional to the energy of the ions passed. This can easily be corrected for in comparisons of peaks at different energies [31], although the counting statistics cannot be normalized. Brongersma and Mul [21] used a more sophisticated analyzer, based on the design of Kuyatt and Simpson [32] in which scattered ions are slowed to a low, fixed energy before entering the analyzer. After analysis, they are re-accelerated to ~5 keV before they strike the detector. Various analyzer designs have been used. Hughes and Rojansky [33] showed that a 127° curved plate analyzer provides re-focussing at the exit slit, in the plane of the analyzer radii. Details of such an analyzer mounted inside the scattering chamber on a rotatable table have been published by Wheatley and Caldwell [34]. Comprehensive evaluation of an 85° curved plate analyzer attached outside the scattering chamber has been described by Suurmeijer [24]. Calculations of trajectories, fringing field distributions, focussing properties, etc. for various types of electrostatic analyzers have been published by Wollnik [35]. Niehus and Bauer [36] have recently reported an interesting combination of ion scattering and Auger electron spectroscopy in which a commercial cylindrical mirror analyzer (CMA) for Auger analysis can be switched in situ to low-energy ion analysis by reversing the polarity.

Counting rates depend critically on proper alignment or aim of the analyzer with respect to the beam spot on the target. Good initial alignment during assembly is necessary, and it is also very helpful to have mechanical means of optimizing the aim with beam on the target [34].

Detectors for low-energy ion scattering are usually electron multipliers of either the channel or Cu—Be dynode types. The detector should be bakeable if it is mounted inside the UHV chamber. Since these multipliers generally exhibit a rapid decrease in detection efficiency for ion energies below ~2 keV, ions exiting from the analyzer aperture should be accelerated accordingly, e.g. by applying the 3000 V multiplier bias with entrance end negative.

Electronic equipment for data collection can be quite simple. The essential components are a coupled pair of well regulated d.c. power supplies, one negative and one positive, for the analyzer, together with a pre-amplifier, main amplifier, digital scaler and timer system to count the pulses from the detector. With these and a steady beam current, one can step the voltage on the analyzer through small energy increments, counting the ion yield for each step, to obtain a spectrum. A beam current integrator—scaler combination can be used to normalize ion dose in case of a fluctuating current, as mentioned previously. A combination of a ratemeter, single channel analyzer, and automatic voltage sweeper with an X—Y plotter is more convenient and more commonly used [20,21,28], while computerized data collection systems can provide facilities such as analyzer voltage scanning, printer and tape output, plotting of spectra, etc. [24].

Depending on the type of work to be pursued, other features which may be required in the scattering chamber include the following: goniometer target holder which allows tilting and azimuthal rotation; target heating by radiation, electron bombardment, or current through the target; ion bombardment cleaning by a separate ion gun; LEED optics; Auger electron spectroscopy; mass spectroscopy for SIMS analysis; and electron emitting filament for neutralizing charge deposited on insulator targets by the ion beam.

III. Ion scattering principles

A. KINEMATICS

An important conclusion from the early work described in the introduction was that the prominent peaks in low-energy scattering spectra result from simple two-body collisions between the incident ions and single lattice atoms, with the host lattice playing little or no part. Attempts to detect deviations from this behavior at energies down to 100 eV have failed [11]. Furthermore, energy losses due to electronic transitions during the collision are too small to be detected in most experimental systems. Therefore, the energy E_1 retained by an ion of mass M_1 with an incident energy E_0 after scattering from a target atom of mass M_2 through an angle θ_L (see Fig. 1) is given by eqn. (1) which is based on conservation of kinetic energy and momentum.

$$\frac{E_1}{E_0} = \frac{M_1^2}{(M_1 + M_2)^2} \left\{ \cos \theta_L + \left(\frac{M_2^2}{M_1^2} - \sin^2 \theta_L \right)^{1/2} \right\}^2 \tag{1}$$

For 90° scattering which is frequently used, this reduces to

$$\frac{E_1}{E_0} = \frac{M_2 - M_1}{M_2 + M_1} \tag{2}$$

Thus, the energy scale becomes a mass scale for target atoms at the surface, with higher energy indicating larger mass.

B. SCATTERED YIELD

Although the identification of the target atoms is relatively simple, requiring a well-defined incident beam energy and accurate energy analysis of the scattered ions, the quantitative determination of the *number* of target atoms is not so straightforward. The scattering yield depends on the differential scattering cross-section $d\sigma(\theta)/d\omega$ for the ion—atom collision and on the probability, P_i, that the scattered particle remains ionized, as well as on the number of target atoms. The scattering cross-section in turn is a function of the interaction potential $V(r)$ between the ion and target atom which is generally not well established at low energies as it is in the MeV region where the simple Coulomb potential, the pre-exponential factor in eqn. (3), applies. Furthermore, ion-fractions, P_i or Y^+/Y, have been reported for only a few systems, and for energies only down to ~ 4 keV. However, if these quantities were known, the scattering yield of ions per unit solid angle, ω, should be given by an expression of the form $dn/d\omega = n_0 NP_i d\sigma(\theta)/d\omega$ where n_0 is the number of particles which strike the target during the measurement, n is the number of scattered particles, N the density of target atoms in the accessible depth, and the other quantities as defined above. The restriction of N to the "accessible depth" of the technique, one or two monolayers, is obviously important; atoms lying deeper will be missed.

If a reasonable interaction potential is known or assumed [37] then the differential scattering cross-section can, in principle, be derived by methods of classical mechanics [37,38]. This may be rather simple or quite complicated depending on the form of the potential chosen. Cross-sections derived from the Bohr screened Coulomb potential have frequently been used in low-energy scattering [11,16,29] in chemical composition analysis. This potential is expressed by

$$V(r) = \frac{Z_1 Z_2}{r} e^2 \exp\left(-\frac{r}{a}\right) \tag{3}$$

in which r is the nuclear separation, $a = a_0/(Z_1^{2/3} + Z_2^{2/3})^{1/2}$ is the electron screening length, and $a_0 = 0.53$ Å is the radius of the first Bohr orbit in the hydrogen atom. Convenient tables from which scattering cross-sections can be derived were published by Everhart et al. [39] and a finer-grained compilation requiring less interpolation has been furnished by Bingham [40], together with the criteria on validity.

The Born—Mayer potential $V(r) = A \exp(-br)$ has been used by Heiland and Taglauer for both composition analysis [41] and surface structure analysis [42,43]. Tables of the A and b parameters for all elements have been calculated by Abrahamson [44]. Since this potential does not contain the Coulomb factor, $Z_1 Z_2 e^2/r$, it is not repulsive enough for close encounters and is valid only for rather "soft" collisions with large impact parameters and small scattering angles. (The impact parameter is the perpendicular distance between the incident trajectory of the ion and the initial position of the target atom, and it is a basic parameter in scattering calculations.) Begemann and Boers [45] used either the Born—Mayer or the Firsov potentials in structure studies by multiple scattering, depending on whether r was greater or less than 1 Å. The testing of various potentials is hampered by uncertainty in neutralization probabilities in the case of electrostatic energy analysis, and should be aided greatly by measurements of neutral yields.

C. ION NEUTRALIZATION

Ion neutralization is a very important phenomenon in low-energy ion scattering, as was pointed out in the introduction to this chapter. In the energy region below 5 keV, the scattered ions which are collected in an electrostatic analyzer constitute only a small fraction of the total yield. Thus, there is inefficient utilization of the scattered yield, and also considerable complication in quantitative interpretation since neutralization efficiency, $(1 - P_i)$ might conceivably depend on various parameters such as ion energy, substrate material, identity of an adsorbed target atom, etc. On the other hand, the most outstanding feature of low-energy ion scattering, its very fine surface selectivity, has generally been attributed to a neutralization effect, in which ions penetrating beyond the first layer or two of atoms at the surface are neutralized much more completely than those which scatter from the surface. The sharply peaked spectra of low-energy ion scattering differ very markedly from those observed at higher energies in which there is generally a broad tail or background at energies below the upper limit determined by elastic scattering from surface atoms. This background tends to obscure the peaks for elements lighter than the substrate atoms. The transition between the two types of behavior is illustrated in Fig. 4, which contains spectra for ^{4}He scattered from polycrystalline gold. Four different incident energies are represented. For E_0 = 25 keV, there is a broad background due to ions which penetrated and lost energy on their inward and outward paths in addition to that lost in the back-scattering collision. This background decreases very rapidly as the incident energy is reduced, and a sharp surface peak develops. To explain such behavior, one is forced to conclude that there is more efficient neutralization of ions which penetrate, as has been supposed, or that the total scattering yield from beneath the surface has decreased in a rather surprising way. It should be noted that the neutralization model differs from the behavior of H and He at higher energies where, according to both theoretical [46,47] and experimental [31] evidence, the neutralization occurs as the particle leaves the surface and is apparently independent of the depth to which the particle has penetrated. The resolution of this question requires information on neutral scattering yields, which ESA spectra do not provide. The technique employed at higher energies cannot be used at low energies

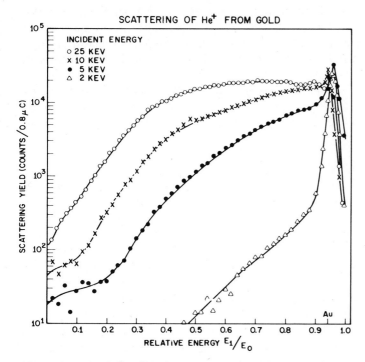

Fig. 4. Spectra of $^4He^+$ backscattered from polycrystalline gold at various incident energies. After Ball et al. [16].

because it utilizes a silicon surface barrier detector whose dead-layer is too thick.

However, recent time-of-flight (TOF) experiments suggest that both possibilities mentioned above actually do contribute to the characteristic shapes of low-energy ion scattering spectra. Energy spectra of scattered neutrals as well as ions were derived from TOF measurements in which both could be detected [48]. In Fig. 5, spectra are shown for He scattered from polycrystalline gold for 2 incident energies, 32 and 8 keV. There is a pair of spectra for each E_0, one spectrum for ions plus neutrals and another, slightly lower, for neutrals alone. For $E_0 = 8$ keV, the scattered yield decreases throughout the low-energy region and the change in shape from 8 to 32 keV resembles the trend of the ion spectra in Fig. 4.

In Fig. 6, spectra are shown for Ar, scattered from gold, also derived from time-of-flight measurements. In this case, there is a sharp surface peak corresponding to single scattering from surface atoms. There is also a double scattering shoulder on the high energy side of the peak, and decreasing yield from the peak down to $E/E_0 = 0.1$ where the upturn is presumably due to recoil gold atoms knocked off by the argon ion bombardment. These Ar neutral spectra also resemble ion spectra for the same E_0 [16], although the

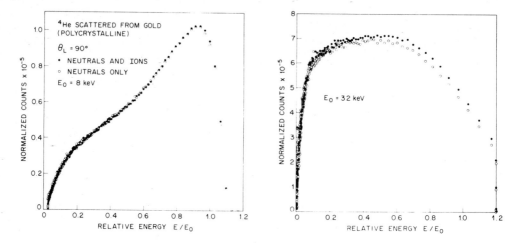

Fig. 5. Energy spectra, derived from TOF measurements, for ^{4}He scattered from poly-crystalline gold at two incident energies, 8 and 32 keV. Neutral spectra were obtained by deflecting ions out of the scattered beam. Data taken from Buck et al. [48].

peaks are not as sharp in the present case. The present spectra are similar to those which Chicherov [49] obtained for 16 keV Ar on Au by stripping or re-ionizing the scattered Ar neutrals in He or H_2 and then analyzing the Ar^+ ions in an ESA. Furthermore, a fall-off in yield at low energy has also been observed for hydrogen by a stripping technique [50]. The shapes of these neutral and neutral plus ion spectra cannot, of course, be explained by a preferential neutralization effect which excludes neutrals from part of the

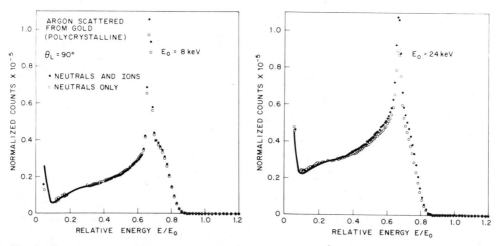

Fig. 6. Energy spectra, derived from TOF spectra, for Ar scattered from polycrystalline gold at two incident energies 8 and 24 keV. After Buck et al. [48].

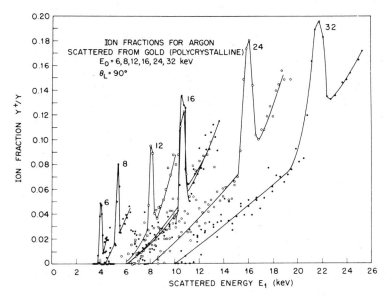

Fig. 7. Ion fractions, (ions)/(ions plus neutrals), for argon scattered from polycrystalline gold, for 6 different incident energies. After Buck et al. [48].

energy distribution. A reasonable explanation seems to be that the low-energy scattering cross-sections are so large, e.g. at 10 keV, 2.2×10^{-18} cm^2 for Ar on Au and 3.3×10^{-19} cm^2 for He on Au, that a significant fraction of the beam is scattered out of the original direction during passage of each layer of atoms.

A calculation [48] based on this assumption, and on several others which are more appropriate for He than for Ar, has been used to illustrate that the beam attenuation effect could cause a fall-off at low energies quite similar to that observed in the He case (Fig. 5). The model also predicts an upturn for higher incident energy, $\geqslant 100$ keV, as is found experimentally. Calculated curves for Ar do not match the experimental as well, mainly owing to neglect of multiple scattering events, but the beam attenuation effect must be even more important in the Ar case.

Nevertheless, although the beam attenuation effect appears to have a strong influence on the spectrum, there is also preferential neutralization of ions which penetrate. In Fig. 7, the ion fraction Y^+/Y (ion yield/ion plus neutral yield) of scattered Ar is plotted against scattered energy, for several incident energies. The curves were derived from energy spectra such as those in Fig. 6. Each curve is peaked at an energy corresponding to single scattering from gold surface atoms ($E/E_0 \sim 0.66$). The high-energy branch of each curve represents ions which were scattered twice, close to the surface, while the lower branch must represent those which penetrated deeper. The peak values of Y^+/Y rise with increasing energy, a trend which is expected for neutralization of ions as they leave the surface, since slower ions are neutralized more efficiently. There is clear evidence of neutralization depending on path length inside the solid: the low-energy branch of the $E_0 = 32$ curve lies below the curves for other incident energies at any common energy and the peak for each curve, representing singly scattered particles, is higher than its double scattering branch even though the double scattered particles have come off with higher energy [51]. If there were no depth effect on neutralization, there should be no peaks and no vertical separation between curves for different incident energies, which is indeed the behavior of light ions at higher energies [31]. Ion-fraction data for He at these energies did not exhibit distinct peaks; however, more structure and clearer indications of depth dependence may be expected to appear at lower energies and with better energy resolution. Ion fractions of singly scattered He are lower ($\sim 0.01-0.04$) over this energy range than those for Ar ($\sim 0.04-0.20$).

Details of the neutralization process, either inside or outside the solid, have not yet been established. Processes which have been considered [11,51, 52] for neutralization outside as the ion leaves are the Auger and resonance transfer mechanisms [53,54] both of which exhibit an exponential dependence of ion-fraction on exit velocity

$$Y^+/Y = e^{-v_0/v_\perp} \qquad (4)$$

in which v_0 is a characteristic velocity and $v_\perp$ the velocity component normal to the surface, which implies an influence of exit angle [51].

An influence of target atom identity has been demonstrated in the case of S adsorbed on Ni [41]. For 1 keV Ne^+ ions neutralization was about 7 times lower for scattering from S atoms than from Ni.

Neutralization remains a major area for further experimental and theoretical investigation. (See note added in proof, p. 99.)

IV. Surface composition analysis

A. CALIBRATION

The questions concerning neutralization efficiencies and scattering cross-sections discussed above make it necessary, for quantitative analysis, to calibrate the technique using standard samples and comparisons with other techniques. The dependence of scattered yield on target atom density is, of course, an important matter. A linear dependence is desirable but not necessarily assured in view of the very shallow depth which is probed. Smith [11] found a linear relation between scattered yield and coverage for a series of Au—Ni alloys. Ball et al. [16] measured scattering yields of He and Ar from gold deposits on silicon whose density had been determined by neutron

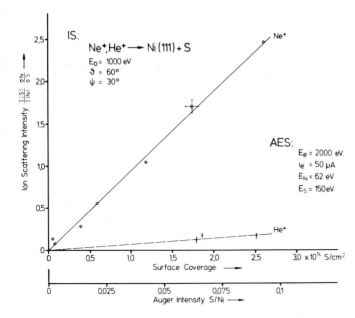

Fig. 8. Relative ion scattering intensities $(I_s/I_{Ni})\sigma_{Ni}/\sigma_s$ as function of sulfur coverage on Ni. The sulfur coverage was determined by AES using the calibration of Perdereau [56]. Scattering cross-sections were calculated with Born—Mayer potentials. After Taglauer and Heiland [41].

activation. A gold detection limit of 7×10^9 atoms/cm²/count was determined on a sample which had 10^{12} atoms/cm² ($\sim 10^{-3}$ monolayer) according to neutron activation results. However, the scattered yield from a sample with 3.6×10^{15} atoms/cm² indicated a sensitivity of 1.6×10^{11} atoms/cm²/count, i.e. fewer counts per gold atom from the sample with four monolayers of coverage. It was proposed that the gold deposit in this case was not a continuous film, but rather consisted of cap-shaped islands thicker than the sampling depth of the beam.

For the cases of S and O adsorbed on Ni at submonolayer coverage, a linear dependence of the scattering intensity on surface coverage was found by Taglauer and Heiland [41,55]. Some of their results are shown in Fig. 8 where the S/Ni peak ratios are plotted against S coverage, which was determined by Auger measurements. The Auger measurements in turn had been calibrated by a radiotracer experiment [56]. The authors concluded from this linearity that the neutralization efficiency was independent of coverage in this range. The detection limits for S and O on nickel were reported as 10^{-3} to 10^{-4} monolayer. One rather surprising feature of the sulfur results is the large difference in S/Ni ratios between Ne and He scattering. It may be due to a difference in shadowing of Ni by S which would be a larger effect in the case of Ne.

Niehus and Bauer [36] have found very substantial departures from linearity for O and Be adsorbed on W. In the case of oxygen, they suggested that the sub-linear increase of signal with coverage could be due to a neutralization effect associated with a change in work function induced by O adsorption.

B. TECHNOLOGICAL APPLICATIONS

Turning to applications of a technological nature, Fig. 9, from work of McKinney and Leys [57], shows qualitative surface analysis of a hot-rolled Fe—Mo—Re alloy, by both $^3He^+$ and $^{20}Ne^+$ beams, and illustrates the advantage of a heavier ion to improve mass resolution. In the $^3H^+$ spectrum, Fe and Mo peaks are resolved but there is only a small shoulder at the Re position, whereas all three peaks are clearly resolved by ^{20}Ne, in accordance with eqn. (2). These authors also point out the characteristic absence of a low-energy background for ^{20}Ne in contrast to the ^3He spectrum. This difference is probably associated with the beam attenuation effect mentioned in section III.C. For light elements such as ^{12}C and ^9Be, ^{20}Ne cannot be used for 90° scattering but mass resolution with He is quite adequate. These authors report higher sensitivity with $^3He^+$ than with $^4He^+$. This is presumably due to the higher velocity and larger ion fraction of ^3He at a given energy.

In Fig. 10, an example [58] is shown in which the activation process of a thermionic cathode was studied by scattering of 1 keV Ne ions. This cathode, of the tungsten "dispenser" type, consists of porous tungsten im-

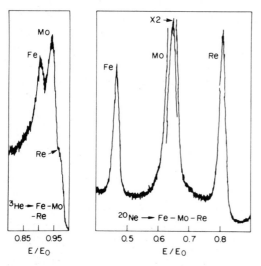

Fig. 9. $^3He^+$ and $^{20}Ne^+$ backscattering from Fe—Mo—Re alloy. Incident energy was 1.5 keV. After McKinney and Leys [57].

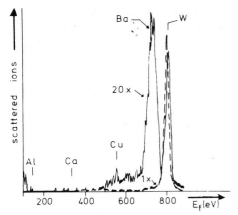

Fig. 10. Energy spectra of $^{20}Ne^+$ ions scattered from Ba-doped tungsten cathode. — — —, Sputter-cleaned cathode before activation; ——— activated cathode at temperature of 1165°. After Brongersma and Schouten [58].

pregnated with a mixture of oxides — Al_2O_3, BaO, and CaO. At room temperature (broken curve), a strong tungsten peak is observed, and little else. After activation by heating to 1165°C, and while the cathode remains at 1165°C, the spectrum shows a Ba peak larger than the W peak, plus small peaks for Cu, Ca, and Al. The Ba which has migrated to the surface is presumably responsible for the increase in electron emission. The copper is an impurity. Both the W and Ba peaks are smaller than the original W peak by a factor of ~ 20. This was attributed to neutralization of scattered Ne^+ ions in the sheath of electrons surrounding the hot cathode.

Other types of technological applications have included analysis of passivated iron—chromium stainless steels [59], oxygen contamination in electroplated metals [30], and depth profiling of TiO_2 (200Å)—Au(175Å)—glass structures [60]. In these and most other technological uses of the technique, results have usually been qualitative or semi-quantitative with the main emphasis on the fine depth selectivity.

V. Surface structure

A. SHADOWING EFFECTS

The surface sensitivity of low-energy ion scattering suggests uses in surface structure work, some of which have been realized while others require further refinement of the technique. Early examples, e.g. CO on Ni, and the polar surfaces of CdS mentioned in the introduction, took advantage of the "shadowing" effect or suppression of a peak for element A because atom B lies above it.

References pp. 100—102

94

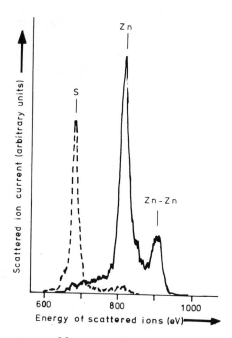

Fig. 11. $^{20}Ne^+$ spectra taken on the two opposite (111) faces of a ZnS crystal. Specular reflection of 1000 eV $^{20}Ne^+$ is used; scattering angle is 45°. After Brongersma and Mul [63].

Recently, a question was raised concerning the absolute identification of polar faces on non-centrosymmetric crystals such as ZnS and CdS. Assignments were originally made by anomalous X-ray scattering [61] and later these were conveniently characterized by etch figures [62]. However, reinvestigation by quantum field theory of the anomalous X-ray criteria cast doubt on the original interpretation by classical theory. The earliest ion scattering results [10] on such a material, CdS, agreed with the older convention, although in the spectra obtained with 2 keV He$^+$ ions the Cd/S ratios (3.2 and 1.7) were greater than unity for both faces of the crystal. A more recent answer to the question, by Brongersma and Mul [63], is shown in Fig. 11 for the case of ZnS. The surface identified by etching as the Zn-terminated face does, indeed, show a large Zn single scattering peak and a smaller Zn double scattering peak, but no evidence of S. The opposite side showed a prominent S peak, a small double scattering hump and only a hint of Zn. The original identification was thus confirmed quite conclusively for ZnS, and also for CdS by He scattering.

Low-energy ion scattering and LEED seem to complement each other rather well. LEED measurements have for some time been used to determine the symmetry and spacing of surface atoms but only recently, and with considerable computational effort, have they given the location of foreign atoms relative to the substrate surface atoms [64]. Heiland and Taglauer [42] used 600 eV He$^+$ scattering in conjunction with LEED to study the surface topography of oxygen on the nickel (110) surface. A well-known LEED pattern (2 × 1) was obtained, for which at least two different surface structures had been proposed. In the ion scattering experiments, the shadowing of Ni atoms by oxygen for several incidence angles was interpreted as supporting a model in which every second place in the top (110) rows of Ni is occupied by an oxygen atom [65]. Measurements of Brongersma and Theeten [66] recently on the (100) Ni surface resolved the spacing of the O atom above Ni at 0.9 Å. In this case, qualitative shadowing effects were not sufficient and a computer simulation technique was employed.

B. DOUBLE AND PLURAL SCATTERING, SURFACE DEFECT ANALYSIS

The surface composition and structure information described thus far is derived mainly from single scattering peaks. However, double and multiple scattering processes may contribute additional features to the spectra, especially in the case of heavy ions and forward scattering angles. The extra

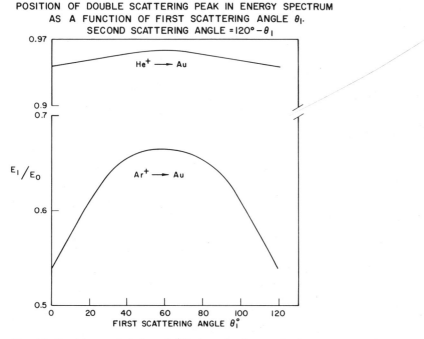

Fig. 12. Positions of Ar$^+$ and ^{4}He$^+$ peaks due to double scattering from gold atoms, with scattering restricted to one plane. After Ball et al. [16].

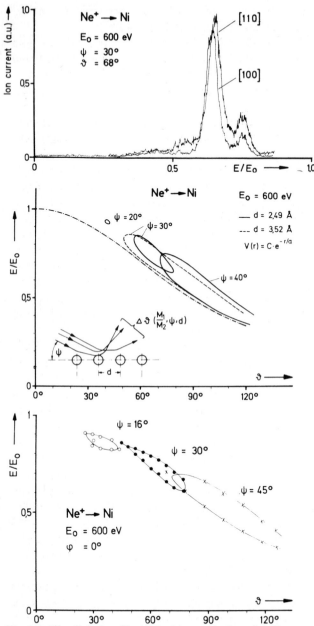

Fig. 13. The "chain effect" in ion scattering. (a) Energy spectra of $^{20}Ne^+$ ions scattered from a Ni(110) surface. The ion trajectories are directed along the [110] or [100] crystal directions. (b) Results of numerical calculations of the "chain effect". The dot-dashed line corresponds to single binary collisions with E_1 given by eqn. (1). (c) Experimental results of scattering along [110] chains, for three angles of incidence, ψ. After Taglauer and Heiland [43].

features may be simple shoulders for polycrystalline targets (Fig. 6) or distinct peaks when a single crystal surface is studied (Fig. 11).

The peaks in the spectra for argon scattering from polycrystalline gold (Fig. 6) have high energy shoulders which result from double scattering near the surface; similar shoulders are observed in ESA spectra [16]. The main peak at E/E_0 = 0.66 results from single scattering through 90° but some ions also reach the detector after scattering from two atoms, an event which has lower but not negligible probability. If it is co-planar scattering, or close to it, the total energy loss will be less than for single scattering through 90°, which can be shown by applying eqn. (2) twice. In Fig. 12 is shown the extent to which the co-planar double scattering energy can exceed the single scattering energy for Ar or ^{4}He scattering from Au ($\theta = 120°$). In the argon case, the spread is significant and scattering cross-sections are large so that double scattering shoulders are easily detectable. Non-co-planar sequences can produce all energies below $0.66E_0$. For He on gold, both the single and double scattering losses are small and so is their difference. Furthermore, scattering cross-sections are small so that a double scattering shoulder would be very hard to detect in this case. Families of spectra for Ne, Ar, Kr, and Xe scattered from polycrystalline gold and copper have been measured [67] as a guide to the amount of interference to be expected from double and multiple scattering in composition analysis.

In scattering from single crystal surfaces, the regular spacing of atoms along surface rows leads to more interesting multiple scattering behavior. The atomic chain model, introduced by Kivilis et al. [68], has received considerable theoretical and experimental attention in recent years. It involves scattering of ions when the planes of incidence and reflection coincide with low index rows of atoms in the surface. Examples of the phenomenon are shown in Fig. 13 for the case of Ne$^+$ scattering from a nickel crystal. In Fig. 13(a), two typical energy spectra are shown for scattering from two different surface rows in the (110) Ni surface. Each spectrum has a major peak at an energy close to that for single binary scattering, and a smaller peak at higher energy which results from two or more successive collisions at smaller angles and with less energy loss. Computer simulations of the sequential scattering process produce double valued curves of E/E_0 vs. scattering angle, as in Fig. 13(b) which may be compared with the experimental results in Fig. 13(c). The dot-dashed curve in Fig. 13(b) represents single binary collisions while the loops show that in the chain model there are two possible combinations of single scattering events which lead to the same total scattering angle but different energies. An interesting feature is that the theory predicts minimum and maximum scattering angles which are caused by mutual shielding of atoms in the chain. That is, not all impact parameters are accessible. Large scattering angles, requiring small impact parameters, are limited by the steering effect of the atom(s) preceding the target atom in the row while small scattering angles are limited by outward deflections by the atoms following. All scattering outside these angles is then prevented. This causes loops of limited range in θ. The angle of incidence ψ influences the size of the loop, as can be seen in the figure. That is, smaller angles of incidence lead to more shadowing. At a large angle of incidence, 40—45°, small scattering angles are cut off, but not large scattering angles since direct hits can now be made. Close spacing of atoms (2.49 Å) along the (110) direction also leads to more shadowing, and consequent shortening of the loop. This is not shown in Fig. 13(c) but it did occur. These experimental results while showing the general behavior do depart appreciably from the numerical calculations, in which the Born—Mayer potential was used. More recent results give better agreement using the Moliere potential [69].

Scattering outside the theoretical angular limits has been attributed [51, 70] to surface defects, i.e. to scattering from atoms which have no neighbors at the proper position to deflect the incoming or outgoing atom. The influence of defects on the atomic chain scattering has been studied extensively by Begemann and Boers [45] who associated large changes in shape of the loops in E vs. ψ curves with upward or downward steps, as well as single vacancies in surface rows. The chain scattering thus appears potentially useful for studying nucleation and growth of films, and surface step effects in catalysis. Consideration of thermal lattice vibrations is needed for refinement of the technique [71].

A more comprehensive review of multiple scattering behavior may be found in a recent paper by Suurmeijer and Boers [72].

VI. Conclusions

Low-energy ion scattering for surface analysis is useful for qualitative and semi-quantitative composition analysis on a wide variety of materials, and is particularly attractive for problems requiring very fine, monolayer, depth selectivity. More general utility in quantitative analysis will require extensive calibration against standards, which has been required also for most other surface analytical techniques, and measurements of neutralization probabilities and scattering cross-sections. Progress has been made in all of these approaches. In surface structure applications, clear-cut solutions to some problems have been made by rather simple shadowing effects, and further progress is being made by numerical calculations. Theoretical and experimental studies of atomic chain scattering offer good reason to expect that this phenomenon will be useful in surface defect analysis. Low-energy ion scattering measurements are compatible with other surface measurements in the same UHV chamber, and complementary information is often forthcoming.

Note added in proof

New insight into the neutralization mechanisms of low energy ion scattering has been provided very recently by Erickson and Smith [73]. They observed oscillations in scattered-ion yield as a function of incident energy for He$^+$ scattered from several elements, including indium whose characteristic behavior is shown in Fig. 14. This phenomenon is attributed to oscillations in neutralization probability similar to those studied in gases by Ziemba et al. [74]. The peaks are more or less evenly spaced when the yield is plotted against $1/v$, which is proportional to interaction time. Although a complete theoretical description is lacking, especially for the high Z ions and atoms, a qualitative explanation is that resonant or quasi-resonant charge exchange occurs and that for sufficiently high velocity the interaction time is so short, $\sim 10^{16}$ sec, that an electron from a neutral target atom has only time to jump to the ion, but not back again. Lower energies and longer interaction times permit additional transfers of the electron back and forth.

The phenomenon has several interesting implications for low energy ion scattering, one of which is illustrated in Fig. 14. There the differences in spacing and number of the peaks for indium in the pure and compound forms suggest that chemical binding information might be obtained from such measurements. These authors also point out that the very distinct differences between curves for Pb and Bi could serve to distinguish between

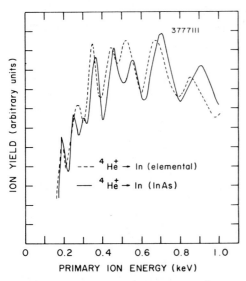

Fig. 14. Oscillations in yield of He$^+$ scattered from indium, as a function of incident energy. $\theta_L = 90°$. The In(InAs) curve has been doubled in amplitude for comparison. After Erickson and Smith [73].

References pp. 100—102

these two elements, whose mass difference is too small to produce a detectable difference in scattered energy.

In addition to He^+, both Ne^+ and Ar^+ have also exhibited oscillations [73,75]. The oscillatory mechanism is evidently not the only one by which low energy ions are neutralized, since a high degree of neutralization does occur even on target materials which produce no oscillations. The other processes mentioned above (Section III.C) having an exponential dependence on $1/v$ must be considered, and also those which occur inside the solid. However, the recognition that oscillatory exchange may occur should clarify the neutralization picture considerably, helping to explain anomalous variations in ion yield. Experimenters should, however, beware of spurious oscillations due to experimental factors such as changes in beam intensity and position on the target.

Acknowledgments

The author is indebted to H.H. Brongersma, P. Mul, W. Schouten and J.B. Theeten; J.T. McKinney and J.A. Leys; D.P. Smith; E. Taglauer and W. Heiland for permission to use figures and data and to H.H. Brongersma, D.L. Malm and W. van der Weg for informative discussions. The hospitality of Professor M.J. Sparnaaij and his group at the Philips Research Laboratory during part of the writing is also gratefully acknowledged.

References

1 C. Brunnee, Z. Phys., 147 (1957) 161.
2 B.V. Panin, Sov. Phys. JETP, 42 (1962) 313; Sov. Phys. JETP, 15 (1962) 215. (1962) 215.
3 E.S. Mashkova and V.A. Molchanov, Sov. Phys. Dokl., 7 (1963) 829.
4 J.M. Fluit, J. Kistemaker and C. Snoek, Physica, 30 (1964) 870.
5 S. Datz and C. Snoek, Phys. Rev. A, 134 (1964) 347.
6 M. Kaminsky, Atomic and Ionic Impact on Metal Surfaces, Springer-Verlag, Berlin, 1965, pp. 237—263.
7 C. Snoek and J. Kistemaker, Advan. Electron. Electron Phys., 21 (1965) 67.
8 E.S. Mashkova and V.A. Molchanov, Radiat. Eff., 16 (1972) 143.
9 D.P. Smith, J. Appl. Phys., 38 (1967) 340.
10 W.H. Strehlow and D.P. Smith, Appl. Phys. Lett. 13 (1968) 34.
11 D.P. Smith, Surface Sci., 25 (1971) 171.
12 S. Rubin, Nucl. Instrum. Methods, 5 (1959) 177.
13 J.A. Davies, J. Denhartog, L. Eriksson and J.W. Mayer, Can. J. Phys., 45 (1967) 4053.
14 W.K. Chu, J.W. Mayer, M.A. Nicolet, T.M. Buck, G. Amsel and F. Eisen, Thin Solid Films, 17 (1973) 1.
15 J.M. Poate and T.M. Buck, in R.B. Anderson and P.T. Dawson (Eds.), Experimental Methods in Catalytic Research, Academic Press, New York, in press.
16 D.J. Ball, T.M. Buck, D. MacNair and G.H. Wheatley, Surface Sci., 30 (1972) 69.
17 L. Wahlin, Nucl. Instrum. Methods, 27 (1964) 55.
18 M. von Ardenne, Tabellen zur Angewandten Physik, B and I, VEB Deutscher Verlag der Wissenschaften, Berlin, 1962, Chap. 3.

19 R.G. Wilson and G.R. Brewer, Ion Beams with Applications to Ion Implantation, Wiley, New York, 1973, pp. 17—115.
20 M. Grundner, W. Heiland and E. Taglauer, I.P.P. Lab Rep., Inst. für Plasmaphysik, Garching, Munich 9, Sept. 8, 1973.
21 H.H. Brongersma and P.M. Mul, Surface Sci., 35 (1973) 393.
22 M. Menzinger and L. Wahlin, Rev. Sci. Instrum., 40 (1969) 102.
23 E.P.Th. Suurmeijer, A.L. Boers, S.H.A. Begemann, Surface Sci., 20 (1970) 424.
24 E.P.Th. Suurmeijer, Thesis, University of Groningen, The Netherlands, 1973.
25 A.O. Nier, Rev. Sci. Instrum., 18 (1947) 398.
26 G. Carter and J.S. Colligan, Ion Bombardment of Solids, Heinemann Educational Books, London, 1968, p. 423.
27 G. Dearnaley, J.H. Freeman and R.S. Nelson, Ion Implantation, North-Holland, Amsterdam, 1973, Chap. 4.
28 R.F. Goff and D.P. Smith, J. Vac. Sci. Technol., 7 (1970) 72.
29 R.E. Honig and W.L. Harrington, Thin Solid Films, 19 (1973) 43.
30 D.L. Malm and M.J. Vasile, J. Electrochem. Soc., 120 (1973) 1484.
31 T.M. Buck, G.H. Wheatley and L.C. Feldman, Surface Sci., 35 (1973) 345.
32 C.E. Kuyatt and J.A. Simpson, Rev. Sci. Instrum., 38 (1967) 103.
33 A.L. Hughes and V. Rojansky, Phys. Rev., 34 (1929) 284.
34 G.H. Wheatley and C.W. Caldwell, Jr., Rev. Sci. Instrum., 44 (1973) 744.
35 H. Wollnik, in A. Septier (Ed.), Focusing of Charged Particles, Vol. II, Academic Press, London, 1967.
36 H. Niehus and E. Bauer, Surface Sci., 47 (1975) 222.
37 I.M. Torrens, Interatomic Potentials, Academic Press, New York, 1972.
38 H. Goldstein, Classical Mechanics, Addison Wesley, Reading, Mass., 1950, Chap. 3.
39 E. Everhart, G. Stone and R.J. Carbone, Phys. Rev., 99 (1955) 1287.
40 F.W. Bingham, Sandia Res. Rep. No. SC-RR-66-506, Clearing house for Fed. Sci. and Tech. Info., NBS, U.S. Dept. of Commerce, 1966.
41 E. Taglauer and W. Heiland, Appl. Phys. Lett., 24 (1974) 437.
42 W. Heiland and E. Taglauer, J. Vac. Sci. Technol., 9 (1972) 620.
43 E. Taglauer and W. Heiland, Surface Sci., 33 (1972) 27.
44 A.A. Abrahamson, Phys. Rev., 178 (1969) 76.
45 S.H.A. Begemann and A.L. Boers, Surface Sci., 30 (1972) 134.
46 B.A. Trubnikov and Yu.N. Yavlinskii, Sov. Phys. JETP, 25 (1967) 1089.
47 W. Brandt and R. Sizmann, Phys. Lett., 37A (1971) 115.
48 T.M. Buck, W.F. van der Weg, Y.-S. Chen and G.H. Wheatley, Surface Sci., 47 (1975) 244.
49 V.M. Chicherov, Sov. Phys. JETP, 16 (1972) 231.
50 R. Behrisch, W. Eckstein, P. Meischner, B.M.U. Scherzer and H. Verbeek, in S. Datz, B.R. Appleton and C.D. Moak (Eds.), Atomic Collisions in Solids, Vol. 1, Plenum Press, New York, 1975, p. 315.
51 W.F. van der Weg and D.J. Bierman, Physica, 44 (1969) 177.
52 A.G.J. de Wit, G.A. v.d. Schootbrugge and J.M. Fluit, Surface Sci., 47 (1975) 258.
53 A. Cobas and W.E. Lamb, Phys. Rev., 65 (1944) 327.
54 H.D. Hagstrum, Phys. Rev., 96 (1954) 336.
55 E. Taglauer and W. Heiland, Surface Sci., 47 (1975) 234.
56 M. Perdereau, Surface Sci., 24 (1971) 239.
57 J.T. McKinney and J.A. Leys, 8th Nat. Conf. Electron Probe Anal., New Orleans, Louisiana, August 13—14, 1973.
58 H.H. Brongersma and W. Schouten, Acta Electron., 18 (1975) 47.
59 J.T. McKinney and R.P. Frankenthal, Nat. Assoc. Corrosion Engineers, Anaheim, California, March 22, 1973.

60 R.F. Goff, 5th Int. Vacuum Cong. Amer. Vac. Soc., Boston, Mass., 1971.
61 D. Coster, K.S. Knol and J.A. Prins, Z. Phys., 63 (1930) 345.
62 E.P. Warekois, M.C. Lavine, A.N. Mariano and H.C. Gatos, J. Appl. Phys., 33 (1962) 690. (Erratum: J. Appl. Phys., 37 (1966), 2203.)
63 H.H. Brongersma and P.M. Mul, Chem. Phys. Lett., 19 (1973) 217.
64 J.E. Demuth, D.W. Jepsen and P. Marcus, Phys. Rev. Lett., 31 (1973) 540.
65 L.H. Germer and A.U. MacRae, J. Appl. Phys., 33 (1962) 2923.
66 H.H. Brongersma and J.B. Theeten, Surface Sci., to be published.
67 S.P. Sharma and T.M. Buck, J. Vac. Sci. Technol., 12 (1975) 468.
68 V.M. Kivilis, E.S. Parilis and N.Yu. Turaev, Sov. Phys. Dokl., 12 (1967) 328.
69 W. Heiland, H.G. Schäffler and E. Taglauer, in S. Datz, B.R. Appleton and C.D. Moak (Eds.), Atomic Collisions in Solids, Vol. 2, Plenum Press, New York, 1975, p. 599.
70 P. Dahl and N. Sandager, Surface Sci., 14 (1969) 305.
71 L.K. Verhey and A.L. Boers, in S. Datz, B.R. Appleton and C.D. Moak (Eds.), Atomic Collisions in Solids, Vol. 2, Plenum Press, New York, 1975, p. 383.
72 E.P.Th. Suurmeijer and A.L. Boers, Surface Sci., 43 (1973) 309.
73 R.L. Erickson and D.P. Smith, Phys. Rev. Lett., 34 (1975) 297.
74 F.P. Ziemba, G.J. Lockwood, G.H. Morgan and E. Everhart, Phys. Rev., 118 (1960) 1552.
75 H.H. Brongersma and T.M. Buck, Surface Sci., to be published.

Chapter 4

SURFACE ANALYSIS BY X-RAY PHOTOELECTRON SPECTROSCOPY

W.M. RIGGS and M.J. PARKER

I. Introduction

The technique of X-ray photoelectron spectroscopy (XPS) is a relative newcomer to the field of surface analysis. Even though the earliest measurements of the electron kinetic energy distribution resulting from X-ray irradiation of solid materials were reported in the early part of this century [1—4], the resolution attainable at the time was insufficient for observation of actual peaks in photoelectron spectra and therefore the accuracy and precision of measurement were far from adequate for real surface analysis. In 1954, a high resolution electron spectrometer for low energy electrons produced by X-ray excitation was operated for the first time by the Swedish group headed by Siegbahn [5]. It was through the use of this instrument to measure the energy spectra of X-ray stimulated electrons that the phenomenon of photo-peaks was first observed. It was quickly recognized that the available precision in locating the photo-peaks provided a far more accurate means for studying atomic orbital energy systematics than those previously available, and the technique was first used for this purpose. It was soon discovered, however, that chemical shift effects occurred, and a classic paper was published in 1958 [6], showing that the chemical difference between copper and its oxides was clearly distinguishable by XPS. The work which immediately followed, also from the Swedish laboratories, elaborated this point and showed that chemical states of non-metallic atoms as well as the oxidation states of metals could be distinguished in many cases. Because of these readily observed and useful chemical effects, the Swedish group coined the name "Electron Spectroscopy for Chemical Analysis" to describe this technique. The corresponding acronym "ESCA" is widely used synonymously with the generic name "X-Ray Photoelectron Spectroscopy" and will be so used in this chapter. In 1966, a publication [7] appeared which pointed out explicitly that the signal obtained in the X-ray photoelectron experiment did arise from material at the sample surface. Although this paper was not quantitatively explicit in terms of the observation depth of the technique, others interpreted data in the paper as evidence that the effective sample thickness for XPS was approximately 100 Å. More recent work has shown that typical sampling depths (more explicitly defined as the electron mean free path) are in the range of 5—25 Å for metals and metal oxides, while values ranging from 40 to 100 Å are common for organic and polymeric materials. Through the late 1960's and into the early part of the 1970's, ESCA was primarily of interest to chemists for its ability to probe sensitively the electron density around atoms in molecules through observation of the chemical shift effect. The research papers which have appeared over the last decade describing details of the chemical shift effect and how this provides information on chemistry of molecular compounds provides an excellent background for interpretation of spectra of monolayer and sub-monolayer quantities of materials present at surfaces. It is the combination of high surface sensitivity and the ability to provide chemical information about species observed at surfaces that is the great strength of ESCA and provides its unique position among surface analysis techniques.

The plan of this chapter is to discuss first the fundamentals of XPS including a basic discussion of the chemical shift effect with illustrations designed to provide qualitative insight into the systematics of the chemical shift. The basic physics will be described in rudimentary terms only, as more sophisticated discussions are available elsewhere [8,9]. A discussion of the escape depth or effective sample thickness in ESCA will also be included in this section. There will then follow a discussion of instrumentation for ESCA, including comparison of the approaches to ESCA instrumentation embodied in the various commercial electron spectrometers. The important subject of sample charging and methods of correcting for sample charge will be discussed. There will also be a discussion of the means for enhancing surface sensitivity when desired and of the information obtainable from ESCA by using sample conditioning experiments such as ion etching. The final section will include illustrative applications of ESCA to such fields as catalysis, polymers, and metals. The objective inherent in this approach is to provide a progressive understanding of the various elements comprising the body of information, which allows complete and detailed surface analysis with this technique. This will be emphasized through the use of specific examples.

To understand where ESCA fits relative to other surface analysis techniques, it is useful to consider five factors: (a) its non-destructive nature, (b) the ability to study plastic and organic surfaces, (c) the factors which govern the sampling depth, (d) the vacuum considerations for doing both applied and fundamental research, and (e) the chemical shift effect. These factors are discussed below, highlighting the similarities and differences which they provide to ESCA compared with other techniques.

ESCA can be considered a non-destructive technique because the X-ray beam used to excite the photoelectron spectrum is relatively harmless to most materials, especially when compared with surface analysis techniques which depend upon ion or electron bombardment of the surface. Thus, for most samples, the material can be observed in the photoelectron spectrometer and then kept for archival purposes or used for other analytical tests. Although some photo-decomposition occurs with some kinds of systems, this phenomenon is relatively rare and almost never prevents useful data from being obtained.

The signal observed with ESCA originates in the sample surface itself whereas, by contrast, ISS depends upon observation of species scattered from the surface and SIMS and ion probe techniques depend on measuring species sputtered from surfaces. This means that ESCA does not have intrinsic depth profiling capability as ion scattering and sputtering techniques do. This can be an advantage or a disadvantage. If depth profile information is required, intrinsic depth profiling capability is a desirable characteristic. However, if one is attempting to measure traces of materials producing weak signals, the ability to signal average on a surface which does not change as a function of experiment time may be invaluable. Of course, it is possible to use sputtering devices in conjunction with ESCA measurement when depth profile information is required. The compromises implied by these alternatives must always be considered when choosing a technique for a particular application.

The ability to obtain detailed chemical information from plastic and organic surfaces is an unparalleled capability of ESCA. Electron bombardment techniques, such as AES, tend to be too destructive of plastic surfaces for use in their analysis. Ion sputtering and ion scattering techniques are rather destructive and do not provide chemical information directly. The electron mean free path of 40—100 Å in polymeric materials provides a surface sensitivity unmatched by other techniques commonly used to obtain information from polymer surfaces, such as attenuated total reflectance (ATR) infrared spectroscopy, where the penetrating depth of the probing radiation is of the order of microns. Thus, for the study of polymer surface treatments which are important in such areas as adhesion, weathering, dyeability, and printability of polymeric materials, ESCA is the only technique with sufficient surface sensitivity and specificity to detect the surface chemistry which dominates the properties of polymer surfaces.

The sample depth in ESCA is governed by the electron mean free path (MFP) in the sample material. Since this is also the effect which governs sample depths in AES, the effective sample observed in both techniques is essentially the same. The MFP is a function both of sample composition and of the kinetic energy of the escaping electrons. Thus, the effective sample thickness may not be exactly the same for photoelectron peaks observed at different points in the spectrum. This factor must be considered for careful quantitative studies.

One of the important characteristics of ESCA not shared by all other techniques is the ability to perform surface analysis measurements at relatively moderate vacuum levels. When electron beam excitation is used, as with AES for example, UHV must be used to prevent formation of carbonaceous deposits on the sample with attendant masking of the surface to be measured. However, the relatively gentle nature of the X-ray beam allows observation of surfaces for many hours at moderate vacuum levels, particularly if the vacuum composition is favorable as is typically the case when ion pumps or other "dry" pumping is utilized. It is, of course, true that atomically clean metal surfaces, for example, cannot be maintained for significant periods of time at pressures above 10^{-9} to 10^{-10} torr. The vast majority of surface analysis work, however, is not done on atomically clean surfaces and therefore does not require UHV for ESCA.

The chemical shift effect is the other characteristic which sets XPS apart from other well known surface characterization techniques. A decrease in electron density in the valence region around an atom in a molecule produces an increase in the binding energy of core level electrons. Thus, binding energy shifts can be readily interpreted in terms of well understood chemical concepts. Although sophisticated theoretical studies of this effect have been carried out, the application of rudimentary chemical intuition and understanding is normally sufficient for interpretation of chemical shifts in ESCA. By contrast, chemical shifts in AES are more difficult to interpret. As yet, no chemical shift information is available at all from ISS. Mass spectrometric fragmentation patterns provide indirect chemical information from ion probe and SIMS techniques since the composition of the ions sputtered from the surface is, of course, related to the composition of the surfaces themselves.

II. Fundamentals

A. X-RAY ABSORPTION

The basic ESCA experiment is shown schematically in Fig. 1 for a free atom in the gas phase. X-ray photons, $h\nu$, from a nearly monoenergetic beam are directed onto the sample. The photons are absorbed by sample atoms with each absorption event resulting in the prompt emission of an electron.

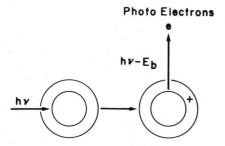

Fig. 1. The physical basis of ESCA.

Electrons from all the orbitals of the atom with a binding energy, E_b, less than the X-ray energy are excited, though not with equal probability. Thus, some peaks are more intense in the spectra than others. Since energy is conserved, the kinetic energy, KE, of the electron plus the energy required to remove it from its orbital to the spectrometer vacuum must equal the X-ray energy. If the X-ray energy is known and the kinetic energy is measured with the electron spectrometer, the binding energy of the electron in the atomic orbital can be obtained. In practical solid state experiments, a correction for the spectrometer work function, ϕ_s, must also be applied, normally as part of the spectrometer calibration procedure. Thus, one obtains $E_b = h\nu - KE + \phi_s$.

B. QUALITATIVE ANALYSIS

Since the atomic structure of each element in the periodic table is distinct from all the others, measurement of the positions of one or more of the electron lines allows the ready identification of an element present at a sample surface. As an example, Fig. 2 shows the ESCA spectrum of tetra-propylammonium difluorodithiophosphate. Each of the elements in this compound produces at least one electron line in the spectrum with the exception of hydrogen, which is not detected in ESCA experiments. In addition, the presence of oxygen indicates the likelihood that the sample was partially oxidized from exposure to air or that it contains some water of

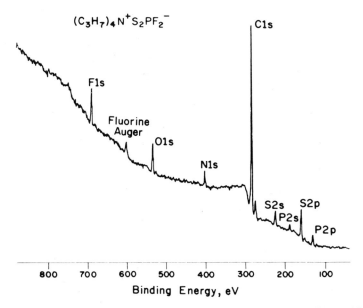

Fig. 2. Survey spectrum illustrating qualitative elemental identification by ESCA.

hydration. Note that elements adjacent to one another in the periodic chart produce electron lines which are well separated from one another so that no ambiguity exists in identification of adjacent elements. Thus, carbon, nitrogen, oxygen, and fluorine from the first period and sulfur and phosphorus from the second period are all easily distinguished from one another.

C. QUANTITATION

The intensity of the signal observed is, of course, a function of the amount of material present. Thus, observation of signal intensities in ESCA spectra can provide semi-quantitative and quantitative analyses of surface material. In addition to the concentration of the element producing the signal, ESCA intensities depend upon the MFP of the electrons and the efficiency of absorption of the exciting X-rays by the sample material. If these factors are well understood or if good standards are available, quantitative analysis can be performed using ESCA. Considering these factors in order, the concentration of elements present at the sample surface is, of course, the quantity usually desired and it can be obtained from the signal intensity if adequate account can be taken of the effects of the other variables.

The MFP of the electron and the way it varies with kinetic energy of the electron and composition of the matrix, is currently a subject of research. Preliminary information providing some insight is, however, available. A compilation of data from many sources of chemical and physics literature on electron MFP in metals is shown in Fig. 3. The MFP's are seen to range from

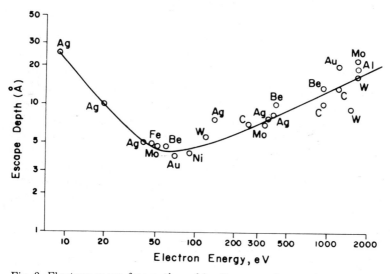

Fig. 3. Electron mean free path vs. kinetic energy in metals.

approximately 5 to about 30 Å. The shape of this curve and the tendency for data points to cluster about it regardless of the material being observed suggests that the kinetic energy of the electron in metals is a major factor governing escape depths. The minimum around 50 eV indicates that scattering interactions of free electrons with other electrons, and with plasmons in solids, tend to maximize near a kinetic energy of about 50 eV. It is probably fortunate that the vast majority of practical ESCA measurements occur on electrons with kinetic energies of above 100 eV; thus, to a first approximation, the electron MFP in ESCA experiments increases linearly with kinetic energy on a log/log plot. The scatter of the points around the line, while undoubtedly due in part to experimental error and differences among the techniques used to make the measurements, is also due in part to intrinsic differences in the ability of the different matrices to scatter electrons . This difference becomes more important when very different materials such as metals, oxides and polymeric materials are compared. The MFP of 1000 eV electrons in typical polymeric materials ranges from 40 to 100 Å, while measurements on oxide systems suggest that electron MFP's in oxides are somewhat longer than they are in metals. A rule of thumb, which gives qualitatively correct impressions, is that MFP tends to be of the order of 5—20 Å in metals, about 15—40 Å in oxides, and about 40—100 Å in polymers. Further insight is currently being developed through research in this field, but the results which are available to date are quite tentative and preliminary. Thus, it would be premature to include them in this discussion since considerable change will undoubtedly occur as the theories describing these effects are refined.

The most important factor in doing relatively quick semi-quantitative estimations with ESCA is the efficiency with which X-rays are absorbed by atoms of the sample. If the cross-section for X-ray absorption of a particular atomic subshell from which an electron is being observed is high, then the sensitivity for this element will be high. Conversely, the lower the absorption cross-section, the less sensitive the element will be for ESCA observation. This accounts for the relative weakness of valence band spectra in ESCA, where the low binding energy electrons interact only weakly with the incoming X-rays. Wagner [10] has shown (Fig. 4) that a systematic variation of elemental sensitivity occurs throughout the periodic chart. In Wagner's work, all sensitivities are referenced to fluorine and most of the measurements are made on fluoride compounds for this reason. Fluorine is assigned an atomic sensitivity of one and all other elemental sensitivities are then plotted relative to this number. The sensitivity of the light elements increases with atomic number to $Z = 12$, where it drops dramatically. This behavior is repeated three more times in the periodic chart and corresponds to the successive observation of 1s, 2p, 3d and 4f electrons as those producing the strongest electron lines in the ESCA spectra. The resemblance of Fig. 4 to an X-ray absorption spectrum leads immediately to the suggestion that the

TABLE 1

Photoelectric cross-sections relative to $C1s$ level for $MgK\alpha_{1,2}$ (1254 eV) radiation

	$1s_{1/2}$	$2s_{1/2}$	$2p_{1/2}$	$2p_{3/2}$	$3s_{1/2}$	$3p_{1/2}$	$3p_{3/2}$	$3d_{3/2}$	$3d_{5/2}$	$4s_{1/2}$	$4p_{1/2}$	$4p_{3/2}$	
1 H	0.0002												
2 He	0.0089												
3 Li	0.059	0.0008											
4 Be	0.200	0.007											
5 B	0.492	0.022											
6 C	1.00	0.047	0.0006	0.0012									
7 N	1.78	0.084	0.0025	0.0049									
8 O	2.85	0.134	0.0073	0.0145									
9 F	4.26	0.199	0.0178	0.0352									
10 Ne	5.95	0.277	0.0381	0.0751									
11 Na	7.99	0.390	0.0714	0.141	0.0059								
12 Mg		0.525	0.121	0.239	0.0261								
13 Al		0.681	0.193	0.380	0.0485	0.0012	0.0023						
14 Si		0.855	0.292	0.573	0.0726	0.0050	0.0097						
15 P		1.05	0.422	0.828	0.0998	0.0129	0.0253						
16 S		1.25	0.590	1.155	0.130	0.0269	0.0527						
17 Cl		1.48	0.810	1.564	0.163	0.0493	0.0963						
18 Ar		1.71	1.06	2.07	0.199	0.0503	0.161						
19 K		1.96	1.37	2.67	0.249	0.122	0.238			0.006			
20 Ca		2.21	1.74	3.39	0.305	0.169	0.330			0.023			
21 Sc		2.46	2.18	4.24	0.356	0.216	0.420	0.002	0.003	0.027			
22 Ti		2.72	2.68	5.22	0.408	0.268	0.521	0.006	0.009	0.031			
23 V		2.98	3.26	6.33	0.462	0.326	0.633	0.014	0.021	0.034			
24 Cr		3.23	3.92	7.60	0.511	0.382	0.740	0.030	0.044	0.014			
25 Mn		3.48	4.63	8.99	0.575	0.460	0.892	0.048	0.071	0.040			
26 Fe		3.70	5.43	10.5	0.634	0.535	1.04	0.079	0.116	0.043			
27 Co		3.92	6.28	12.2	0.693	0.616	1.19	0.122	0.179	0.045			
28 Ni		4.16	7.18	13.9	0.753	0.701	1.36	0.181	0.265	0.048			
29 Cu		4.38	8.18	15.9	0.805	0.779	1.50	0.268	0.390	0.019			
30 Zn		4.55	9.29	18.0	0.873	0.882	1.70	0.365	0.532	0.052			
31 Ga			10.6	20.5	0.945	0.993	1.92	0.485	0.708	0.074	0.006	0.011	
32 Ge				21.2	1.02	1.11	2.15	0.631	0.920	0.094	0.018	0.034	
33 As					1.10	1.24	2.40	0.802	1.17	0.114	0.037	0.071	
34 Se					1.18	1.37	2.65	1.00	1.46	0.134	0.064	0.123	
35 Br					1.26	1.50	2.92	1.24	1.80	0.156	0.100	0.191	
36 Kr					1.35	1.64	3.20	1.50	2.19	0.178	0.144	0.276	
37 Rb					1.43	1.79	3.48	1.81	2.63	0.209	0.187	0.361	
38 Sr					1.52	1.93	3.78	2.15	3.14	0.242	0.230	0.445	
39 Y					1.61	2.08	4.09	2.54	3.70	0.273	0.268	0.521	C
40 Zr					1.70	2.24	4.40	2.97	4.33	0.305	0.307	0.596	C
41 Nb					1.79	2.39	4.71	3.45	5.01	0.333	0.340	0.661	C
42 Mo					1.89	2.54	5.03	3.97	5.77	0.364	0.379	0.739	C
43 Tc					1.98	2.69	5.36	4.54	6.60	0.397	0.419	0.818	C
44 Ru					2.07	2.84	5.68	5.17	7.51	0.429	0.460	0.899	C
45 Rh					2.15	2.98	6.00	5.84	8.48	0.463	0.501	0.981	C

$4f_{5/2}$	$4f_{7/2}$	$5s_{1/2}$	$5p_{1/2}$	$5p_{3/2}$	$5d_{3/2}$	$5d_{5/2}$	$5f_{5/2}$	$5f_{7/2}$	$6s_{1/2}$	$6p_{1/2}$	$6p_{3/2}$
0.006											
0.021											
0.025											
0.029											
0.013											
0.014											
0.015											
0.016											
0.016											

TABLE 1 (continued)

	$1s_{1/2}$	$2s_{1/2}$	$2p_{1/2}$	$2p_{3/2}$	$3s_{1/2}$	$3p_{1/2}$	$3p_{3/2}$	$3d_{3/2}$	$3d_{5/2}$	$4s_{1/2}$	$4p_{1/2}$	$4p_{3/2}$	4
46 Pd					2.24	3.12	6.33	6.58	9.54	0.494	0.538	1.06	(
47 Ag					2.33	3.25	6.64	7.36	10.7	0.531	0.586	1.15	(
48 Cd					2.40	3.39	6.96	8.22	11.9	0.571	0.636	1.26	(
49 In					2.48	3.51	7.27	9.13	13.2	0.611	0.689	1.37	(
50 Sn					2.54	3.62	7.58	10.1	14.6	0.653	0.743	1.48	1
51 Sb					2.60	3.71	7.86	11.1	16.1	0.696	0.799	1.60	1
52 Te					2.67	3.79	8.14	12.2	17.7	0.741	0.856	1.73	1
53 I					2.75	3.87	8.37	13.3	19.3	0.785	0.913	1.86	1
54 Xe					2.83	3.95	8.64	14.5	21.1	0.831	0.971	1.99	1
55 Cs					2.84	4.04	8.94	15.8	22.9	0.877	1.03	2.12	1
56 Ba						4.10	9.26	17.0	24.8	0.924	1.09	2.26	2
57 La						4.06	9.52	18.2	26.5	0.971	1.15	2.40	2
58 Ce							9.67	19.7	28.6	1.00	1.18	2.49	2
59 Pr							9.75	21.1	30.7	1.04	1.22	2.59	2
60 Nd								22.6	32.9	1.07	1.26	2.70	2
61 Pm								24.3	35.3	1.10	1.30	2.81	
62 Sm								26.1	37.9	1.14	1.34	2.91	
63 Eu								28.2	40.9	1.17	1.37	3.01	
64 Gd								24.3	43.4	1.20	1.41	3.13	
65 Tb									20.8	1.22	1.43	3.21	
66 Dy										1.25	1.45	3.30	4
67 Ho										1.27	1.47	3.39	4
68 Er										1.29	1.49	3.48	4
69 Tm										1.31	1.50	3.56	
70 Yb										1.32	1.51	3.64	4
71 Lu										1.34	1.52	3.73	4
72 Hf										1.36	1.53	3.83	
73 Ta										1.38	1.54	3.93	
74 W										1.39	1.55	4.03	
75 Re										1.41	1.55	4.13	
76 Os										1.42	1.55	4.24	
77 Ir										1.43	1.55	4.34	
78 Pt										1.44	1.54	4.45	
79 Au										1.45	1.53	4.55	
80 Hg										1.45	1.52	4.65	
81 Tl										1.46	1.50	4.75	
82 Pb										1.46	1.47	4.86	
83 Bi										1.44	1.45	4.96	

X-ray absorption cross section for the sub-shell being observed is the factor which dominates the observed sensitivity. The scatter observed about the lines in Fig. 4 for sodium ($Z = 11$), as an example, is due to matrix effects. Note that these matrix effects can produce errors as large as ±50%. Scofield [11] has published an extensive list of X-ray absorption cross-sections for the sub-shells of electrons in the elements of the periodic chart and has specifically utilized X-ray energies of 1254 eV and 1487 eV to produce X-ray

$4f_{5/2}$	$4f_{7/2}$	$5s_{1/2}$	$5p_{1/2}$	$5p_{3/2}$	$5d_{3/2}$	$5d_{5/2}$	$5f_{5/2}$	$5f_{7/2}$	$6s_{1/2}$	$6p_{1/2}$	$6p_{3/2}$
		0.018									
		0.046									
		0.063	0.006	0.011							
		0.076	0.017	0.031							
		0.090	0.033	0.063							
		0.104	0.054	0.104							
		0.117	0.080	0.156							
		0.132	0.112	0.218							
		0.152	0.140	0.278					0.005		
		0.174	0.166	0.334					0.017		
		0.193	0.189	0.382	0.019	0.027			0.021		
0.069	0.088	0.189	0.180	0.365					0.018		
0.126	0.161	0.196	0.186	0.378					0.018		
0.200	0.257	0.202	0.191	0.390					0.018		
0.296	0.378	0.208	0.196	0.400					0.018		
0.416	0.531	0.213	0.201	0.412					0.019		
0.562	0.718	0.219	0.205	0.422					0.019		
0.693	0.887	0.235	0.223	0.465	0.022	0.031			0.022		
0.949	1.21	0.228	0.211	0.440					0.019		
1.20	1.52	0.232	0.214	0.449					0.019		
1.49	1.89	0.237	0.216	0.457					0.019		
1.82	2.31	0.240	0.219	0.464					0.019		
2.20	2.78	0.244	0.220	0.471					0.019		
2.63	3.33	0.247	0.222	0.478					0.019		
3.05	3.87	0.261	0.237	0.519	0.021	0.029			0.023		
3.50	4.45	0.271	0.252	0.562	0.054	0.074			0.026		
3.99	5.08	0.290	0.268	0.606	0.098	0.136			0.029		
4.92	5.75	0.306	0.283	0.651	0.152	0.212			0.031		
5.08	6.46	0.321	0.299	0.697	0.217	0.303			0.033		
5.67	7.22	0.337	0.314	0.743	0.293	0.410			0.033		
6.30	8.03	0.350	0.324	0.774	0.431	0.593					
6.97	8.87	0.366	0.340	0.825	0.508	0.709			0.017		
7.68	9.79	0.380	0.353	0.877	0.619	0.865			0.017		
8.43	10.8	0.397	0.368	0.935	0.707	0.997			0.040		
9.22	11.8	0.413	0.383	0.996	0.804	1.14			0.051	0.004	0.008
10.0	12.8	0.430	0.398	1.06	0.900	1.29			0.059	0.011	0.023
10.9	14.0	0.446	0.412	1.12	0.997	1.44			0.068	0.021	0.046

absorption coefficients relevant to ESCA measurements. His data have been normalized to a relative atomic sensitivity of 1.00 for carbon, and the resulting relative sensitivities for magnesium X-rays have been tabulated as shown in Table 1. The normalization is the work of Swingle [12]. The use of this table as a guide to the relative atomic sensitivity of the elements agrees well both with Wagner's data in Fig. 4 and also with empirical experience. For example, the table shows the relative sensitivity of oxygen to carbon to be

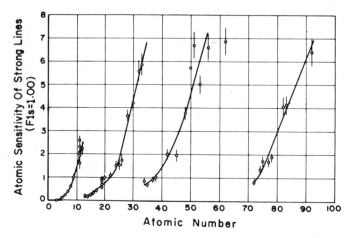

Fig. 4. Relative sensitivity of the elements. After Wagner [10].

approximately 3.0, which is in agreement to within ±20% with the observed ratios from a wide variety of pure compounds, such as succinic acid, poly-ethylene terephthalate, carbonates, and other organic and inorganic com-pounds. Thus, if observed peak intensity ratios in ESCA spectra are cor-rected according to the relative sensitivity ratios in this table, useful element abundance ratios can be obtained with far less complexity than if a complete analysis must be made which takes into account the variation of electron MFP with matrix and kinetic energy. An example of this can be found in the applications section on organic surfaces, where the fluorine-to-carbon peak intensity ratio in Teflon having the empirical formula of CF_2 is shown to be 8.2. If this value is corrected for the relative atomic sensitivity of fluorine to carbon, given in the table as 4.26, an atom abundance ratio of 1.94 is obtained, which is in good agreement with the known actual abundance ratio of 2.0.

With appropriate standards, the technique can be used to make other kinds of quantitative measurements, such as the thickness of very thin films. An example is the measurement of the thickness of an amide layer on a polyethylene film, also discussed in the applications section under organic surfaces. Most elements can be detected at 0.1 at.% abundance with 10—15 min data acquisition times. In favorable cases, the detection limits have been pushed below one atom in 100,000.

III. Chemical shifts

The bulk of the published ESCA research in the last fifteen years has been directed toward understanding the systematic shift in peak positions resulting from changes in the chemical structure and oxidation state of chemical compounds. This work provides a valuable resource to chemists engaged in surface characterization since chemical shifts observed from species at surfaces can be interpreted using the published fundamental work as a basis. Only a few examples of this kind of basic information will be discussed here and the reader is referred to primary references such as the work of Siegbahn [8] and the research literature for more detailed information pertinent to specific chemical systems.

A. ORGANIC STRUCTURAL INFORMATION

The spectrum obtained from polyethylene terephthalate, a polyester widely used in packaging and photographic films and synthetic textile fibers, is shown in Fig. 5. Examination of the structural formula reveals that there are three fundamentally different kinds of carbons in the compound: benzene ring carbons, carboxyl group carbons, and those in the CH_2 group bound to the oxygen of the terephthalic acid unit. Each of these carbons is in a different electrostatic environment and therefore exhibits a different chemical shift, producing carbon $1s$ peaks in different positions in the spectrum. Carbon in the most electronegative environment (the ester carboxyl group) appears at the highest binding energy. This is because the electronegative oxygen atoms withdraw electron density from the valence and bonding orbitals of the carbon atom, thereby reducing the screening of the core electrons from the nuclear charge and increasing their binding energy. Thus, in

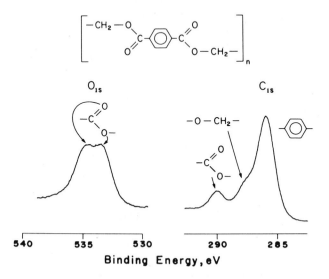

Fig. 5. Carbon $1s$ and oxygen $1s$ spectra of polyethylene terephthalate showing a chemical shift related to molecular structure.

general, atoms with highly electronegative substituent groups can be expected to exhibit higher binding energies than the same atoms bound to groups with lower electronegativity. Accordingly, the —OCH$_2$— carbon is at an intermediate binding energy between carboxyl and ring carbons, and the ring carbons exhibit the lowest binding energy. The different chemical environment of the two kinds of oxygen atoms in this molecule is also reflected in the oxygen 1s spectrum as seen in Fig. 5. Here, the carboxyl oxygen appears at a higher binding energy than the oxygen atom bridging the terephthalate and glycol carbons.

As would be expected from the above qualitative discussion, substitution of highly electronegative elements, such as fluorine, will induce the largest chemical shifts. This is illustrated by the series of fluoropolymers ranging from polyethylene to polytetrafluoroethylene, which differ through the series by the sequential substitution of fluorine for hydrogen in the monomer unit. The carbon 1s spectra of three members of this series are shown in Fig. 6 [13]. Inspection of the spectra reveals that substitution of fluorine does cause the carbon 1s peak to shift to significantly higher binding energies. The second member of the series, polyvinylfluoride, exhibits two peaks in a one-to-one intensity ratio as expected from the presence of one monofluorinated carbon and one carbon with no fluorine atoms in the monomer

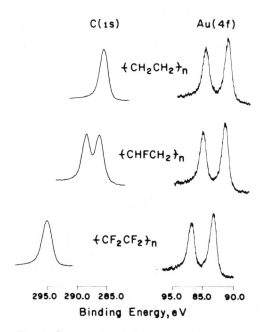

Fig. 6. Spectra from fluorinated polymers showing chemical shifts in carbon 1s induced by fluorine substitution. The Au 4f spectra are from vapor deposited gold utilized for charging correction.

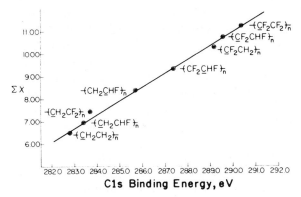

Fig. 7. Carbon 1s peak positions in fluoropolymers correlated with substituent electronegativity.

unit. The maximum shift is observed for the fully fluorinated polytetra-fluoroethylene. The gold 4f spectra also shown in this figure are from small amounts of gold deposited on the polymer surface to make accurate correction for sample charging effects. Charging effects and the means for correcting them are discussed elsewhere in this chapter.

These chemical shifts can be understood in more than a purely qualitative fashion as is illustrated in Fig. 7, where the charge-corrected carbon 1s binding energies have been plotted versus the calculated electronegativity of the substituents bound to the carbon. The observed correlation suggests that in well-understood systems the accurate measurement of peak positions can be used to identify definitively chemical species present at organic surfaces. Some specific examples of using this kind of information in the surface analysis of polymers is given in Section VI.A.1.

B. INORGANIC STRUCTURAL AND CHEMICAL INFORMATION

The spectra of aluminum, palladium, and tantalum oxides, shown in Figs. 8—10, illustrate that a chemical shift to higher binding energy will be observed for most metals when oxidized. The shift is frequently of the order of 1 eV per unit change in oxidation state, as in the examples shown. The aluminum spectrum is from a metal sample cleaned by abrasion in the laboratory atmosphere. Thus, the peak at the higher binding energy is due to the overlayer of aluminum(III) oxide and the sharper peak at lower binding energy is from the aluminum substrate below the approximately 15 Å thick oxide layer. The palladium spectrum is from a mixture of palladium(II) oxide and palladium metal, while the tantalum spectrum is from a tantalum foil sample with tantalum(V) oxide on the surface. In the case of the palladium and tantalum spectra shown, the 3d and 4f spectra are doublets; therefore, the presence of two distinct chemical species produces a doublet of

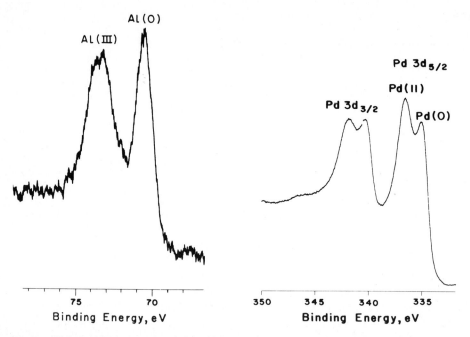

Fig. 8. Aluminum $2p$ spectrum from aluminum metal with passivating oxide layer showing chemical shift of Al(III) relative to Al(0).

Fig. 9. Palladium $3d$ spectra of mixture of Pd(0) and Pd(II) oxide illustrating chemical shift.

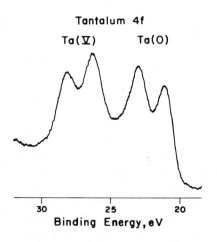

Fig. 10. Tantalum $4f$ spectra from tantalum foil with surface oxide.

TABLE 2

Useful chemical shifts in X-ray excited Auger spectra

| | $2p$ | $3d$ | Auger apparent binding energy | | Auger transition |
			Mg anode	Al anode	
Mg	49.8			300.8	
MgO	51.2			307.2	
Al	72.8			93.5	KLL
Al$_2$O$_3$	75.4			100.2	
Zn		9.9	261.0	494.0	
ZnO		10.7	265.6	498.6	
Ga		18.5	185.3	418.3	
Ga$_2$O$_3$		19.7	190.8	423.8	
Ge		29.4	108.5	341.5	
GeO$_2$		33.2	116.3	349.3	LMM
As		42.05		261.85	
As$_2$O$_3$		45.8		268.6	
Se		55.5		179.7	
Na$_2$SeO$_3$		58.3		185.2	
Ag		374.0	895.2	1128.2	
Ag$_2$SO$_4$		374.2	899.2	1132.2	
Cd		411.55	869.7	1102.7	
CdO		411.65	873.4	1106.4	
In		451.8	843.3	1076.3	
In$_2$O$_3$		452.3	846.9	1079.9	MNN
Sn		492.9	815.3	1048.3	
SnO$_2$		494.6	820.4	1053.4	
Sb		537.3	789.2	1022.2	
Sb$_2$O$_3$		539.2	793.7	1026.7	
Te		583.7	761.9	994.9	
TeO$_2$		587.2	767.1	1000.1	

doublets. In the case of oxidation of tantalum(0) to tantalum(V), the chemical shift is so large that the entire tantalum $4f$ doublet due to the oxidized species falls to the left (higher binding energy) of the tantalum(0) peaks. In cases such as these, the presence and identity of oxidized species of metals at surfaces is easy to determine. There are cases known for metals, such as tin and silver, where the chemical shift upon oxidation is too small to be analytically useful. Fortunately, in virtually all of these cases, the most prominant X-ray excited Auger line, which, of course, is also visible in the ESCA spectrum, exhibits an analytically useful chemical shift between the oxidized and

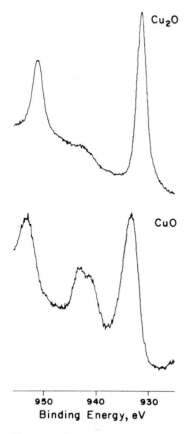

Fig. 11. Copper $2p$ spectra of Cu(I) and Cu(II) oxides showing presence of "shake-up" satellite only on Cu(II).

reduced species. A tabulation of a group of such metallic elements is shown in Table 2. The photoelectron chemical shift and the Auger chemical shift are tabulated for the elements and selected representative oxidized compounds of these elements [14].

The spectra of copper(I) and copper(II) oxides shown in Fig. 11 exhibit an analytically useful shift of the peak position to approximately 1.6 V higher binding energy upon oxidation from copper(I) to copper(II). Another interesting and useful feature is also observed. The large subsidiary peak at about 943 eV in the copper(II) oxide is referred to as a "shake-up" peak and appears only in the spectrum of the copper(II) species. This type of subsidiary peak is due to a process where the primary photoelectron on leaving the atom causes the excitation of a valence electron to an unfilled higher energy level, thus removing some energy from the photoelectron and making it appear at apparently higher binding energy in the spectrum. At least in the

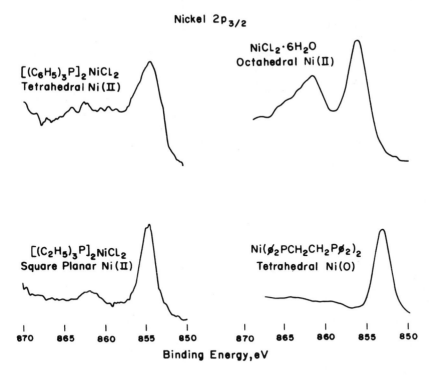

Fig. 12. Nickel $2p_{3/2}$ spectra from organometallic compounds. Intense "shake-up" satellites are only observed from paramagnetic species.

first row transition metals, the shake-up transition appears to involve promotion of an electron from a ligand orbital to an empty metal $3d$ orbital. Such peaks only appear at significant strengths in paramagnetic as opposed to diamagnetic compounds, thus providing another piece of valuable chemical information. This is clearly illustrated in the case of the nickel $2p$ spectra in Fig. 12. Here, the tetrahedral nickel(II) and octahedral nickel(II) compounds, whose spectra are shown at the top of the figure, are both paramagnetic species, and both show intense shake-up transitions. Spectra of the diamagnetic planar nickel(II) and tetrahedral nickel(0) compounds, shown in the lower half of the figure, do not exhibit significant shake-up peaks. Thus, the presence or absence of these peaks can be used in many cases as a diagnostic tool to determine the magnetic character and, by implication, the chemical structure and oxidation states in inorganic and organometallic compounds. While it is certainly not recommended that ESCA be used as a substitute for magnetic measurements in characterizing bulk inorganic species, the technique can be invaluable in characterizing inorganic species at surfaces where magnetic measurements may be impossible to make.

References pp. 157—158

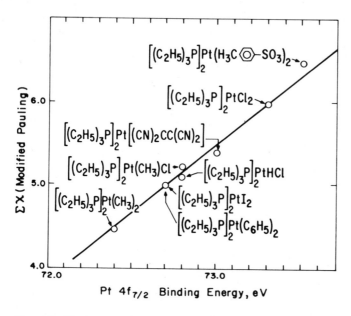

Fig. 13. Platinum $4f_{7/2}$ peak positions in organometallic compounds correlated with substituent electronegativity.

As with organic compounds, the substitution of groups with different electronegativities causes subtle changes in the chemical shift in inorganic species and can be used to study the details of the electronic environment of species at surfaces. This is illustrated in Fig. 13 for a series of platinum(II) compounds where the calculated electronegativity of the substituents on the metal atom is plotted versus the charge-corrected binding energies observed for core levels of the metal itself. A linear correlation is observed which is good enough to distinguish among quite a number of different inorganic species [15].

This kind of basic chemical information is available in the ESCA literature and may be utilized as a guide in studying complex organometallic and inorganic chemistry at surfaces in a variety of practical systems involving, for example, catalysis, lubrication, adhesion and corrosion. While these chemical effects do not produce chemical shifts large enough to be as useful in structural studies as, for example, the chemical shift effects in nuclear magnetic resonance, they are nevertheless valuable in providing supplemental chemical information about materials present at surfaces. Furthermore, the existing ESCA literature, which is growing at a rapid rate, provides a wealth of background information to aid the surface scientist in interpreting observed phenomena.

IV. Instrumentation

A. INTRODUCTION

The basic elements of an X-ray photoelectron spectrometer are depicted in Fig. 14. The functions of the spectrometer are to produce intense X-radiation, to irradiate the sample to photoeject core electrons, to introduce the ejected electrons into an energy analyzer, to detect the energy-analyzed electrons, and to provide a suitable output of signal intensity as a function of electron binding energy. There are a number of commercial instruments presently available which accomplish these functions, each having a different approach for design of the source, energy analyzer and detector. Some of the characteristics of these systems are listed in Table 3 [16,17].

B. X-RAY SOURCES

A basic X-ray source includes a heated filament and a target anode. A potential is applied between the filament and the anode to accelerate electrons emitted from the filament toward the target. Electron bombardment of the target causes emission of X-rays which are characterized by a continuum (termed Bremsstrahlung radiation) upon which is superimposed dis-

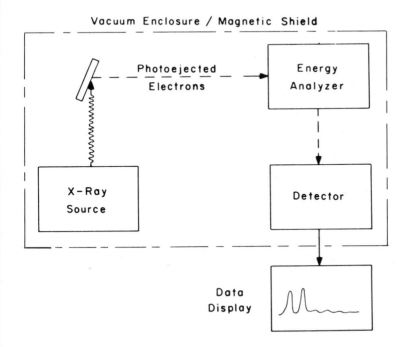

Fig. 14. The basic elements of an X-ray photoelectron spectrometer.

References pp. 157—158

Characteristics of commercial ESCA instruments

	AEI ES-200	Du Pont 650B	Hewlett–Packard ESCA-5950A	McPherson ESCA-36	PEI	Vacuum generators ESCA-3
Energy analyzer						
Type	Hemispherical, retarding lens	Energy filter, retarding field	Hemispherical, retarding lens	Hemispherical	Cylindrical mirror, retarding field	Hemispherical, retarding lens
Scan mode	Retarding and hemisphere voltages	Retarding voltage	Retarding voltage	Hemisphere voltage	Mirror voltage	Retarding voltage
X-Ray source						
Anode	Mg, Al, Cr, Cu	Mg, Al	Al[a]	Mg, Al, Cu	Al	Mg, Al
Sample						
Standard size (cm)	0.5 × 1.5	0.63 or 1.11	0.7 × 1.0	1 × 2	1 × 2	Not available
Temperature range (°C)	−150 to +350	−180 to +500[b]	−180 to +300[b]	−190 to +400[b]	Ambient	−160 to +300[b] or ambient to +600
Preparation[c]	IE, SD	IE, SD	IE, SD	IE, SD	IE, SD	IE
Detector						
Type	Channel electron multiplier	Channel electron multiplier	Multichannel detector and vidicon	Channel electron multiplier	Channel electron multiplier	Channel electron multiplier
Vacuum						
Pump	Diffusion	Ion	Ion	Turbomolecular	Ion	Diffusion
Base pressure (torr)	$10^{-7\,d}$	$10^{-8\,d}$	10^{-8}	$10^{-7\,d}$	10^{-10}	10^{-10}
Data acquisition						
Analog	Yes	Yes	No	No	Yes	Yes
Digital	Optional computer	Optional MCA[e]	MCA[e], optional computer	Computer	No	Optional MCA[e]

[a] Dispersed K_α radiation.
[b] Accessory sample probe.
[c] IE = sample ion-etching accessory; SD = evaporative sample-deposition accessory.

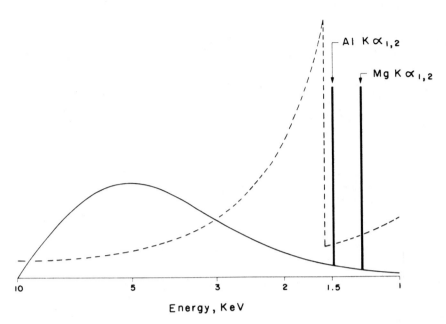

Fig. 15. The intensity distribution from an aluminum or magnesium anode X-ray source at 10 kV (solid curve); the mass absorption coefficient of aluminum as a function of energy (broken curve) is superimposed.

crete wavelengths of varying intensity as illustrated in Fig. 15. Continuous radiation occurs as the bombarding electrons decelerate and interact with the electrons of the target. The continuum has a maximum energy limit which corresponds to the maximum energy of the exciting electrons, while the intensity maximum (disregarding the "characteristic" wavelength peaks) occurs at an energy corresponding roughly to 50% of the energy maximum. The overall intensity of the continuum is directly proportional to the filament current, potential applied between the filament and anode, and the atomic number of the target material; however, the relative distribution of the continuum remains virtually constant. Electrons are also ejected from inner orbitals of target atoms during the bombardment process. The target atoms thus excited are unstable and undergo relaxation processes; this involves a re-arrangement of orbital electrons from outer orbitals to fill vacancies in the inner orbitals. With each such transition, the atom moves to a less energetic state and radiation is emitted at an energy corresponding to the difference in energy between the initial and final states. Characteristic radiation from the target is produced only when the electrons from the filament have sufficient energy to eject the inner electrons of the target atoms. The threshold energy varies with atomic number [18].

For the purpose of producing a photoelectron spectrum, the X-ray source will ideally produce monochromatic radiation having sufficient energy to excite core electrons of all the elements in the periodic table. A monochromatic source is desirable because the width of the incident X-ray line contributes to the width of the resultant photoelectron line; the width of the photoelectron peaks in turn affect the resolution of the electron spectrometer, that is, the ability of the system to distinguish between closely spaced peaks. (In XPS, resolution is normally given as the Full line Width at Half Maximum height, FWHM.) In commercial electron spectrometers, characteristic radiation emitted from magnesium or aluminum anodes is normally used as the desired near-monochromatic radiation. The K_α doublet of magnesium has an energy of 1253.6 eV with a composite line width approximately 0.7 eV FWHM; the K_α doublet of aluminum has an energy of 1486.6 eV with a composite line width of approximately 0.85 eV FWHM. The K_α lines of the heavier elements are more energetic sources of excitation, but they also have broader line widths; for example, the copper K_α line width is about 2.6 eV FWHM, and the molybdenum K_α line width is about 6 eV FWHM. Moreover, at higher energies, the $\alpha_{1,2}$ doublets in the X-ray spectrum separate, and this gives rise to confusing artifacts in the photoelectron spectrum. Thus, X-radiation from the heavier elements is normally undesirable for electron spectroscopy.

X-ray sources are constructed in a variety of ways, but they share basic elements. The filament is heated by means of a current which produces a high density electron cloud around the filament due to the thermionic emission. Electrons in this cloud are accelerated towards the anode by the potential applied between the anode and filament. An electrostatic shield may be interposed to eliminate a line-of-sight from the filament to the target and thus minimize deposition of filament material onto the target. The conversion of high voltage electrons into X-rays is an inefficient process (only about one percent of the total applied power emerges as useful radiation). It is therefore necessary to cool the anode to dissipate that portion of the energy which appears as heat. For ESCA experiments, the generated X-rays pass through a window and impinge on the sample surface. This window prevents the entry of scattered electrons from the X-ray source into the sample chamber. In X-ray photoelectron spectrometers, an aluminum window, which is normally used with either an aluminum or a magnesium target, purifies the X-ray spectrum by absorbing Bremsstrahlung radiation above approximately 1600 eV. As an X-ray beam passes through the window, the beam is attenuated by an amount dependent upon the thickness and the density of the window material; furthermore, X-rays of different energies are attenuated by differing amounts by the same window material. Thus, it is desirable to have the window fabricated from very thin material to maximize the photon flux at the sample surface. The ability to use the absorption properties of an aluminum window to advantage is illustrated in Fig. 15 for aluminum and magnesium anodes. The absorption characteristics of aluminum are shown superimposed on the emission spectrum. The K absorption edge of aluminum is approximately 70 eV above the aluminum K_α doublet and about 300 eV above the magnesium K_α doublet. Therefore, the aluminum window effectively filters the Bremsstrahlung radiation at higher energies without greatly attenuating either the aluminum or the magnesium K_α lines.

The geometrical relationship of the anode to the sample in the photoelectron spectrometer is critical to the efficiency of using the X-rays produced. As the anode—sample distance is decreased, higher photon flux results at the sample surface. In fact, for a point source of X-rays, the photon flux to a sample of fixed size varies inversely with the square of the distance between them. Larger anode area also results in higher photon flux, provided the sample surface completely intercepts the cone of radiation from the anode. Geometry varies from instrument to instrument, but the planar anode, which is most commonly employed, is shown in Fig. 16. If the takeoff angle of the exciting radiation is low relative to the plane of the anode face, the anode appears to be a line source when viewed from the sample position. Planar anodes are common in demountable X-ray sources; this type of assembly permits rapid changing of anodes.

The geometry of a commercially used X-ray source, designed for maximum efficiency, is illustrated in Fig. 17. The source incorporates an annular anode which subtends a large solid angle and is closely coupled to the sample surface. The anode surface is a conical section and extends through 180° in a plane parallel to the sample surface. The result is that the anode partially "surrounds" the sample, and the central portion of the sample surface is approximately equidistant from all points on the anode face. This type of geometry produces a very high photon flux at the sample, contributing to high spectrometer sensitivity.

If the energy width of the excitation source can be reduced, a corresponding reduction in the line width of the resulting photoelectron peak will be obtained. Therefore, one of the objectives of instrument design is to effectively monochromatize the excitation source. True source monochromatization can be achieved through crystal dispersion of the generated X-rays [19], as shown in Fig. 18. The X-ray monochromator uses the properties of a

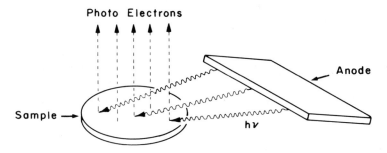

Fig. 16. The geometry of a planar anode X-ray source.

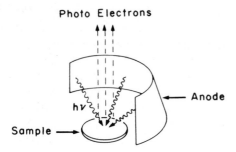

Fig. 17. The geometry of an annular anode X-ray source.

spherically curved crystal to diffract and "focus" radiation diverging from the X-ray source. The crystal, of course, does not focus X-rays in the optical sense but rather diffracts them so they converge at the sample surface. The X-ray source anode, the crystal and the sample are located along the locus of a circle, with the crystal being tangent to the circle. The radius of curvature of the crystal is equal to the diameter of the circle, and the target and sample are so placed on the circle that they both make the same angle relative to a line normal to the crystal face. With this configuration, only radiation which satisfies the Bragg diffraction law will be reflected and brought to a focus at the sample [20]. In crystal diffraction, less than 1% of the radiation incident on the crystal is reflected to the sample. This loss in signal can be partially compensated by using three crystals, as illustrated in Fig. 18, to increase the X-ray flux at the sample surface and by using a large-area detector. (See Section IV.D.)

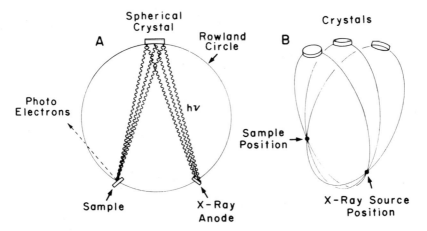

Fig. 18. Monochromatizing the X-ray source. (a) The basic elements of a crystal mono-chromator; (b) the three-crystal geometry utilized in a commercial application of this technique.

C. ELECTRON ENERGY ANALYZERS

The function of the energy analyzer is to measure the number of photo-ejected electrons as a function of their energy. An analyzer may be either magnetic or electrostatic, but in either case there are certain requirements imposed upon it. The analyzer must be placed in a suitable vacuum environment, with pressure less than about 10^{-5} torr to minimize electron scattering through collision with residual gas molecules. Since electrons are influenced by stray magnetic fields (including the earth's magnetism), it is necessary to cancel these fields within the enclosed volume of the analyzer. The degree of cancellation necessary depends on the type of analyzer being employed; for a spatially dispersive instrument, stray magnetic fields are typically reduced to a level of about 10^{-4} gauss; for a nondispersive instrument, it is sufficient to reduce stray magnetic fields to about 10^{-3} to 10^{-2} gauss. Stray field cancellation is accomplished in a magnetic dispersive instrument by using three sets of mutually perpendicular compensative coils. In an electrostatic analyzer, either dispersive or nondispersive, stray field cancellation is accomplished by using Mu metal shielding. Electrostatic analyzers are used in all commercial spectrometers now available because of the relative ease of stray field cancellation.

Three types of analyzer geometry are presently used in commercial instrumentation. They are the dispersive spherical sector capacitor, the dispersive cylindrical mirror and the nondispersive energy filter and are shown schematically in Figs. 19—21.

The geometry of the spherical sector analyzer (Fig. 19) is similar to a

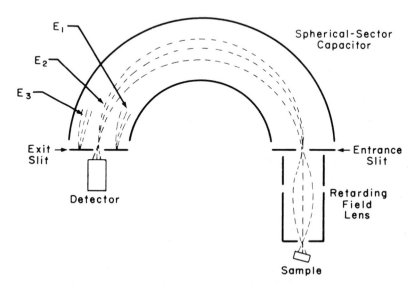

Fig. 19. The geometry of a spherical-sector analyzer.

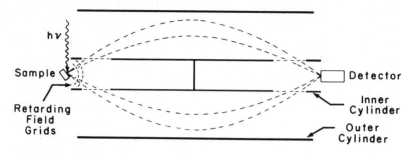

Fig. 20. The geometry of a cylindrical mirror analyzer.

prism and lens system. Electrons with different energies will be separated as they traverse the electric field; electrons with the same energy but diverging from each other as they enter the analyzer will be brought to a focus at the exit slit. Because of its symmetry, the spherical sector analyzer is space focusing, that is a point source is reimaged as a point (ignoring aberrations) [21]. Because this analyzer has a well defined focal plane, it lends itself to multidetector techniques (Section IV.D). A spectrum can be produced by scanning the voltage applied to the spherical plates of the analyzer such that electrons having successive energy values are allowed to pass through the exit slit and reach a detector. Alternately, a fixed voltage is applied to the ana-

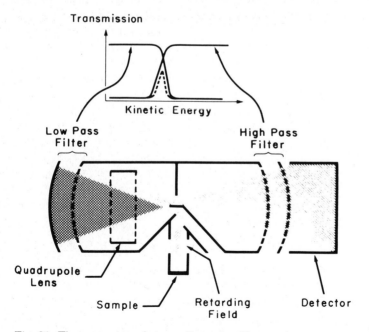

Fig. 21. The geometry of a non-dispersive filter analyzer.

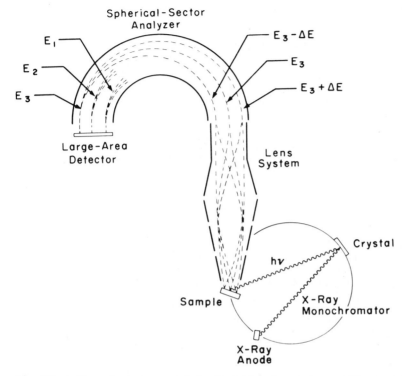

Fig. 22. A dispersion compensated spherical-sector analyzer with monochromatized X-ray source.

lyzer and the retarding field is scanned; photoelectrons are slowed down by the retarding field, and only those electrons that match the sector voltage will reach the detector. The former scanning mode produces constant resolving power throughout the spectrum, but peak widths therefore change. Retardation gives constant peak widths and better sensitivity to electrons of low initial kinetic energy.

In Fig. 22 is a schematic representation of a spherical sector analyzer designed for use with a monochromatized X-ray source [19]. With monochromatization, the energy of the incident X-ray photons varies across the sample by the width of the X-ray line (about 1 eV). Thus, the photoelectrons from a given energy level in the sample will also have an energy spread, and the higher energy electrons will follow trajectories through the analyzer with larger radii than the lower-energy electrons. To compensate for this, the analyzer and lens are so arranged that low-energy electrons enter the analyzer closer to the center of curvature than high-energy electrons, and so all of the electrons from one energy level in the sample are brought to a single focus at the exit focal plane of the analyzer. The four-element lens retards the photoelectrons and also focuses an image of the sample, with constant magnification, at the entrance to the analyzer.

A cylindrical mirror analyzer consists of two coaxial cylinders with a negative potential applied to the outer cylinder. The sample and detector are located along the common axis of the two cylinders. The potential applied between the inner and outer cylinders creates a cylindrical retarding field. Entrance and exit slots are cut around the periphery of the hollow inner cylinder; the planes of these slots are both parallel to each other and normal to the coaxis of the cylinders. To prevent field distortion, a metal mesh is used to bridge the slots. The axial symmetry of the cylindrical mirror produces a space focusing analyzer [22]. The basic elements of a cylindrical mirror analyzer, as applied to XPS, are shown schematically in Fig. 20. Photoejected electrons are decelerated by a retarding field (formed by spherical grids), pass through the entrance slot, are deflected by the radial field between the cylinders, pass through the exit slot, and are detected. For a given set of analyzer conditions, only monoenergetic electrons will traverse the system and be detected. A baffle in the inner cylinder prevents the passage of electrons through the cylinder from sample to detector. A spectrum can be produced either by scanning the voltage applied to the outer cylinder or by scanning the retarding field while maintaining a fixed mirror potential. The commercial adaptation of this analyzer utilizes tandem cylindrical capacitors to improve resolution; electrons traverse two analyzing fields before being detected.

The nondispersive analyzer shown schematically in Fig. 21 uses two spherically symmetric retarding fields [23]. The first field encountered by the photoelectrons is used in the reflection mode as a low-energy pass filter; the second field is used in a transmission mode as a high-energy pass filter. There is a small energy overlap between the low-energy pass and high-energy pass filters which yields a narrow transmission function as illustrated. Photoelectrons from the sample are decelerated by a retarding field to match their energy to that of the analyzer. A low resolution parallel-plate energy analyzer, at a 45° angle with respect to the incident electron beam, deflects electrons within a selected energy range through the entrance aperture to the analyzer toward the low-energy pass filter. Electrons with kinetic energy at or less than the nominal potential on this filter are reflected and focused back through a central aperture in the analyzer; electrons with greater energy strike the electrostatic mirror surface and are scattered from the electron beam. The portion of the electron beam reflected by the low pass filter is aligned by a quadrupole lens for transmission through the aperture. Those electrons having kinetic energies that exceed the nominal potential of the high-energy pass filter are transmitted to the detector; lower-energy electrons are reflected back into the body of the analyzer and are not detected. A spectrum is produced by scanning the retarding field while the analyzer is maintained at a fixed pass energy.

D. DETECTORS

The detector commonly used in commercial instrumentation is a channel electron multiplier. The exception to this is the system illustrated in Fig. 22 which employs a large-area detector.

The channel electron multiplier is an electrostatic device which employs a continuous dynode surface (a thin-film conductive layer on the inside of a tubular channel) requiring only two electrical connections to establish the conditions for electron multiplication. The output of the multiplier is a series of pulses that are fed into a pulse amplifier/discriminator and then into a digital-to-analog converter (DAC), a multi-channel analyzer (MCA), or a computer.

Instruments having a well defined focal plane can employ multidetector arrays. The large-area detector to be described is an example of such a detector and is used in commercial instruments with X-ray monochromators; this use of multidetector elements partially compensates for the low X-ray intensity obtained from the monochromator. Photoelectrons reaching the exit plane of the analyzer impinge on an array of channel electron multipliers. With this arrangement, a 10 eV range (or larger) of electron energies can be detected simultaneously. An electron which impinges on a multiplier generates an amplified pulse of electrons on the output side which is then accelerated and strikes a phosphor screen mounted on a glass window in the vacuum system. A corresponding light pulse is emitted by the screen and detected by a vidicon mounted outside the window. This camera transforms the light pulse into a signal which is transferred to the memory of a multi-channel analyzer; the MCA also monitors the scan voltage of the camera and uses this information to sort the pulses into its memory according to the energies of the electrons.

E. VACUUM SYSTEM

The vacuum system has two requirements: it must maintain the sample chamber and analyzer at a pressure so the photoejected electrons have a long MFP relative to the internal dimensions of the spectrometer, and it must reduce the partial pressure of reactive residual gases to a level that will not unduly contaminate the sample surface and interfere with the study of that surface. Thus, a major consideration in vacuum system design is vacuum composition as well as the ultimate vacuum level. The type of pump, the materials from which the sample chamber is constructed, and the design of the baffling between the pump and sample chamber determine the cleanliness of the system.

Three types of pumping system are used in commercial instruments. The most common type is an oil diffusion pump backed by a mechanical fore-pump. Oil backstreaming from this type of pump can cause contamination of the sample chamber, but proper cryogenic baffling can eliminate, or at least minimize, this problem. One manufacturer uses a turbomolecular pump backed by a rotary mechanical forepump. The third type of system, which uses a getter-ion pump backed by a sorption forepump, is acquiring increased popularity for clean surface studies because of the absence of contamination from pump fluids. If the spectrometer is equipped with a sample pretreatment chamber, it is separated from the sample chamber by a suitable valve and pumped separately from the sample chamber and analyzer. If sample pretreatment takes place in the analyzer sample chamber, then differential pumping is employed to maintain the sample chamber as a separate vacuum region from the analyzer.

F. SAMPLE HANDLING

The sample may be either a solid with a regularly defined surface or a powder and may be a conductor, semiconductor, or non-conductor. Powders may be pelletized, compressed into a suitable holder, or dusted onto a carrier such as double-sided adhesive tape. The choice of sampling technique will be dictated by the form of the sample and the type of sample holder used in a particular electron spectrometer. It is desirable but not essential to have the sample in electrical contact with the spectrometer. Samples that are insulated from the spectrometer can acquire a positive charge at their surface during irradiation because photoelectrons are being ejected. This so-called "charging effect", which is described in Section V.A, may lead to an apparent shift in binding energy. The degree of this effect is a function of X-ray source and sample chamber design. In most instruments, there is a sufficient number of background electrons in the sample chamber to either neutralize the surface charge or stabilize it at a small value. However, there is a marked reduction in the number of background electrons when a monochromatized X-ray source is used, and the charging effect can be large; in this case, the surface of the sample can be neutralized by using an auxiliary low-energy electron source, such as a "flood gun", located in the sample chamber.

The initial study of a sample surface normally takes place with the sample in an "as received" condition. However, all commercial instruments provide a means for cleaning or chemically modifying a sample surface without exposure to the ambient atmosphere between such treatment and the subsequent examination of the surface. Sample preparation may be performed either directly in the sample chamber or in a pretreatment chamber, depending upon the design of the spectrometer. Some common sample surface pretreatments are etching via an ion bombardment gun mounted in the sample chamber, deposition of a metallic thin film via an evaporator mounted in the sample chamber, exposure to a controlled reactant atmosphere, and exposure to non-ambient temperatures. A glove bag or glove box mounted at the entrance to the sample chamber is a simple but effective way for transferring samples from sealed containers to the spectrometer under an inert atmosphere.

G. DATA ACQUISITION AND PROCESSING

Commercial instruments acquire data digitally and present them in either an analog or a digital mode. Analog operation is the most direct and simplest method of acquiring and presenting data. The output of the pulse amplifier / discriminator is fed to a digital-to-analog convertor which drives the Y-axis of an X—Y recorder. The X-axis of the recorder is driven by a voltage which is proportional to the binding energy and synchronous with the scan function of the analyzer. The resulting record is a spectrum of counting rate vs. binding energy.

Digital systems utilize either a multi-channel analyzer or a computer to acquire and process data. Both devices employ a digital processor with either core or solid state memory, data and address registers, and control circuits; the data acquisition and reduction routines in the MCA are hard wired, while the computer utilizes software routines. Digital counting of the output of the discriminator measures the number of pulses per unit time. Each pulse adds a count to an address (channel) in memory; pulses are accumulated in that channel for a time interval determined by an internal clock. At the end of the time interval, the address register is advanced to the next channel, and pulses are accumulated at this new address. This process continues until all channels have been addressed. If the system is operating in a repetitive scan mode, this sequence will automatically begin again. The analyzer scan function is synchronized with the processor clock; thus, each address in memory is associated with a binding energy increment as well as a time interval. Synchronization of the scan of the analyzer with the processor not only serves to label the memory address but also results in the coherent addition of subsequent data to each channel. The stored or processed data may be displayed on an oscilloscope, plotted on an X—Y recorder, or punched onto paper tape for subsequent manipulation or storage. Some commercial electron spectrometers also display the spectrum on an oscilloscope during data acquisition so that the operator can observe the incoming signal. A multi-channel analyzer or computer permits limited data manipulation. Among the routines that may be provided are digital smoothing, transfer of data from one portion of memory to another, addition or subtraction of spectra, background suppression, normalization of spectra, presentation of derivative and integral curves based on the basic spectrum, and simultaneous display of spectra in different memory regions for visual comparison on an oscilloscope. A computer may additionally provide the software routine to permit the automatic sequential scanning of a number of binding energy ranges and routines for peak deconvolution and curve fitting.

A digital processor permits the operator to make multiple scans of a selected binding energy region to enhance the signal-to-noise of the spectrum. Repetitive scanning is usually used for enhancing of weak signals. However, signal strength depends, all other things being equal, upon the design characteristics of the electron spectrometer (X-ray source, analyzer, and detector). Thus, a single scan made on one instrument may provide as much usable information as multiple scans made on another less sensitive instrument. For comparable scan times, a recorder with suitable input filter networks is as effective as a digital processor in rejecting white noise, interference, and impulse noise. Distortion, which is essentially the introduction of foreign frequency components onto the desired signal, is a greater problem with a digital processor than a recorder. Distortion is usually caused by the interaction of the signal with various modulating and processing devices and is often encountered in pulse measuring systems. On the other hand, drift poses a problem for the recorder, particularly when scanning at very slow rates. With either technique, there is a limit to the amount of noise that can be rejected, particularly if the noise spectrum exhibits some periodicity or coherence that coincides with the desired signal. Also with either acquisition technique, signal enhancement is purchased at the expense of increased measurement time. Either device may be used where strong signals are encountered, but a digital processor has definite utility for the detection of trace impurities or insensitive elements.

V. Some experimental variables

A. CHARGING EFFECTS

During X-ray irradiation, nonconductive samples can undergo a change in surface potential. This surface charge is normally positive, and retards the photoejected electrons. This decrease in the kinetic energy of the photoelectrons results in higher apparent binding energies than the true values [24] and therefore limits the amount of chemical information obtainable from the data. The mean potential at which the surface charge stabilizes depends on the electron emission from the sample, the photoejected electrons from the X-ray window and sample chamber walls that impinge on the sample, and electron conduction in the sample surface [25]. The magnitude of the charge effect will vary from instrument to instrument since the instrumental parameters of X-ray flux, X-ray window surface condition, and sample compartment geometry influence the net flow of electrons from the sample surface [25,26].

B. CHARGE COMPENSATION

The object of charge compensation is the determination of absolute binding energies. Compensation may generally be obtained by using (1) charge neutralization, or at least reducing the charge to less than 0.1 eV, (2) an internal standard, or (3) an external standard.

One technique for charge neutralization is to flood the sample environment with low-energy electrons; the source of these electrons can be a filament mounted in the sample compartment. In practice, the photoelectron line for a surface species, such as carbon, is monitored while the filament current is adjusted. As filament current is increased progressively, the FWHM of the observed peak will decrease, reach a minimum, and eventually start to increase again as the sample surface becomes negatively charged. In this latter case, the negative surface charge serves to accelerate the photoelectrons rather than retard them. The peak observed on a charged surface may be broadened because of the variation in the charge across the sample surface or because of a charge gradient into the surface. The relative importance of these two effects is not clear from existing published data. The rationale for using this technique is that a minimum line width corresponds to zero surface charge; it assumes that the observed species is in electrical equilibrium with the sample and accurately tracks the charge on the sample. This last assumption is one of the keys to the successful application of this technique; in the absence of electrical equilibrium, the binding energies measured for other surface species will be in error.

For using an internal standard, a species is introduced into the sample matrix or substrate [27]. A known binding energy of the reference element is assumed to be the same for all samples being compared. An example is the study of organometallic compounds with the assumption that the binding energy of organic carbon is invariant. Since chemical change clearly influences binding energies, care and chemical intuition must be exercised in using this approach.

In the third approach, an external standard is used to compensate for surface charge. The reference material is typically a vacuum deposited noble metal or carbon contamination on the sample surface [24,26,28,29]. The use of carbon as a reference is attractive since this ubiquitous material is present on all but the most meticulously cleaned samples. However, the various sources of contamination may not all produce carbon with the same $1s$ binding energy. Moreover if the sample surface is chemically active, the interaction of the carbon with this active surface might also shift the binding energy of the carbon. Ideally, the external standard should have a well defined binding energy that is insensitive to the environment, does not react with the sample surface, and is in electrical equilibrium with the sample. In many applications the noble metals best satisfy these criteria.

Due to the ease of sample preparation and chemical inertness of the reference material, vacuum deposition of a noble metal is often preferred over other charge correction methods. In systems where a noble metal may react with the surface (i.e. with some cyanide and halide salts [30]) the vacuum deposition technique may not be applicable. Gold is generally used as the reference material because it is the least reactive of the noble metals and the easiest to vaporize.

If an accurate determination of binding energy is required, the most effective approach to charge compensation is to use a combination of methods. The use of vapor deposited gold together with flood electrons is one such approach [31]. As previously described, filament current is increased until the width of the carbon 1s peak is at a minimum; at this value of filament current, the apparent binding energy of the gold peak must appear at the calibration value for the spectrometer (i.e. the binding energy for gold in electrical equilibrium with the spectrometer). If these two conditions for carbon and gold are not simultaneously satisfied, it indicates improper sample preparation.

Whether charge correction and binding energy determinations are to be approximate or exact, the binding energies of suitable calibration lines must be known. To minimize measurement error, the calibration line should be as close as possible to the line of interest; the use of a pair of calibration lines, which bracket the line of interest, will provide increased measurement accuracy. Some suitable calibration lines (in eV) are as follows [26]: Pt $4f_{7/2}$ = 71.0; Au $4f_{7/2}$ = 83.8; Cu $3s$ = 122.9; C $1s$ (graphite) = 284.3; Pd $3d_{5/2}$ = 335.2; Ag $3d_{5/2}$ = 368.2; Ag $3p_{3/2}$ = 573.0; Cu $2p_{3/2}$ = 932.8.

C. DEPTH PROFILING VIA ION ETCHING

As noted elsewhere in this chapter, ESCA is a relatively non-destructive technique compared with other methods of surface analysis. No surface species are removed during the ESCA measurement, and the soft X-ray source used for excitation avoids many of the problems associated with thermal degradation of sensitive materials. Sample composition remains constant during analysis, allowing extensive characterization of the first 10—20 Å of a surface without significant time-dependent effects. However, for certain applications, a significantly broader scope of information can be obtained by coupling ESCA measurements with destructive techniques such as ion etching. Through a repetitive, sequential process of ESCA measurement followed by ion etching, depth profiling data can be obtained.

In Fig. 23, the results of a depth profiling study of a zirconium coating on aluminum metal are shown [32]. The relative atomic abundances of the observed elements (obtained by normalizing the individual peak height for X-ray absorption cross-section relative to the carbon 1s level) are plotted versus etching time. In this example, Ar^+ ions were used to sputter the sample surface at a rate of the order of 10's of angstroms per minute. Thus, the data show that the zirconium and a fluorine contaminant are located in an extremely thin layer at the surface and are not uniformly distributed in the bulk of the material. The profile of the fluorine 1s line shows that this

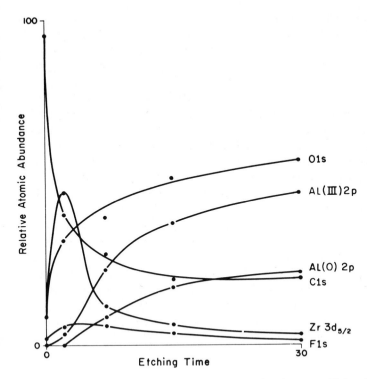

Fig. 23. Depth profile of zirconium-coated aluminum, utilizing ESCA combined with ion etching.

contaminant is associated with the zirconium coating. There are other points of interest. For example, the carbon contamination on the sample surface dominates the spectrum initially but is readily removed under ion bombardment. The Al(0) and Al(III) lines associated with the substrate are not visible initially but increase in intensity as overlayers are removed, thus suggesting that the substrate initially was completely covered by the zirconium coating. It may be further observed that an aluminum oxide is present at the interface between the bulk aluminum and the zirconium coating.

Depth profiling data obtained with ESCA appears to be in substantial agreement, both qualitatively and quantitatively in terms of elemental analysis, with data obtained by AES [33,34]. Though AES has been used extensively for depth profiling, this technique does not normally provide information about the chemical state of the surface elements; in this respect therefore, ESCA can provide more information. The work of Coad and Cunningham [33] shows that there are two additional factors favoring ESCA. Photoelectron peak intensities for different elements can be readily normalized and correlated by using theoretically based and empirically verified relative intensity factors, whereas Auger peak intensities are correlated from a theoretical basis requiring the application of calculated correction factors to the observed intensities. Further, the measurement of ESCA peak intensities is straightforward since the number of electrons observed, $N(e)$, is obtained directly from the observed peak areas. Auger intensities, however, are normally estimated from the peak-to-peak amplitude of the derivative signal and are subject to significant errors if the peak shape changes.

Caution should be exercised in chemical interpretation of ESCA data following ion bombardment since the ion etching process itself can cause changes to the surface being examined. For example, the bombardment of an oxide surface by Ar^+ ions can cause chemical reduction of the surface to a lower oxide or to the metal [34,35]. This process can be advantageously used to observe intermediate oxidation states as a higher oxide is reduced to its metallic form. However, in those cases where the true distribution of the oxidation states is desired, care must be taken in using ion etching.

The successful use of ion etching as a complimentary tool to surface analysis by ESCA requires an understanding of the physico-chemical effects that are associated with ion bombardment. These effects include differential sputtering of the surface species, diffusion and migration of constituent elements between the surface and the substrate, chemical reactions between the surface species and the incident ions, and gross heating of the surface due to the transfer of energy from the incident ions to the sample [35].

D. GRAZING ANGLE ESCA

As the angle of electron exit relative to the sample surface is decreased, the effective sample area is increased while the effective sample depth is proportionately decreased; this has the effect of increasing the peak intensities of surface species [36]. The practical implication of this observation is that by tilting the sample at an angle to the spectrometer, it is possible to enhance the surface sensitivity of ESCA and thus help distinguish surface phenomena from substrate properties. While this has greatest application in the area of polymer analysis, where the MFP tends to be long, it is also applicable to the study of very thin layers such as oxides on metals.

The reason for enhancement of surface sensitivity is shown diagrammatically in Fig. 24, while Fig. 25 provides a quantitative tabulation of the degree of enhancement for several values of electron exit angle, θ [37]. In Fig. 24, M is the MFP of a photoelectron, D is the depth of sample observed, and θ is the electron exit angle. Since the MFP is constant for a given material and electron energy, decreasing θ results in smaller values of D. Thus, the relative intensities observed for surface species, as compared to sub-surface species, will be enhanced in the ESCA spectrum as the exit angle is decreased.

Examples of the application of grazing angle ESCA are shown in Figs. 26 and 27 [37]. Freshly cleaned aluminum rapidly forms a 10—15 Å layer of

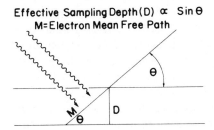

Effective Sampling Depth(D) ∝ Sin θ
M = Electron Mean Free Path

Fig. 24. Diagramatic presentation of the parameters influencing effective sampling depth.

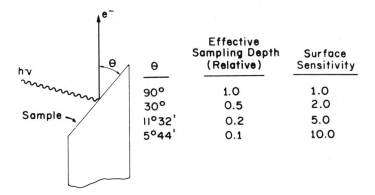

θ	Effective Sampling Depth (Relative)	Surface Sensitivity
90°	1.0	1.0
30°	0.5	2.0
11°32'	0.2	5.0
5°44'	0.1	10.0

Fig. 25. Theoretical values of effective sampling depth and relative surface sensitivity for various values of take-off angle.

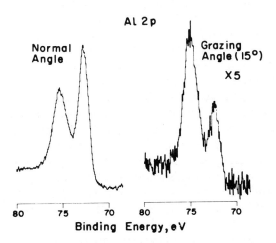

Fig. 26. ESCA spectra of freshly cleaned aluminum at normal and grazing exit angles.

References pp. 157—158

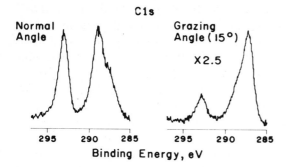

Fig. 27. ESCA spectra of surface contamination on polyvinylidene fluoride at normal and grazing exit angles.

aluminum(III) oxide when exposed to atmospheric conditions. The aluminum $2p$ peaks can be readily observed for both the oxidized aluminum in the overlayer and the metallic aluminum in the substrate. The left-hand portion of Fig. 26 shows the spectrum at a 90° exit angle while the right-hand spectrum is for a 15° exit angle. The overall spectral intensity is somewhat decreased because of foreshortening of the sample as viewed from the spectrometer; however, there is a marked relative enhancement of the signal from the surface material. This is a clear indication that the oxide is present as a very thin layer at the sample surface. In Fig. 27, the application of the grazing angle technique to an organic material is illustrated. The left side of the figure shows the carbon $1s$ spectrum obtained at a 90° exit angle from a sample of polyvinylidene fluoride "as received". This spectrum shows the two peaks due to the $-CF_2-$ and $-CH_2-$ carbons in the substrate polymer as well as a shoulder on the right side of the polymer $-CH_2-$ peak due to surface contamination. The same sample at a 15° exit angle is shown on the right side of this figure . The increase in the relative intensity of the peak arising from surface contamination, which now dominates the spectrum, is an unambiguous indication that the contaminant is a surface species.

VI. Applications

A. ORGANIC SURFACES

Many surfaces of high scientific and technological interest are organic. These include polymer surfaces in the form of both films and fibers, rubber surfaces, and organic coatings such as lubricants and microencapsulating materials. Investigation of these kinds of surfaces ranges from inquiry into their fundamental nature and behavior to applied research in adhesion, dyeability, surface oxidation, and other surface reactions as they relate to industrial processes or products. The ability to measure the elemental composition and the chemical nature of these surfaces has application in answering a wide variety of questions on subjects such as printing on polymer films, heat sealing of packaging films, oxidation of rubber, and the nature of reactions occurring at polymer surfaces. ESCA is an absolutely unique tool for application in these areas. Other techniques which traditionally have been used for polymer surface analysis include ATR infrared spectroscopy, which is very specific for organic functional groups at surfaces but has a sampling depth of the order of several thousands of angstroms at best. Techniques such as contact angle measurements provide physical characterization sensitive to the uppermost molecular layer of the surface but are incapable of providing detailed chemical information about surfaces. Other modern surface characterization techniques depending upon electron beam or ion beam excitation tend to be destructive of organic surfaces and cannot provide as much chemical information as ESCA. The examples which follow, while not all-inclusive, will serve to suggest the broad range of applicability of ESCA to polymer surface characterization.

1. Reactions at polymer surfaces

ESCA is the central analytical technique in a study of the surface chemistry of fluoropolymers [38]. In Fig. 28, a typical example from this work,

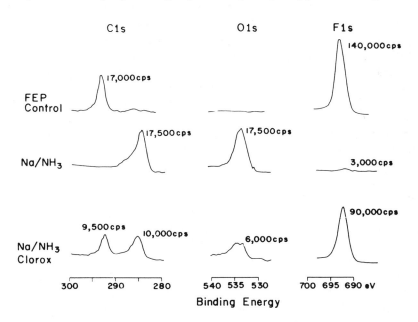

Fig. 28. Carbon, oxygen and fluorine spectra showing chemical effects of surface treatments on FEP Teflon.

References pp. 157—158

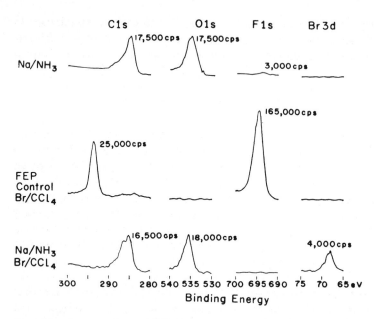

Fig. 29. Spectra from FEP Teflon surfaces showing bromination of carbon—carbon double bonds in Na/NH$_3$ treated sample.

the kind of chemical detail available through ESCA is shown. The spectra of the control sample show a carbon 1s peak at high binding energy (as expected for a fluoropolymer) and an intense fluorine spectrum. After treatment with sodium/ammonia, which is commonly used for promoting adhesion on fluorocarbons, the carbon spectrum is changed from that of a fluorocarbon to that of a hydrocarbon with some oxidized components. Oxygen on the treated surface verifies that oxidiation is one of the byproducts of this treatment. The nearly complete absence of fluorine suggests that the reaction at the surface extends to a layer of the order of 50—100 Å thick, thereby obscuring from observation the fluorocarbon substrate which remains. The change in water contact angle from high values (non-wettable) to low values (wettable) is consistent with the ESCA spectra which show that the treated surface no longer has fluorocarbon character. Subsequent treatment of the sodium/ammonia etched surface with chlorox reduces the thickness of the oxidized hydrocarbon surface, as shown by the reappearance of an intense fluorine peak and a carbon peak at high binding energy concurrent with the reduction in oxygen peak intensity and the retention of low water contact angles.

In a further aspect of this work, the presence of unsaturated carbon—carbon bonds in the surface is demonstrated (Fig. 29). Treatment of the sodium/ammonia etched surface with bromine in carbon tetrachloride brominates double bonds at the surface; the appearance of bromine 3d spectra is shown in this figure. The change in appearance of the carbon spectrum on bromination of carbon—carbon double bonds is consistent with a different chemical shift of the brominated carbons relative to the same carbons before bromination.

2. Wool fiber surfaces

ESCA can be utilized to study surface chemistry of wool fiber and its relationship to the end use properties of wool [39,40]. Examination of the carbon, oxygen and sulfur spectra of wool, before and after exposure to corona and low temperature discharge, shows that the carbon and oxygen spectra do not change significantly. However, the appearance of a peak at high binding energy in the sulfur spectrum indicates oxidation of the cystine sulfur in protein molecules in the wool structure occurs on exposure to either kind of discharge. Comparing the binding energies of these sulfur peaks with model compounds shows this sulfur to be in the +6 oxidation state, probably an $R{-}SO_3H$ group. Further study of wool with fluorocarbon coatings grafted to the fiber surfaces by a discharge treatment revealed that the intensity of the observed fluorine peak is correlated with the ability of the treated fiber to reject oil and therefore to be stain resistant.

3. Lubricants at surfaces

The presence, amount and type of lubricant present at various kinds of surfaces including organic surfaces such as polymers and inorganic surfaces such as metals, is a subject of widespread importance. One example of the application of ESCA in this area is the measurement of a high molecular weight amide used as a lubricant at the surface of polyethylene films [41]. In this application, films with different thicknesses of the amide lubricant

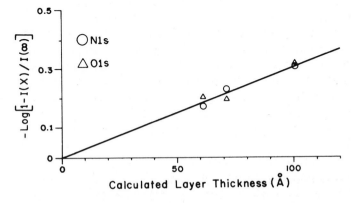

Fig. 30. Signal intensity correlated with thickness of polyamide layer on polyethylene.

are obtained and the ESCA signals from the nitrogen and oxygen atoms in the amide molecule are measured and plotted as shown in Fig. 30. The thickness of the amide layer correlates nicely with the intensity of the observed ESCA signal. Thus, measurement of such an intensity can be used as an analytical method of assaying the amount of lubricant present on the film surface.

4. Ion exchange at surfaces

Processes have been developed [42] which allow grafting of organic functional groups capable of ion exchange, such as acrylic acid, to otherwise inert polymer surfaces, such as polyethylene, polypropylene or polyesters. Acid—base reactions occurring with such surfaces form salts with metal ions such as silver, barium, calcium, lead and mercury and have been used as the basis of a method to assay for the amount of acrylic acid bound to the grafted surface quantitatively [43]. Conversely, ion exchange surfaces can also be utilized to scavenge metals from aqueous solutions at sub-parts-per-million levels. The signal intensity from the metal ion on the resulting surface can then be used as an analytical measure for the presence and amount of metal which was in solution, assuming that the exposure time and conditions are well controlled [44].

B. INORGANIC SURFACES

ESCA applications to important inorganic surfaces encompass the areas of catalysis, corrosion, and electrochemistry, as well as the composition of fracture interfaces, surface composition and composition profiles of semiconductor devices, passivation of metal surfaces and others. Possibly the most important of these areas where ESCA has begun to be extensively applied is in catalysis research (Section VI.C). The literature describing ESCA research in most of the other areas of application to inorganic surfaces is sparse, in part because other techniques such as Auger, ion microprobe and ion scattering have more traditionally been applied to study of these areas. Since recent developments make ESCA a fast technique (as easy to use as techniques such as Auger), applications where oxidation state information is desired will increasingly utilize ESCA as the method of choice.

1. Metal/metal oxide films

Among the few published applications to date on inorganic surfaces is a study of interactions in multi-component metal and metal oxide thin films [45]. In this work, changes in the oxidation states in alloys and at interfaces are examined using ESCA. The oxidation of a Nichrome surface, for example, results in the chromium component being selectively oxidized to chromic oxide while nickel remains in the metallic state. When this surface is overlayed with a 20 Å thick aluminum film, the chromic oxide is reduced to chromium and aluminum is oxidized to aluminum oxide in the region near the interface between the two metals.

2. Electrode passivation

The passivation of a copper–nickel alloy exposed to sodium chloride solutions at different electrochemical potentials is another study involving ESCA [46]. The composition of the film after exposure to the passivation potential indicates nickel oxide is at the surface and gives a spectrum similar to that obtained from a pure nickel specimen passivated in the same environment. However, at anodic potentials below the passivation potential, the surface layer is found to contain copper and nickel in the same ratio as exists on the surface of the alloy.

Another example is the study of the chemical and electrochemical oxidation of important metallic electrode materials, such as palladium [47]. In this electrochemical study, both PdO and PdO_2 are observed at the electrode surface. The exact species present and the distribution among these species is a function of the electrode potential. As might be expected, higher potentials are required to produce higher oxidation states.

3. Chemisorption on metals

The adsorption of carbon monoxide on molybdenum and tungsten films has been studied by ESCA. In this case, it is possible to distinguish the so-called alpha and beta forms of adsorbed CO from a form designated the gamma form [48]. Conclusions about the chemical nature of the adsorbed species can frequently be drawn from such work.

4. Airborne particulates

The characterization of the chemical states of sulfur in pollution aerosols is another application of ESCA [49]. In this study, seven separate chemical species of sulfur are found in atmospheric particulate samples collected in the Los Angeles and San Francisco areas. Species are identified as SO_3, SO_4^{2-}, SO_2, SO_3^{2-}, and two kinds of sulfides. Sulfides are found to be the dominant species, although in some samples the reduced forms of sulfur are present in concentrations comparable to the sulfate.

5. Glass surfaces

The chemical composition and reactions which can occur at glass surfaces are also the subject of study by ESCA [50,51]. Migration of components, such as calcium, in glass is observed to be a function of the temperature to which the glass is exposed. Also, leaching experiments show that materials such as aluminum at the glass surface are depleted on exposure of the surface to acid solutions. By reaction of glass surfaces with silylizing agents having appropriate organic functional groups, ion exchange can be made to occur at the glass surfaces and traces of material, such as lead, can be scavanged from an aqueous solution and observed with ESCA.

C. CATALYSIS

The field of catalysis offers some excellent examples of industrial interest in ESCA since it can provide precisely those data which are of interest in the study of commercially important catalysts. Some of the reasons for studying heterogeneous catalysis are to determine (1) why a catalyst has failed, (2) catalytic mechanisms with the hope of increasing the yield of a process, (3) the optimum time at which a partially spent catalyst should be reactivated, and (4) the mechanism by which a catalyst is poisoned with the intent of lengthening the catalyst's life. These studies all have practical significance as well as scientific interest.

In attacking these problems, ESCA can provide data about the elemental composition of a surface, determine the chemical states of the surface atoms (e.g. their degree of oxidation), and be used to determine the molecular structure of the surface (e.g. the type and relative abundance of functional groups associated with species adsorbed at the surface). However, the most important information derived through ESCA are the differences in the character of the surfaces between active and inactive catalysts and the difference in the surface of an active catalyst before and after a catalytic reaction. This type of information, combined with physical and chemical data obtained from other sources plus records of actual catalyst performance, can provide insight into the behavior of catalytic systems as well as act as a guide in the development of fabrication procedures for catalysts.

An excellent example of the use of ESCA in the field of catalysis is a study of the influence of irradiation on the properties of a silver catalyst used in the conversion of ethylene to ethylene oxide [52]. The emphasis of this study is on the yield—irradiation relationship rather than on a simple activity—irradiation effect. This catalytic process produces water and carbon dioxide as byproducts so the yield of ethylene oxide, and not merely total ethylene conversion, is a measure of catalyst performance. Pre-irradiation of the silver catalyst with gamma rays produces a substantial improvement in ethylene oxide yield. ESCA scans of the catalyst, shown in Fig. 31, identify calcium as an impurity that is present only following irradiation. Additional experiments with supported silver catalysts, prepared with and without calcium salt addition and evaluated for ethylene oxide yield before and after gamma ray exposure, verify that ethylene oxide yield enhancement

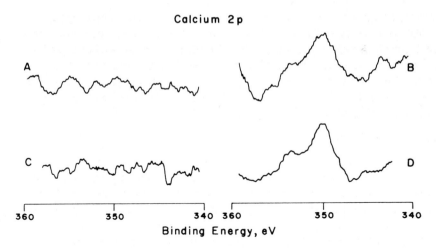

Fig. 31. The effects of irradiation of a silver catalyst used in the conversion of ethylene to ethylene oxide. (a) ESCA scan for Ca for an unirradiated catalyst; (b) ESCA scan for Ca for an irradiated catalyst; (c) ESCA scan for Ca following the reduction and reoxidation of a previously irradiated catalyst; (d) ESCA scan for Ca following re-irradiation of a reduced and reoxidized irradiated catalyst.

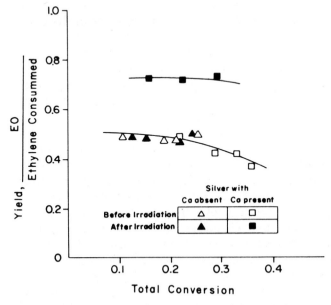

Fig. 32. The yield of ethylene oxide versus total ethylene conversion for a silver catalyst, with and without calcium.

References pp. 157—158

is correlated with the presence of surface calcium. The yield data from these experiments are shown in Fig. 32.

The preceeding example illustrates how ESCA can furnish an essential clue to catalyst behavior by providing direct and positive evidence of elements on the catalyst surface. In the following study, ESCA plays a more substantial role in defining the qualitative difference between catalysts of known high and low activity. Brinen and Melera [53] studied hydrogenation catalysts composed of rhodium supported on charcoal. The spectra for rhodium foil, rhodium sesquioxide (Rh_2O_3), and catalyst samples are shown in Fig. 33. The spectra of the foil and sesquioxide establish a chemical shift between the metal and the oxide forms. Though the three catalysts all have the same bulk chemical analysis, the spectra clearly indicate differences in surface chemistry. The spectra for catalysts A and B are similar and exhibit multiple rhodium species ranging from the metal form to the sesquioxide form, with the oxide form dominating; A and B are high-performance catalysts. The spectrum of C, a low-performance catalyst, shows that metallic rhodium is the dominant species at the surface. Thus in this system, good catalyst activity is associated with a high oxide-to-metal ratio, and it is possible to determine

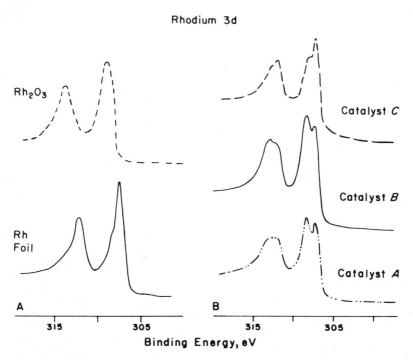

Fig. 33. ESCA spectra of three catalysts and rhodium foil versus rhodium sesquioxide. Catalyst A and catalyst B exhibit high activity while catalyst C exhibits low activity.

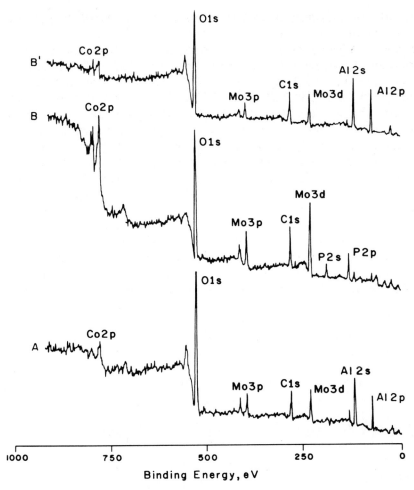

Fig. 34. ESCA spectra of an alumina-supported cobalt molybdate catalyst for removal of sulfur from oil. Spectrum A is from a catalyst of low activity, spectrum B is from a catalyst of high activity, and spectrum B′ represents the bulk composition of the catalyst.

the probable performance of a fresh catalyst from an examination of its ESCA spectrum.

The use of ESCA as a predictive tool is further illustrated by a study of an alumina-supported cobalt molybdate catalyst for removal of sulfur from oil [54]. In Fig. 34, the spectra for two experimental catalysts A and B, which were prepared by the same procedure but exhibit different activities, are shown. The bulk composition of these two catalysts is also identical; spectrum B′, which simultates the bulk catalyst, was obtained from a pulverized sample.

Certain interesting features are immediately apparent from these spectra.

Relative to the alumina support, the catalyst of high activity, B, has a greater concentration of phosphorus, molybdenum and cobalt at the surface than does the catalyst of lower activity, A. In particular, catalyst B exhibits a very high relative concentration of phosphorus at the surface. This suggests that phosphorus is possibly a promotor in this system; further experiments can be constructed to establish what role, if any, phosphorus plays in this reaction process. In contrast, the spectrum of catalyst A closely resembles that of the bulk catalyst B', exhibiting a large aluminum signal and virtually no phosphorus signal. The aluminum signal can be attributed to the exposed surface of the alumina support, which suggests that either there is insufficient catalyst on the surface of the support or there is ineffective catalyst dispersion.

The preceding showed ESCA can be used as a tool in the detailed examination of the fabrication procedure of a catalyst to identify those steps in the process which produce the desired end characteristics of the catalyst. Carrying the investigation one step further, the active catalyst can be examined following exposure to the reaction system as illustrated in Figs. 35 and 36. These figures compare the molybdenum spectra for molybdenum trioxide (MoO_3) and molybdenum disulfide (MoS_2) with fresh and used catalysts. The presence of sulfide in the spectra of the used catalyst suggests that the catalyst has performed its required function of removing sulfur from the oil sample.

Another application is illustrated by the study of tungsten oxide (WO_3) supported on gamma-alumina (γ-Al_2O_3) [55]. The results obtained using ESCA are in agreement with those obtained using various other techniques, with ESCA providing additional information on the valence states of the surface tungsten. In Fig. 37, the changes in chemical state arising from exposure of the WO_3 surface to a hydrogen atmosphere at various temperatures are shown. The spectra clearly show the reduction of W(VI) to W(0). A similar experiment with WO_3 supported on γ-Al_2O_3 is illustrated in Fig. 38; there is no detectable reduction observable thus supporting other evidence for the formation of the tungstate $Al_2(WO_4)_3$ at the surface of the calcined catalyst.

A practical example is the study of auto exhaust catalysts [56]. In Fig. 39 and Table 4, the results of this examination of copper oxides and alumina-supported copper catalysts are given. The sample numbers in the table correspond to the spectrum numbers in the figure so spectrum I is for $CuAl_2O_4$ and II is for CuO. Spectrum III is similar to CuO but the spectra for IV and V resemble $CuAl_2O_4$. As shown in the table, the calcination temperature required to convert CuO to $CuAl_2O_4$ decreases as the effective surface area of the alumina support increases; moreover, the presence of $CuAl_2O_4$ correlates with increased physical stability of the catalyst. Also shown in Table 4 are data obtained by X-ray absorption edge spectroscopy. This latter technique provides information similar to that obtainable using ESCA except that the ESCA data shown here can be collected in about one hour while the

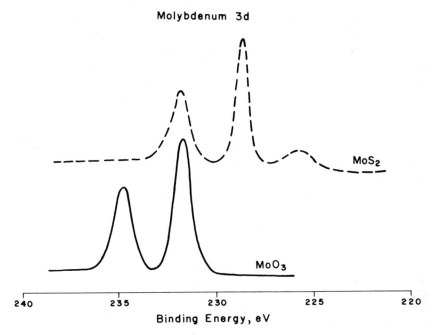

Fig. 35. ESCA spectra of MoO$_3$ versus MoS$_2$; the peak at ca. 226 eV in the MoS$_2$ spectrum is due to the sulfur 2s line.

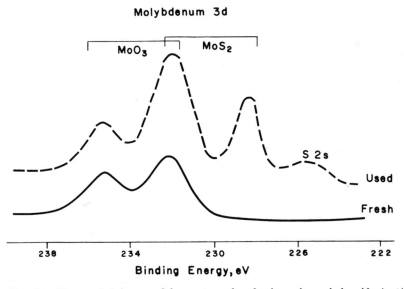

Fig. 36. The molybdenum 3d spectra of a fresh and used desulfurization catalyst; the spectrum of the used catalyst exhibits both oxide and sulfide species.

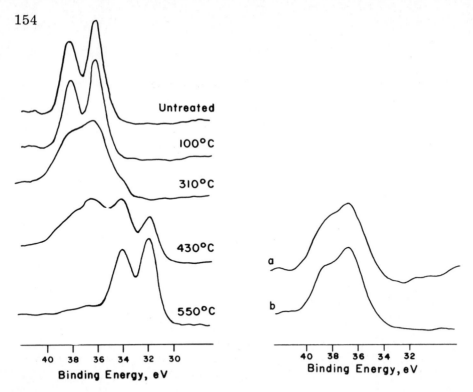

Fig. 37. ESCA spectra of unsupported WO$_3$ treated for two hours with hydrogen at different temperatures.

Fig. 38. The tungsten 4f spectra of γ-Al$_2$O$_3$ supported WO$_3$. (a) After calcination for 2 h at 550° C in air; (b) after subsequent reduction for 2 h at 550° C in hydrogen.

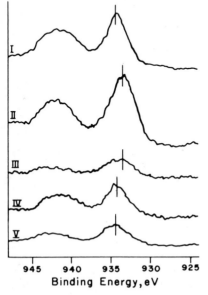

Fig. 39. The copper 2$p_{3/2}$ spectra of copper oxides (I and II) versus Al$_2$O$_3$-supported copper catalysts (III, IV and V). (Sample numbers refer to Table 4.)

TABLE 4

ESCA binding energy shifts

Sample no.	Copper (wt. %)	Supporting surface area (m^2/g)	Calcination temperature (°C)	Shift of binding energy from Cu^0 value (eV)	Shift of K-absorption edge from Cu^0 value (eV)
I	$CuAl_2O_4$			3.20	7.94
II	CuO			2.25	3.86
III	10.3	72	500	2.35	4.50
IV	10.3	72	900	3.10	7.73
V	8.8	301	500	3.30	7.73

X-ray absorption data require about 30 h of instrument time.

Finally, an example of the application of ESCA technique to a chemical process problem is illustrated in Figs. 40 and 41 [57]. The purpose of this examination of a charcoal-supported palladium catalyst is to determine the cause of the catalyst deactivation. The reaction process is the reduction of a

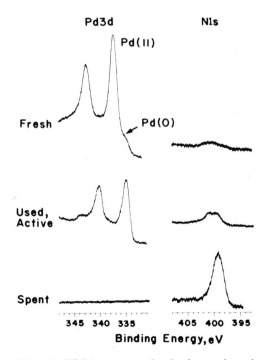

Fig. 40. ESCA spectra of a fresh, used, and spent charcoal-supported palladium catalyst.

References pp. 157—158

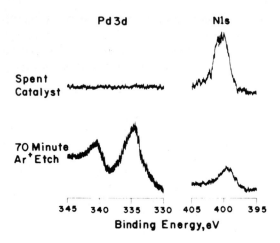

Fig. 41. ESCA spectra of a spent charcoal-supported palladium catalyst before and after ion etching.

nitrogen-containing organic compound. X-ray fluoresence analysis showed no difference in the total palladium content between an active and a spent catalyst. Therefore, poisoning of the catalyst surface by either sulfide or chloride compounds was suspected. However, ESCA spectra show these poisons to be absent, and further that an active catalyst exhibits a definite palladium signal while the spent catalyst shows no trace of palladium on the surface. Conversely, nitrogen spectra show a relatively low intensity signal from the active catalyst and a relatively high intensity signal from the spent catalyst. These results support the hypothesis that catalyst deactivation is caused by the adsorption of a nitrogen-containing byproduct of the reaction on the catalyst surface, thus blocking the active palladium sites. Additional evidence supporting this hypothesis was obtained by ion etching the spent catalyst as shown in Fig. 41; the palladium signal is restored and the nitrogen signal is reduced.

VII. Summary

The various experiments which have been described above demonstrate the utility of ESCA in the study of organic and inorganic surfaces and catalysis; the scope of the technique ranges from basic research to routine testing. Specifically in the field of catalysis, it is possible to investigate directly the chemical composition of the surface and determine the chemical differences, if any, between catalysts of high and low activity. The bonding between atoms can be very important in catalysis, affecting the stability, activity, and selectivity of a catalyst; ESCA frequently provides the only means for studying this bonding. The ESCA technique can be used to examine directly all elements except hydrogen; it can detect these elements whether the sample is crystalline or amorphous. ESCA can be expected to find increasing application in many areas of science and technology requiring detailed information on surface composition and chemistry.

References

1 H. Robinson and W.F. Rawlinson, Phil. Mag., 28 (1914) 277.
2 H. Robinson, Proc. Roy. Soc., Ser. A, 104 (1923) 455.
3 H. Robinson, Phil. Mag., 50 (1925) 241.
4 M. de Broglie, Compt. Rend., 172 (1921) 274.
5 K. Siegbahn, Alpha, Beta, and Gamma-Ray Spectroscopy, North-Holland, Amsterdam, 1965.
6 C. Nordling, E. Sokolowski and K. Siegbahn, Ark. Fys., 13 (1958) 483.
7 K. Larsson, C. Nordling, K. Siegbahn and E. Stenhagen, Acta Chem. Scand., 20 (1966) 2880.
8 K. Siegbahn, et al., ESCA, Almquist and Wiksells, Uppsala, 1967.
9 D.A. Shirley, Advan. Chem. Phys., 23 (1972) 85.
10 C.D. Wagner, Anal. Chem., 44 (1972) 1050.
11 J.H. Scofield, Lawrence Livermore Laboratory Rep. UCRL-51326, 1973.
12 R.S. Swingle, Du Pont Experimental Station, Wilmington, Delaware, private communication, 1974.
13 C.R. Ginnard and W.M. Riggs, Anal. Chem., 44 (1972) 1310.
14 C.D. Wagner, private communication.
15 W.M. Riggs, Anal. Chem., 44 (1972) 830.
16 C.A. Lucchesi and J.E. Lester, J. Chem. Educ., 50 (4) (1973) A205, and 50 (5) (1973) A269.
17 Manufacturers' literature.
18 R. Jenkins and J.L. De Vries, Practical X-Ray Spectrometry, Philips, Eindhoven, 1969.
19 K. Siegbahn, D. Hammond, H. Fellner-Feldegg and E.F. Barnett, Science, 176 (1972) 245.
20 L.S. Birks, X-Ray Spectrochemical Analysis, Interscience, New York, 1959.
21 B. Wannberg, U. Gelius and K. Siegbahn, J. Phys. E, 7 (3) (1974) 149.
22 H.Z. Sar-el, Rev. Sci. Instrum., 38 (1967) 1210.
23 J.D. Lee, Rev. Sci. Instrum., 44 (1973) 893.
24 W.P. Dianis and J.E. Lester, Anal. Chem., 45 (1973) 1416.
25 M.F. Ebel and H. Ebel, J. Electron Spectrosc., 3 (1974) 169.
26 G. Johansson, J. Hedman, A. Berndtsson, M. Klasson and R. Nilson, J. Electron Spectrosc., 2 (1973) 295.
27 J.J. Ogilvie and A. Wolberg, Appl. Spectrosc., 26 (1972) 401.
28 D.J. Hnatowich, J. Hudis, J.L. Perlman and R.C. Ragaini, J. Appl. Phys., 42 (1971) 4883.
29 C.R. Ginnard and W.M. Riggs, Anal. Chem., 46 (1974) 1306.
30 D. Betteridge, J.C. Carver and D.M. Hercules, J. Electron Spectrosc., 2 (1973) 327.
31 P. Citrin, Bell Laboratories, Murray Hill, New Jersey, personal communication, 1974.

158

32 Application Data Bulletin ES-0010, E.I. du Pont de Nemours and Company (Inc.), Instrument Products Division, Wilmington, Delaware, 1974.
33 J.P. Coad and J.G. Cunningham, J. Electron Spectrosc., 3 (1974) 435.
34 W.E. Baitinger, N. Winograd, J.W. Amy and J.A. Munarin, Proc. IEEE Int. Reliability Phys. Symp., Las Vegas, Nevada, 2—4 April 1974.
35 K.S. Kim, W.E. Baitinger, J.W. Amy and N. Winograd, Proc. Int. Conf. Electron Spectrosc., Namur, Belgium, 16—19 April 1974, Elsevier, Amsterdam.
36 W.A. Fraser, J.V. Florio, W.N. Deglass and W.D. Robertson, Rev. Sci. Instrum., 44 (1973) 1490.
37 Application Data Bulletin ES-0009, E.I. du Pont de Nemours and Company (Inc.), Instrument Products Division, Wilmington, Delaware, 1974.
38 D.W. Dwight and W.M. Riggs, J. Colloid Interface Sci., 47 (1974) 650.
39 M.M. Millard, Anal. Chem., 44 (1972) 828.
40 M.M. Millard, K.S. Lee and A.E. Pavlath, Text. Res. J., 42 (1972) 307.
41 Application Data Bulletin ES-0008, E.I. du Pont de Nemours and Company (Inc.), Instrument Products Division, Wilmington, Delaware, 1974.
42 A. Bradley, Chem. Technol., (1973) 232.
43 M. Czuha and W.M. Riggs, 167th ACS Nat. Meet., Los Angeles, April 1974.
44 M. Czuha and W.M. Riggs, Pac. Conf. Chem. Spectrosc., San Francisco, October, 1974.
45 N. Winograd, W.E. Baitinger, J.W. Amy and J.A. Munarin, Science, 184 (1973) 565.
46 L.D. Hulett, A.L. Bacarella, L. LiDonnici and J.C. Griess, J. Electron Spectrosc., 1 (1972) 169.
47 K.S. Kim, A.F. Gossman and N. Winograd, Anal. Chem., 46 (1974) 197.
48 S.J. Adkinson, C.R. Brundle and M.W. Roberts, J. Electron Spectrosc., 2 (1973) 105.
49 N.L. Craig, A.B. Harker and T. Novakov, Atmos. Environ., 8 (1974) 15.
50 G.D. Nichols, D.M. Hercules, R.C. Peck and D.J. Vaughan, Appl. Spectrosc., 28 (1974) 219.
51 D.M. Hercules, L.E. Cox, S. Onisick, G.D. Nichols and J.C. Carver, Anal. Chem., 45 (1973) 1973.
52 J.J. Carberry and G.C. Kuczinski, Chem. Technol., (1973) 237.
53 J.S. Brinen and A. Melera, J. Phys. Chem., 76 (1972) 2525.
54 J.S. Brinen, Proc. Int. Conf. Electron Spectrosc., Namur, Belgium, 16—19 April 1974, Elsevier, Amsterdam.
55 P. Biloen and G.T. Pott, J. Catal., 30 (1973) 169.
56 J.L. Ogilvie, A. Wolberg and J.S. Roth, Chem. Technol., (1973) 567.
57 Application Data Bulletin ES-0007, E.I. du Pont de Nemours and Company (Inc.), Instrument Products Division, Wilmington, Delaware, 1974.

Chapter 5

AUGER ELECTRON SPECTROSCOPY

A. JOSHI, L.E. DAVIS and P.W. PALMBERG

I. Introduction

During the last few years, the technique of Auger electron spectroscopy (AES) has emerged as one of the most widely used analytical techniques for obtaining the chemical composition of solid surfaces. A high sensitivity for chemical analysis in the 5—20 Å region near the surface, a rapid data acquisition speed, and the ability to detect all elements above He mark the basic advantages of this technique. The Auger spectrum provides reliable quantitative information and in many cases the status of chemical bonding.

While AES was initially used purely as a research technique, it is now a routine analytical tool. Its range of applications comprises numerous fields, such as semiconductor technology, metallurgy, catalysis, mineral processing, and crystal growth. The range of these applications is growing rapidly and the fundamental mechanisms in AES are becoming better characterized. Several reviews describing the technique and some of its applications have appeared recently [1]. This chapter, while dealing briefly with the principles involved in the Auger process, is primarily aimed at reviewing recent developments in experimental techniques and applications.

The Auger effect was discovered by Auger [2] in 1925 while working with X-rays. The technique of using electron-excited Auger electrons to identify surface impurities was suggested by Lander [3] in 1953. The high sensitivity of the technique was not realized, however, until 1968 when Harris [4] demonstrated the use of differentiation of the energy distribution, $N(E)$ vs. E, curves to obtain the Auger spectra in the present familiar form. The technique became highly popular when Weber and Peria [5] in 1967 demonstrated that the readily available LEED optics were suitable Auger electron spectrometers. Most present day Auger spectrometers employ the cylindrical mirror analyzer [6] which was shown by Palmberg et al. [7] to increase greatly the speed and sensitivity of AES. Quite often, AES is used in conjunction with inert gas ion sputtering to obtain compositional variations as a function of depth [8]. In more recent developments [9], finely focused electron beams with conventional deflection and rastering capability are utilized to obtain two-dimensional compositional analysis of surfaces. In combination with inert gas ion sputtering the scanning system makes three-dimensional elemental imaging possible. While other sources of excitation, such as X-rays [10] and ions [11], can also be used to excite Auger electrons, electron sources are the most widely used. The applications discussed in Section V are limited to electron-excited AES.

II. Fundamentals

The fundamental mechanisms involved in AES are ionization of atomic core levels by the incident electron beam, the radiationless Auger transition, and the escape of the Auger electron into the vacuum where it is detected with an electron spectrometer. The Auger electrons are manifest as small peaks in the total energy distribution function $N(E)$, such as that in Fig. 1, which is produced from a silver target by a 1 kV incident electron beam. The Auger peaks are evident in the $N(E)$ function, but become more pronounced by electronic differentiation which removes the large background consisting mainly of backscattered primary electrons and inelastically scattered Auger electrons. The $N(E)$ function also includes a low energy peak, corresponding to ejected lattice electrons, a pronounced peak at 1 kV consisting of elastically reflected primary electrons, and several small peaks corresponding to characteristic energy losses of reflected primary electrons.

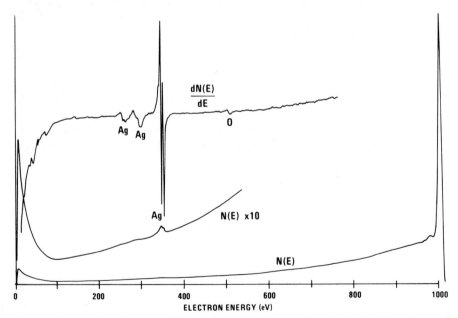

Fig. 1. Energy distribution $N(E)$ and dN/dE for a silver target with primary beam of electrons at 1000 eV.

A. THE AUGER PROCESS

The Auger process can be understood by considering the ionization process of an isolated atom under electron bombardment. When an incident electron with sufficient primary energy, E_P, ionizes a core level, the vacancy is immediately filled by another electron, as depicted by an $L_1 \rightarrow K$ transition in Fig. 2. The energy ($E_K - E_{L_1}$) from this transition can be released in the form of characteristic X-rays (the basis for X-ray fluorescence spectroscopy) or be transferred to another electron, e.g. in the L_2 level, which is ejected from the atom as an Auger electron. The measured energy of the electron is approximately equal to $E_K - E_{L_1} - E_{L_2} - \phi_A$, where ϕ_A is the work function of the analyzer material. This ejection process is termed the $KL_1 L_2$ Auger transition. Several such transitions ($KL_1 L_1$, $KL_1 L_2$, $M_2 M_4$-M_4 ...) can occur with various transition probabilities. The Auger electron energies are characteristic of the target material and independent of the incident beam energy.

It is obvious from the above discussion that at least two energy states and three electrons take part in an Auger process. Hence, H and He atoms cannot give rise to Auger electrons. Similarly, isolated Li atoms having a single electron in the outermost level cannot give rise to Auger electrons. In a solid, however, the valence electrons are shared and the Auger transitions of the type KVV occur involving the valence electrons of the solid. Thus, the Auger electrons from Li have been observed in various compounds containing Li.

The kinetic energy of Auger electrons originating from a WXY transition can be estimated from the empirical relation

$$E_{WXY} = E_W(Z) - E_X(Z) - E_Y(Z + \Delta) - \phi_A$$

The work function of the analyzer material is represented by ϕ_A and Z is the atomic number of the atom involved [1(d)]. The binding energy values may be obtained from the X-ray emission energy tables. The term Δ is introduced because the energy of the final doubly ionized state is somewhat larger than the sum of the energies for individual ionization of the same levels. Experi-

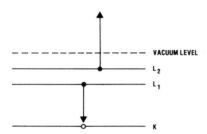

Fig. 2. Energy level diagram depicting relaxation by an L_1 level electron dropping into the K level and emission of an L_2 electron as an Auger electron.

References pp. 218—222

mental values for Δ lie between 1/2 and 3/4. Another expression [12] for estimating Auger transition energies is

$$E_{WXY} = E_W(Z) - 1/2\ \{E_X(Z) + E_X(Z+1)\} - 1/2\ \{E_Y(Z) + E_Y(Z+1)\} - \phi_A$$

Very few calculations have been attempted [13,14] to determine the Auger energies from first principles. Difficulties involved in these calculations relate to the coupling schemes governing the transition from a singly ionized to a doubly ionized state of the atom. The calculations generally involve [15] the magnetic interaction fields produced by the spins and the orbital angular momenta of the partially filled shells in addition to the other electronic interactions (coulomb and exchange). The magnetic interactions are dominant for core electrons of heavy elements, and the electronic configuration follows the *jj* coupling scheme. LS coupling dominates for light

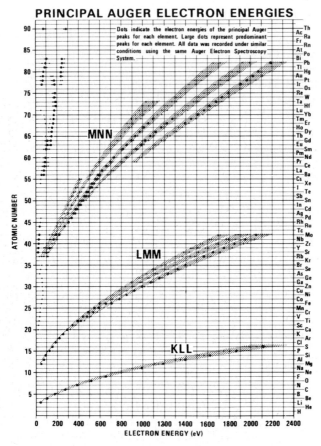

Fig. 3. Most prominent Auger transitions observed in AES.

elements and for the transitions involving low energies. It is necessary to add a spectroscopic term (e.g. $^1P_1$) to the WXY notation to characterize the transitions fully. These coupling schemes govern the energies as well as the transition probabilities. While the energies obtained by such calculations are normally not very accurate, recent calculations by Shirley [16] obtained by the introduction of relaxation energy into the intermediate coupling scheme led to very accurate KLL Auger energies.

The most pronounced Auger transitions observed in AES are given in Fig. 3. These involve electrons of neighboring orbitals, e.g. KLL, LMM, MNN, NOO, MMM, and OOO families. The Auger electron emission and X-ray emission are competing processes in bringing the initially excited atom to a low energy state, but under the conditions normally employed in AES (observed Auger peaks generally involve electron binding energies less than 2.5 keV), the probability of X-ray emission is negligible [1(f),10]. As indicated in Fig. 3, the most prominent Auger peaks result from KLL transitions for elements with $Z = 3$—14, LMM transitions for elements with $Z = 14$—40, MNN transitions for elements with $Z = 40$—79, and NOO transitions for heavier elements. For convenience, the Auger peak energy is commonly identified by the maximum negative excursion in the $dN(E)/dE$ peak. Complete Auger spectra for most elements and some compounds are available in the literature [17].

B. AUGER ELECTRON ESCAPE DEPTH

The high surface sensitivity of AES is due to the limited mean free path of electrons in the kinetic energy range 20—2500 eV. Auger electrons, which lose energy through plasma losses, core excitations, or interband transitions, are removed from the observed Auger peaks and contribute to the nearly uniform background upon which the Auger peaks are superimposed. Since phonon losses are small compared with the natural width of Auger peaks, they do not affect the Auger escape depth. Hence the Auger yield is not dependent on sample temperature.

The Auger electron escape depth is generally determined empirically by depositing atomically uniform overlayers on metallic substrates, while monitoring the diminution of Auger peaks from the substrate. The Auger peaks decay exponentially with overlayer coverage which is consistent with an exponential dependence of escape probability on the depth of the parent atom. A compilation of data from a variety of experimenters has been used to generate the "universal" escape curve shown in Fig. 4, which indicates that the escape depth is not strongly dependent on the matrix. This behavior is expected because the prominent loss mechanisms involve valence band excitations and the valence electron density is not a widely varying function of material. As discussed in Section IV, the small variation of escape depth with material greatly simplifies quantitative analysis.

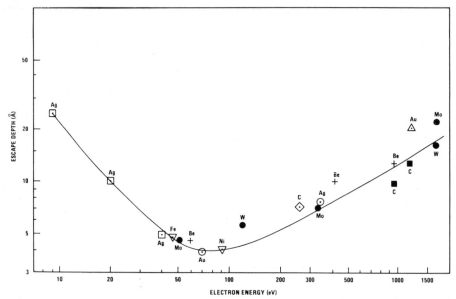

Fig. 4. Auger electron escape depths vs. energy for various materials. ●, M.L. Tarng and G.K. Wehner, Phys. Elect. Conf., 1972; ◻, D.E. Eastman, P.E.C. Alburquerque, 1972; ▽, J.W.T. Ridgeway and D. Haneman, Surface Sci., 24 (1971) 451, 26 (1971) 683; ⊙, P.W. Palmberg and T.N. Rhodin, J. Appl. Phys., 39 (1968) 2425; ◇, K. Jacobi and J. Hölz, Surface Sci., 26 (1971); △, Y Baer et al., Solid State Commun., 8 (1970) 1479; ■, R.G. Steinhardt et al., Proc. Int. Conf. Electron Spectrosc., Asilomar, 1971, p. 557; +, M.P. Seah, Surface Sci., 32 (1972) 703. (Modified version of figure compiled by J.C. Tracy, General Motors Corporation.) After Palmberg [44].

C. CORE LEVEL IONIZATION PROBABILITIES BY ELECTRON IMPACT

Since the Auger transition probability and Auger electron escape depth are independent of the incident electron beam energy, E_P, the dependence of the Auger peak amplitude on E_P is governed completely by the ionization cross-section of the initial core level. Ionization occurs primarily by the incident electrons during their initial passage through the escape depth region (~5—20 Å thick), but backscattered primary electrons can also contribute to the Auger yield when the incident beam energy is substantially greater than the binding energy of the core level involved. In gaseous specimens, where the contribution of scattered primary electrons is negligible, the ionization cross-section vs. E_P characteristics may be determined directly by measuring the Auger electron intensity as a function of primary energy. Hence the contribution of backscattered primary electrons to the Auger yield may be assessed by comparison of characteristic curves from gaseous and solid samples. Vrakking and Meyer [18] have measured the L-shell ionization cross-sections, σ, for gaseous S, P, and Cl, and compared these results with Auger yield vs. E_P curves on solids to evaluate the backscattering factor for silicon.

DuCharme and Gerlach [19] have measured the cross-section behavior as a function of E_P by measuring the KLL Auger current emitted by a monatomic layer of C, N, O, and Na atoms covering a W(100) surface. As E_P/E_K (E_K is the binding energy of the core level) increases from unity, σ rises rapidly from zero to a maximum at $E_P/E_K \simeq 3$. A further increase in primary energy over a large range does not cause a substantial variation in σ. The measurements were compared with theoretical calculations adopted from a first-Born approximation theory [20] of Burhop and the Coulomb—Born theory formulated by Rudge and Schwartz [21]. While the latter theory generally gave a better fit with the data, the measured values were generally higher for $E_P > 2.5\ E_K$. Similar results were obtained for L-shell cross-sections estimated [19] by studying the LMM Auger transitions of S, Cl, Ti, and Cu.

The contribution to σ from backscattered electron excitation is required in these experiments to evaluate single event cross-sections. The calculated energy distribution of the backscattered primaries within the escape depth region is theoretically very complex. Gerlach and DuCharme [22] have used simplifying assumptions to estimate that the backscattering contributions for C on a W(100) substrate from secondary events is about 25% for 1200 eV ($E_P/E_K = 4.23$), normally incident primary electrons. Calculations made by Bishop [23] predict an increasing contribution to ionization with both increasing atomic number and primary energy. The backscattering contribution to ionization is expected to show relatively small variation with the angle of incidence of the primary beam [10]. Experimental backscattering contributions of 10% for Be [24] at $E_K = 1.0$ keV, 20% for Si [25] at $E_P/E_K = 10$, and 25% for Ag [25] at $E_P/E_C = 4$ have been reported.

An estimate of the possible Auger current, I_A, can be made for C on W from the formula, $I_A = \sigma I_P N$, where $\sigma = 3.2 \times 10^{-19}$ cm^2 for $E_P \doteq 1136$ eV [22], $I_P = 10\ \mu$A, and $N = 10^{15}$ carbon atoms/cm^2. The result is an Auger current of 3.2×10^{-9} A or an Auger yield of 3.2×10^{-4}.

The acceptance angle and the transmission of the analyzer further reduces this quantity. For a 90% reduction, the magnitude of the detected current is 3.2×10^{-10} or 2×10^9 electrons/sec.

D. MATRIX EFFECTS

Since Auger electron emission frequently involves valence electrons and always involves the binding energy of core levels, the line shape as well as the most probable energy is strongly influenced by chemical environment. If strong (e.g. ionic) chemical bonding of two or more atoms occurs, the core electron levels may shift several electron volts. In ionic bonding, electron levels of electronegative elements are shifted to lower binding energies, and those of electropositive elements are shifted to higher binding energies due to the net charge transfer. A corresponding chemical shift in Auger electron kinetic energy is observed. When the electron density of states in the valence band is altered with chemical environment, a change in shape of Auger peaks from transitions involving valence electrons is observed.

In practice, measurable and frequently large shifts in Auger energies corresponding to changes in chemical environment are observed. Szalkowski and Somorjai [26] have found a linear energy shift of ~ 0.6 eV for the $L_3 M_{2\,3} M_{2\,3}$ Auger peaks from vanadium oxides as a function of oxidation number. Conversely, energy shifts of valence electrons from the $L_3 M_{2\,3} V$ transition could not be related to the oxidation states. This result suggests that energy shifts of Auger peaks involving valence electrons reflect more the redistribution of the density of states in the valence band than core level energy shifts.

Wagner and Biloen [27,28], using ESCA, have found large energy shifts in peaks from core level (i.e. excluding the valence band) Auger transitions in metal oxides and nonconducting salts. A shift of 7.0 eV was reported [28] for the $KL_{2\,3} L_{2\,3}$ transition for silicon compared with oxidized silicon and 5.4 eV for sodium in NaF, respectively. The corresponding photoelectron binding energy shifts for the above levels ($2P_{3/2}$) were 3.9 eV for oxidized silicon and, essentially, no shift for sodium. The large differences between Auger peak shifts and photoelectron binding energy shifts appear to be due to extra atomic relaxation (polarization) energies arising primarily from polarization of the electron clouds of adjacent atoms toward the core level vacancies. Calculations made by Kowalczyk et al. [29] predict that the extra atomic relaxation energy can cause an energy shift of approximately 10 eV for $LM_{4\,5} M_{4\,5}$ copper and zinc Auger electrons in metallic copper and zinc. Calculations have not been made for relaxation energies in different chemical environments, but it is likely that relaxation is the dominant factor in explaining large Auger peak shifts for core level transitions. Wagner [28] has made use of the difference in Auger and photoelectron energies to characterize the chemical state of oxides and compounds since, in most cases, this is more easily measured than a shift in photoelectron binding energy.

For some time, spectra of valence Auger electrons [1(h), 30—33] have been used to characterize the chemical state of solid surfaces empirically. Figure 5 (top) is an Auger spectrum of a silicon wafer after oxidation. The low energy (60—100 eV) Auger distribution is composed of LVV transition electrons and is characteristic of SiO_2 (75 eV). In this case, the Si peak associated with SiO_2 has undergone both a change in shape as well as a shift in energy. After sputtering, Fig. 5 (bottom), the spectrum is characteristic of elemental silicon with a peak energy of 91 eV.

An interesting example [30] of characterizing surface chemical states using AES is shown in Fig. 6 for the KLL carbon Auger spectra of CO on W(112), $W_2 C$, graphite, and diamond. Hybridization of L shell electrons and,

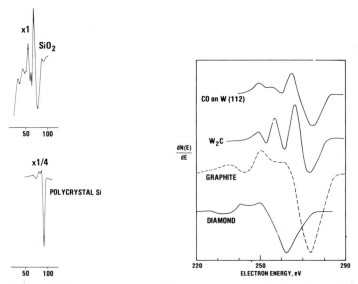

Fig. 5. LVV Auger spectra of SiO_2 and pure polycrystalline silicon.

Fig. 6. Spectra of the carbon KLL Auger transitions for CO on W(112), W_2C, graphite, and diamond. After Haas et al. [30].

thus, changes in the transition probability density is suggested as the cause for the differences in peak shapes.

The shape of the Auger electron line is not a direct measure of the density of states, but rather is a reflection of the transition probability density. Thus, first principle calculations to determine density of states from Auger spectra, or vice versa, is a complex problem. Amelio [34] compared deconvoluted valence spectra of silicon with the theoretical density of states. The comparison of reduced experimental data and theoretical calculations showed considerable agreement in the major maxima and minima features of the density distribution.

Hagstrum [35] and Becker and Hagstrum [36] by deconvoluting Auger line shapes have determined Auger transition probability density functions for the chalcogenides of nickel from ion neutralization spectroscopic studies. They have shown that features in the probability functions correspond to molecular orbital structure characteristic of surface chemical bonds. Sickafus [37] has shown that it is possible to apply similar treatments to electron-stimulated Auger electron line shapes. Specifically, good agreement with ion neutralization spectroscopic results was found for clean nickel and Ni(110)—C(2 × 2)S.

The fine structure apparent in the Auger spectrum is not always entirely

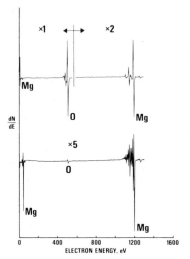

Fig. 7. AES and characteristic loss structure of vacuum-cleaved MgO (top) and evaporated Mg film on cleaved MgO (bottom).

due to the valence band electron density distribution. Auger electrons escaping the surface may lose discrete amounts of energy through plasmon excitations, core level ionization, and interband excitations and appear as distinguishable peaks. In the Auger spectrum these peaks will be apparent at characteristic energies less than the Auger electron energy. Because plasmon losses dominate in Auger fine structure, only these excitations will be discussed.

Plasmons are created by energetic electrons losing discrete amounts of energy to excite collective oscillations of valence electrons under the influence of positive atomic cores [38]. The characteristic frequency ω_p corresponding to the loss energy can be expressed classically as $(4 \pi n e^2/m)^{1/2}$ where n is the electron density, e is the electronic charge, and m is the mass. From this expression, it can be seen that the plasmon loss energy is characteristic of a specific solid and will change according to chemical environment. The large differences in the Auger fine structure from Mg and MgO is shown in Fig. 7.

Mularie and Peria [39] showed that some of the fine structure below the LVV Si Auger transition energy at 91 eV matched energy loss peaks due to plasmon excitations. The 91 eV silicon peak and fine structure were compared to plasmon loss peaks produced by elastically scattered electrons. The fine structure from pure silicon apparent at 73 eV in Fig. 5 (bottom) is attributed to a bulk plasmon loss.

III. Experimental methods

AES instrumentation in its simplest form involves a vacuum system, an electron gun for target excitation, and an electron spectrometer for energy analysis of the emitted secondary electrons. Because the AES technique is surface sensitive, an essential requirement is a UHV system capable of an ultimate vacuum in the 10^{-10} torr range. If every incident gas particle sticks to a surface, a monolayer is adsorbed in about one second at 10^{-6} torr. Even at a vacuum of 10^{-10} torr, an appreciable fraction of a monolayer of C and O can adsorb on active surfaces in a period of 30 min. Contamination from the vacuum system environment is obviously most critical for highly reactive metal surfaces where the sticking coefficient for most residual gases is near unity.

Surfaces which are passivated through exposure to air have a very low sticking coefficient, and hence the vacuum requirements are much less stringent. A vacuum in the low 10^{-8} torr range is usually sufficient for analysis on passivated surfaces. As discussed in Section III.C, the vacuum requirements are also relaxed for profile measurements in which the surface is continuously eroded by an argon ion beam during Auger analysis.

In addition to the basic requirements of an electron gun and an electron spectrometer, most AES systems presently in use employ an ion gun for in-depth profiling measurements, a manipulator for positioning the sample, a means for precisely locating the area of analysis, and frequently an attachment for performing in situ fracture or cleavage of specimens. In this section no attempt will be made to review the numerous variations of AES apparatus; only apparatus which has found widespread use will be considered.

A. ELECTRON ENERGY ANALYSIS

In the development of AES as a practical tool for surface analysis, several types of electron energy analyzers have been employed. These include the $127°$ cylindrical analyzer used by Harris [4] in his pioneering work in AES, the LEED—Auger apparatus first employed by Weber and Peria [5], and the cylindrical mirror analyzer [6,7]. Because of its superior signal-to-noise capability, the cylindrical mirror analyzer (CMA) is used almost exclusively for modern AES apparatus.

The two parameters which characterize electron velocity analyzers are the transmission, T, and the luminosity, L, where $L = \iint T(x,y)\,dxdy$. T is the probability that an electron of proper energy originating from a point (x,y) on the sample will be transmitted through the analyzer. The output signal, I, is then given by

$$I = \iint j(x,y)\, T(x,y)\,dxdy \tag{1}$$

where $j(x,y)$ is the spatial distribution of Auger electrons emitted from the target surface. When the target is excited uniformly over the region where T has appreciable value, eqn. (1) reduces to $I = jL$, where j is the Auger current density emitted over the acceptance area of the analyzer. In this case, it is the luminosity rather than transmission which is the proper figure of merit for the instrument.

When the diameter of the electron beam used to excite the target is very small (typically $<100~\mu m$) compared with the region over which T has appreciable value, eqn. (1) reduces to $I = TJ$, where J is the total Auger current emitted from the sample. Therefore, the highest signal levels will be obtained from an instrument with the highest possible transmission rather than the highest luminosity which is a primary consideration in ESCA apparatus. It is for this reason that the CMA, which is characterized by an unusually large transmission factor, is used almost exclusively in Auger apparatus.

A schematic diagram which illustrates the operation of a CMA for obtaining Auger spectra is shown in Fig. 8. The cylindrical mirror analyzer includes an internal electron gun mounted with its optical axis coincident with the CMA symmetry axis. The electron beam is focused to a fine point on the surface of the specimen which is positioned at the source point of the CMA. Electrons ejected from the point of excitation move radially outwards until they pass through a grid-covered aperture on the inner cylinder. A negative potential applied to the outer cylinder directs electrons with specific energy

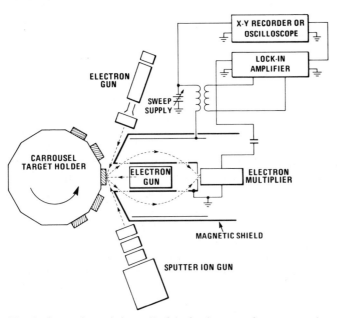

Fig. 8. Operation of the cylindrical mirror analyzer as an Auger spectrometer.

back through a second aperture on the inner cylinder and finally through a small exit aperture on the CMA symmetry axis. The pass energy, E, of the CMA is proportional to the potential applied to the outer cylinder. The range of electron energies, ΔE, which are transmitted is determined by the resolution R given by $R = \Delta E/E$. For commercial Auger spectrometers, R is typically 0.5%.

The secondary electron energy distribution, $N(E)$, is generated by plotting the output of the electron multiplier vs. the negative voltage applied to the outer cylinder. The $dN(E)/dE$ function is generated by superimposing a small a.c. voltage on the outer cylinder voltage and synchronously detecting the in-phase signal from the electron multiplier with a lock-in amplifier. The Y-axis of the X—Y recorder is then proportional to $dN(E)/dE$, and the X-axis is proportional to the kinetic energy of electrons emitted from the specimen.

B. SIGNAL-TO-NOISE CONSIDERATIONS

The sensitivity of AES is determined by the signal-to-noise ratio that can be obtained with the instrumentation. The detectable limit of concentration is limited by shot noise associated with the background current upon which the Auger peaks are superimposed. The background current consists predominantly of primary electrons which are backscattered through the surface a second time after loss of various amounts of energy through electronic excitation in the bulk. The total backscattered current with energy greater than 50 eV is typically 30% of the primary beam current. Assuming a uniform distribution of the backscattered current throughout the energy range from zero to E_P, the background current in the transmitted signal from a CMA is given by

$$I_B = 0.3\ T I_P\ \Delta E/E_P \tag{2}$$

where I_P is the incident beam current and ΔE is the energy window of the analyzer. The transmitted Auger current is given by

$$I_A = \beta I_P\ T\ \Delta E/\gamma\ \ (\Delta E \ll \gamma)\ \text{ or }\ \beta I_P\ T\ \ (\Delta E \gg \gamma) \tag{3}$$

where β is the Auger yield for the Auger transition of interest and γ is the inherent width of the Auger peak. In the shot noise limit, the equivalent noise current is given by [40]

$$I_N = (2\ I_B eB)^{1/2} \tag{4}$$

where e is the electronic charge, B is the measurement bandwidth, and I is the current exiting from the analyzer. In the shot noise limit, the Auger signal-to-noise ratio is derived from eqns. (2)—(4).

$$I_A/I_N = (\beta/\gamma)\ (I_P T\Delta E\ E_P/0.6\ eB)^{1/2}\ \ \ \ \Delta E \ll \gamma$$

$$I_A/I_N = \beta(I_P\ T E_P/0.6\ \Delta EeB)^{1/2}\ \ \ \ \Delta E \gg \gamma \tag{5}$$

References pp. 218—222

In eqn. (5) it is evident that the maximum signal-to-noise ratio is achieved when the energy window of the analyzer is approximately equal to the natural width of the Auger peak. Typical parameters are $I_P = 10\ \mu A$, $E_P = 5\ kV$, $T = 0.1$, $\beta = 10^{-4}$, $\gamma = 3\ eV$, and $B = 1\ Hz$. Under these conditions, the maximum signal-to-noise ratio is 1.3×10^4, which means that the limit of detectability is approximately 77 ppm. The optimum signal-to-noise ratio is normally not achieved, however, because ΔE and γ are not matched. A typical detection limit is 1000 ppm.

From the above considerations it is apparent that the energy window of the analyzer should match the natural width of Auger peaks as closely as possible over the energy range 0—2000 eV. Since Auger peaks are typically 3—10 eV wide, an energy resolution of about 0.5% is about optimum for the best signal-to-noise performance averaged over the entire range. Higher energy resolution can yield fine structure in Auger peaks which is not evident with a resolution of 0.5%, but will result in a significant reduction in signal-to-noise performance for many Auger peaks, especially in the low energy range.

C. THIN FILM ANALYSIS

In combination with ion etching, the inherent surface sensitivity of AES may be utilized to profile the elemental concentration with depth in thin films. As illustrated in Fig. 9, the ion beam forms a crater which is large

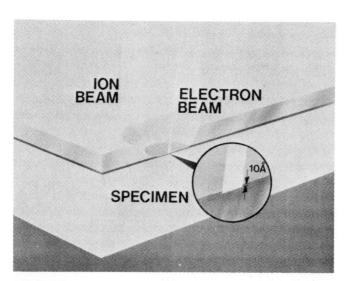

Fig. 9. Illustration of simultaneous ion beam milling and Auger analysis for thin film profiling.

compared with the diameter of the electron beam probe. In-depth profiling is accomplished by continuously sensing the elemental composition of the crater bottom during sputter erosion. Simultaneous ion bombardment has little effect on the Auger measurement, since the number of secondary electrons created by the ion beam is small compared with the secondary electron current created by the electron beam. The static pressure of 5×10^{-5} torr Ar required to operate the sputter ion gun also has a negligible effect on the Auger spectrometer operation.

A major advantage of the simultaneous ion etching and AES detection method over a sequential process in which the ion etching is stopped during the Auger measurement is that surface contamination is greatly reduced. In typical profile measurements, the surface is sputter eroded at a rate of several atomic layers per second. If the partial pressure of active gases can be maintained at 10^{-7} torr or below, the maximum arrival rate for surface contaminants is 0.1 monolayers/sec. Under these conditions, the concentration of surface contaminants cannot exceed a few percent of an atomic layer. When active gases are removed from the static pressure of Ar by a liquid nitrogen cooled Ti sublimation pump, the surface concentration of residual gas impurities can be maintained at a considerably lower level.

In early depth profile measurements [8], the data were acquired by repetitively scanning the pass energy of the CMA over a range sufficiently large to encompass all Auger peaks of interest. The variation in amplitude of the Auger peaks of interest was ascertained from a time base recorder output. The technique is generally quite inefficient, because only a small portion of the scanned energy range contains Auger peaks. In addition, the amplitude vs. sputtering time curves were plotted manually. When the in-depth concentration profiles of only a few elements are of interest, a multiplexing system greatly reduces data acquisition time and provides a more convenient display of the data. An illustration of a multiplexing technique is given in Fig. 10. Energy windows are adjusted to encompass only the peaks of interest, namely Si, O, Cr, and Ni. These energy windows are repetitively scanned in the sequence 1, 2, 3, 4, 1, 2, 3, 4, etc. As each energy window is scanned, the peak-to-peak amplitude of the corresponding Auger peak is electronically measured and plotted automatically on the Y-axis of an X—Y recorder, when the upper limit of the energy range is reached. The displacement of the X-axis is proportional to the elapsed sputter etching time. In multiplexing schemes, it is also useful to vary the measurement sensitivity independently for each channel. A direct recording of the O, Cr, Ni, and Si Auger peak amplitudes vs. etching time for a nichrome film is given in Fig. 11.

Several factors affect the depth resolution of the AES/ion etching profiling technique. These include ion beam uniformity across the sampled area, the Auger electron escape depth, and sample inhomogeneity. Loss of depth resolution from ion beam non-uniformity is negligible if the ion beam is large compared with the electron beam. The contribution from the Auger escape

174

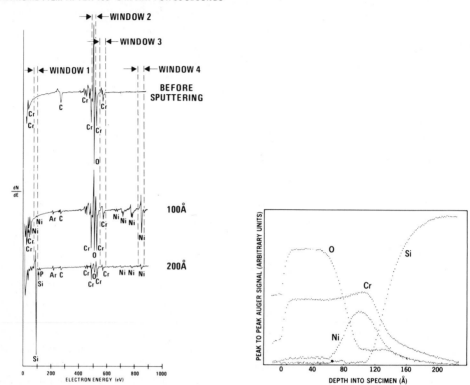

NICHROME FILM AFTER 450° C IN AIR FOR 30 SECONDS

Fig. 10. Auger spectra from nichrome thin film before sputtering and after etching to depths of 100 and 200 Å. Energy windows indicate regions scanned in the multiplexing technique.

Fig. 11. In-depth profile of nichrome thin film on silicon.

depth is only 5—20 Å and remains constant with sputtered depth. Sample inhomogeneity is the most important factor limiting depth resolution and can occur as a non-uniform thickness of thin films over the sampled area, as polycrystalline grains which are comparable in dimension with the thickness of the film, and as precipitation of constituent elements into particles. A non-uniform film thickness obviously limits the depth resolution to a value less than the percent uniformity of the film thickness over the sampled area. The depth resolution is determined by the relative sputtering rates of the precipitates and the matrix in films containing precipitates, and by a variation in sputtering yield with grain orientation in polycrystalline films. In general, the depth resolution of the AES/ion etching technique is limited to about 10% of the sputtered depth, but can approach 3% in optimum cases.

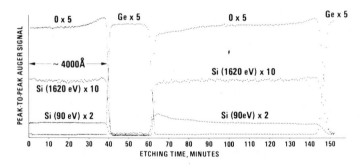

Fig. 12. In-depth profile of optical coating.

The best in-depth resolution that has been achieved occurs in amorphous films. As an example of the type of resolution which has been achieved, Fig. 12 illustrates the sputter profile from an optical film in which the in-depth resolution is about 3% of the sputtered depth.

D. SCANNING AUGER MICROSCOPY

When combined with a finely focused electron beam for target excitation, AES can be used to perform two-dimensional surface elemental analysis. The first attempt of this type was made by MacDonald [9] who combined a CMA with a commercial SEM and demonstrated point analysis [9] as well as elemental maps [41] with sub-micron spatial resolution. The second approach to scanning Auger microscopy was made by Pocker and Haas [42] who installed a field emission electron microscope gun on a standard UHV AES system. A severe limitation in their system was a poor signal-to-noise performance because of unstable beam current from field emission sources. A third approach to scanning Auger microscopy [43] involves optimization of an electrostatic gun inside of the CMA. The primary advantage of this approach is greatly simplified sample alignment.

A schematic diagram of a scanning Auger system [43] which utilizes an internal electrostatic gun is shown in Fig. 13. The electrostatic gun contains four electron optical components: a triode source with a tungsten filament, a condenser lens, an objective lens, and deflection plates. The electron gun control provides the necessary focusing voltages for the lenses and the power for the filament. The scanning system control provides the necessary voltages for deflecting the focused electron beam on the specimen, the deflection voltages for the record display and storage monitor, and the TV interface electronics. Electronics required to monitor and display the target current, secondary electron current, and the Auger signal are included in the scanning system control.

References pp. 218—222

SCANNING AUGER MICROPROBE
BLOCK DIAGRAM

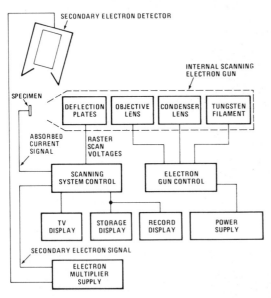

Fig. 13. Block diagram of electron gun and controls used for scanning Auger microscopy.

The electron gun operation is similar to that in a conventional SEM. Two lenses demagnify the crossover formed by the triode source, and the deflection plates raster the beam on the sample. At a minimum beam size of about three microns, the beam current is in the range of 0.1 μA. By relaxing the condenser lens voltage, the beam current can be increased up to about 10 μA with a corresponding increase in beam size. The magnification of the scanning system is controlled by the voltage amplitude applied to the deflection plates.

This scanning Auger system [43] can be used to perform point Auger analysis with a spatial resolution less than three microns or to obtain a two-dimensional mapping of the concentration of a selected surface element. The usual low energy secondary electron or absorbed current displays are used to monitor the surface topography and locate the areas of interest on the sample. To obtain an elemental map (Auger image), the intensity of the display is controlled by the magnitude of the selected Auger peak. The most negative excursion in the differentiated Auger spectrum is taken as a measure of the Auger current.

To demonstrate the operational modes of a scanning Auger system, MacDonald [43] has performed three-dimensional analysis of a microelectronic circuit. The TV display is first used to locate the areas of interest on the circuit chip. Absorbed current micrographs (Fig. 14) are then recorded on

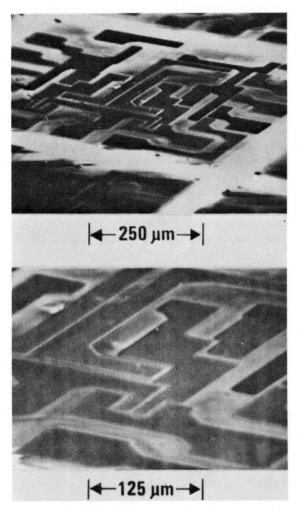

Fig. 14. Absorbed current micrographs of an integrated circuit (top) and an area on the integrated circuit that includes a transistor (bottom). I_P = 0.4 μA, E_P = 5kV.

the record display. The dark areas on these micrographs are metal interconnections consisting of a Au/Mo thin film composite; lighter areas are P/Si. Point analysis is accomplished by slowly reducing the scan voltage on the electron gun deflection plates to zero, while maintaining the area of interest at the center of the TV display. The surface composition of the SiO_2 layer is determined by recording an Auger spectrum as shown in Fig. 15.

A two-dimensional elemental map of the surface is obtained by setting the pass energy of the electron spectrometer at the negative excursion of the Auger peak of interest, and using the output of the lock-in amplifier to

178

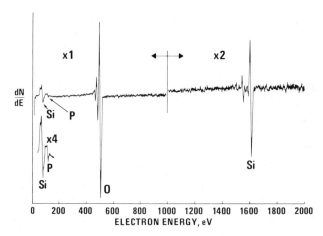

Fig. 15. Auger spectrum for a point analysis of phosphorus-doped SiO$_2$ on an integrated circuit.

intensity modulate the record display as the electron beam is rastered across the sample. Auger images for the elements Si, O, P, and Au are given in Fig. 16. Bright areas on these Auger images correspond to a high concentration of the element analyzed. The recording time for these Auger images was one minute using a beam current of 0.4 μA and a beam voltage of 5 kV.

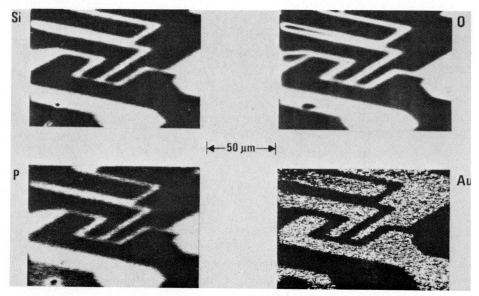

Fig. 16. Auger images of Si, O, P, and Au on an integrated circuit transistor.

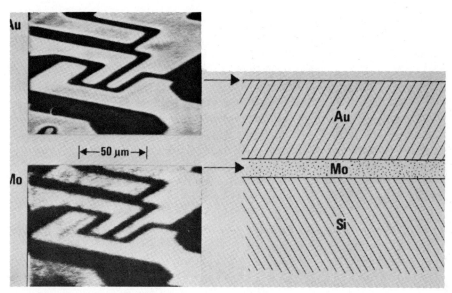

Fig. 17. Three-dimensional analysis of a Si—Mo—Au thin film showing the Au Auger image, the Mo Auger image, after sputter etching into the Mo layer, and a schematic of the composite structure.

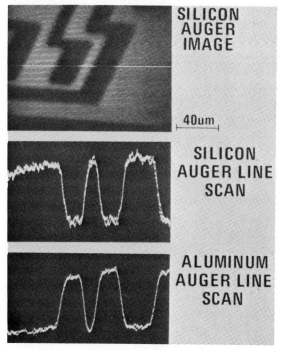

Fig. 18. Absorbed current micrograph of a transistor with a line scan marker (top), a Si Auger line scan, and an Al Auger line scan (bottom).

References pp. 218—222

Three-dimensional analysis using a combination of scanning Auger microscopy and sputter etching is illustrated in Fig. 17. The scanning Auger image for Au was obtained before sputtering to delineate the Au electrode pattern. Ion sputter etching was then performed while monitoring the Mo Auger peak. After sputtering one micron of Au from the sample, the Mo peak appeared, the sputter etching was terminated, and the Mo Auger image shown in Fig. 17 was recorded. These data demonstrate that the Mo layer, which serves to enhance adhesion between the Au electrode and the Si substrate, is uniformly distributed.

A more quantitative measure of the distribution of a particular element on a surface can be obtained by performing a line scan as illustrated in Fig. 18. In this approach, the electron beam is scanned across the integrated circuits as shown in the upper micrograph of Fig. 18, while the amplitude of the Al or Si Auger peak is used to Y-modulate the record display. In the lower portion of Fig. 18, the Y deflection is a direct measure of the concentration of Al.

IV. Quantitative analysis

In practical applications of AES, semiquantitative determination of the elemental composition is often sufficient. AES has been extremely successful in failure analysis, for example by simply using comparative analysis, i.e. demonstrating a difference in the surface composition between good and bad devices. In the evaluation of new processes and chemical characterization of new materials, however, it is frequently important to obtain accurate quantitative information. In this section the fundamental aspects and ultimate capability of AES for quantitative analysis are examined.

A. BASIC MECHANISMS AND ABSOLUTE MEASUREMENTS

To achieve absolute quantitative analysis in AES, it is necessary to relate the Auger current from a specific element to the population density of that

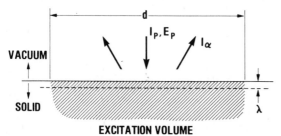

Fig. 19. Schematic representation of source volume of Auger electrons and the primary excitation volume.

element in the surface region. The Auger electron source volume generated by an incident electron beam of diameter d and energy E_P is depicted schematically in Fig. 19. The emitted Auger current produced by a WXY Auger transition in element α can be expressed as [44]

$$I_\alpha(\mathrm{WXY}) = \int\limits_\Omega \int\limits_{E_W}^{E_P} \int\limits_0^\infty I_P(E,Z)\, \sigma_\alpha(E,E_W)\, N_\alpha(Z)\, \gamma_\alpha(\mathrm{WXY}) \exp\left[(-Z/\lambda)\, d\Omega dEdZ\right] \tag{6}$$

where $I_P(E,Z)$ is the excitation flux density, $\sigma_\alpha(E,E_W)$ is the ionization cross-section of the core level W, $N_\alpha(Z)$ is the atomic density of element α at a depth Z from the surface, $\exp(-Z/\lambda)$ is the Auger electron probability for escape, and $\gamma_\alpha(\mathrm{WXY})$ is the WXY Auger transition probability factor. For simplicity, two-dimensional homogeneity of the elemental composition parallel to the surface has been assumed.

To simplify eqn. (6), three-dimensional homogeneity of the chemical composition will be assumed for the region in which the escape probability has significant value. It is also useful to separate the excitation flux density into two components, $I_P(E,Z) = I_P + I_B(E,Z)$, where $I_B(E,Z)$ is the excitation flux due to backscattered primary electrons. With these assumptions, the detected Auger current can be expressed as [44]

$$I_\alpha(\mathrm{WXY}) = I_P T N_\alpha \gamma_\alpha(\mathrm{WXY})\, \sigma_\alpha(E_P,E_W)\, \lambda(1 + R_B) \tag{7}$$

where R_B is defined as the backscattering factor, and T is the transmission of the analyzer.

To carry out accurate quantitative analysis from first principles utilizing eqn. (7), the ionization cross-section, the Auger yield, and the backscattering factor must be known to the desired degree of accuracy. In addition, the absolute Auger current must be accurately measured. A further complication is surface roughness which generally reduces the Auger yield relative to a flat surface [45]. Since these requirements are generally not met for routine Auger analysis, quantitative analysis using first principles is not considered practical at the present time. In the following sections, two methods which offer considerable possibilities for quantitative AES are discussed.

B. MEASUREMENTS WITH EXTERNAL STANDARDS

In this method, Auger spectra from the specimen of interest are compared with that of a standard with a known concentration of the element of interest. The concentration of element α in the test specimen (N_α^T) can be related to that in the standard (N_α^S) using eqn. (7).

$$N_\alpha^T/N_\alpha^S = (I_\alpha^T/I_\alpha^S)\,(\lambda^S/\lambda^T)\,\left[(1 + R_B^S)/(1 + R_B^T)\right] \tag{8}$$

An important advantage of this method is that ionization cross-section and Auger yield data are not required, and the Auger current is reduced to a relative measurement. When the test sample composition is similar to that of the standard, the escape depth and backscattering factor are also removed from eqn. (8) and quantitative analysis reduces to a relative Auger current measurement. It is, of course, essential that the two measurements be made under identical experimental conditions.

When the composition of the standard is not similar to that of the test specimen, careful consideration must be given to the influence of matrix on both the backscattering factor and the escape depth. Also, a variation in matrix which alters the shape of the Auger peak invalidates the use of the Auger peak-to-peak height in the differentiated Auger spectrum as an accurate, relative measure of Auger current [46].

As discussed in Section II.C, estimates of the backscattering factor can be obtained by comparing Auger yield vs. E_P curves with theoretical ionization cross-section vs. E_P curves or with Auger yield vs. E_P data from gaseous specimens where the backscattering factor has negligible value. As expected, experiments of this type [18] demonstrate that the backscattering factor increases with the weight density of the matrix and the primary beam energy, relative to the binding energy of the atomic core level involved. An obvious method of reducing the backscattering factor is to operate E_P near the ionization threshold, but this procedure also reduces the sensitivity through a reduction in ionization cross-section. A more practical solution is the development of formulas based on empirical data derived from the type of experiments mentioned above. It seems probable that an empirical backscattering factor of the form $R_B(E_P, E_w, D)$, where D is the weight density of the matrix, could be sufficiently accurate for quantitative analysis.

An elegant experiment which clearly illustrates the influence of backscattering on the Auger yield has been reported by Tarng and Wehner [47]. They deposited Mo uniformly on W while monitoring various Auger peaks from each element. As indicated in Fig. 20, the 120 eV Mo Auger signal rises to a value about 20% greater than that from pure Mo at an overlayer thickness of about eight monolayers. This overlayer thickness is slightly larger than the escape depth, but small compared with the penetration depth of the incident electron beam. Under these conditions, the backscattering factor is predominately determined by the heavier W substrate; the backscattering factor for the 120 eV Mo peak is 20% greater for a W matrix than for a Mo matrix.

The influence of the matrix on escape depth is not well established. As discussed in Section II.B, available Auger data appear to lie close to a "universal" escape depth curve, but the accuracy of the data is not sufficient to rule out matrix effects of sufficient magnitude to preclude quantitative analysis. More precise measurements on the matrix dependence of the escape depth are required for improving the quantitative capability of AES.

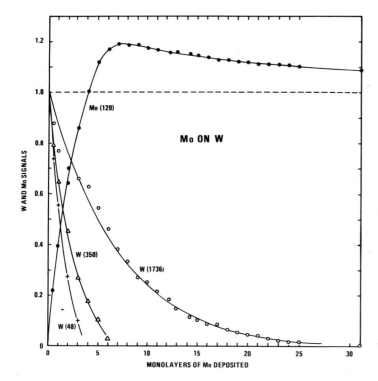

Fig. 20. Mo and W Auger signals as a function Mo overlayer coverage on W substrate. After Tarng and Wehner [47].

C. MEASUREMENTS WITH ELEMENTAL SENSITIVITY FACTORS

A less accurate but highly useful method for quantitative analysis involves the use of elemental sensitivity factors. Assuming that meaningful sensitivity factors can be assigned to all elements, the atomic concentration of element X can be expressed as [17,44]

$$C_X = (I_X/S_X)/\sum_\alpha (I_\alpha/S_\alpha) \qquad (9)$$

where S_α is the relative sensitivity of element α. The use of matrix-independent sensitivity factors obviously neglects variations in the backscattering factor and escape depth with material, and hence the method is normally only semiquantitative. Two important advantages of this method are the elimination of standards and insensitivity to surface roughness. The latter advantage is realized because all Auger peaks are uniformly affected by surface topography.

A first approximation to sensitivity factors can be obtained from Palmberg et al. [17] in which Auger spectra of pure elemental standards were obtained under identical experimental conditions. Specifically, the instrumental sensitivity is given for each spectrum and was calibrated using a silver specimen just prior to obtaining each spectrum. This means that the spectra given can be conveniently used to obtain elemental sensitivity factors relative to Ag for use in eqn. (9). Since the data were taken in the conventional $dN(E)/dE$ mode, it is necessary to base the sensitivity factors on the peak-to-peak height which is only valid when the peak shape is invariant with matrix [46].

D. EXPERIMENTAL RESULTS

To test the validity of the elemental sensitivity factor approach, an Inconel alloy was prepared for a ductile fracture inside a UHV system [48]. The Auger measurement included examination with a scanning Auger system to locate the incident beam on a region which was free from intergranular fracture surfaces. The Auger spectrum in Fig. 21 demonstrates that the only surface impurity in detectable quantity was carbon. Sensitivity factors based on available data [17] were applied to the MNN peak-to-peak heights of Ti, Fe, Cr, and Ni and the atomic concentrations determined from eqn. (9). In Table 1, the calculated atomic concentration levels from ductile fracture measurements of stainless steels and Inconel are compared with spectroscopically determined bulk concentrations. The experimental values derived from AES are in good agreement with the spectroscopically determined bulk values and are not affected appreciably by sputtering with argon ions. The agreement is expected since the Auger peak shapes remain nearly constant in

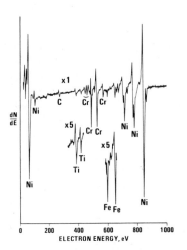

Fig. 21. Auger spectrum from ductile fracture surface of Inconel.

TABLE 1

Comparison of bulk atomic concentrations and values calculated from peak-to-peak height measurements obtained from ductile fracture surfaces [48]

Alloy	Element	$S_x{}^a$	Atomic concentration		
			Bulk atomic absorption	Surface AES	After sputtering AES
304 Stainless	Cr	0.31	0.205	0.24	0.23
	Fe	0.21	0.702	0.68	0.67
	Ni	0.26	0.093	0.09	0.10
316 Stainless	Cr	0.31	0.200	0.18	0.22
	Fe	0.21	0.656	0.71	0.67
	Ni	0.26	0.127	0.10	0.10
	Mo	0.25	0.018	0.01	0.02
600T Inconel	Cr	0.31	0.177	0.20	0.21
	Fe	0.21	0.084	0.07	0.07
	Ni	0.26	0.736	0.72	0.71
	Ti	0.44	0.003	0.01	0.01

[a] Relative elemental sensitivity factors S_x obtained directly from Palmberg et al. [17].

metal alloy matrices, and the similarity of the alloy constituents ensures a minimal variation in the backscattering factor and escape depth. Also, surface roughness to first-order approximation will have an equivalent effect on all signals.

In the case of ionic compounds such as oxides, however, the peak shape of the metallic component is often significantly altered relative to the pure metal. Also, the variation in matrix is expected to produce significant changes in the escape depth and backscattering factor. In Fig. 7, the Auger spectra of vacuum cleaved MgO and evaporated Mg on cleaved MgO were compared to demonstrate characteristic differences in Auger fine structure [44]. If the Auger yield is independent of matrix and the peak-to-peak technique is a valid measure of the Auger current, the density of Mg in the pure metal must be only half that in MgO. The actual atomic density of magnesium in MgO is 1.24 times as large as elemental magnesium. Since backscattering factors for Mg and MgO should be small and not significantly different, the difference must be attributed to either a longer escape depth in MgO or errors in the Auger current measurement caused by a change in peak shape.

Care must be used to eliminate electron beam artifacts when making standardized measurements on oxides, fluorides, chlorides, etc. with intense

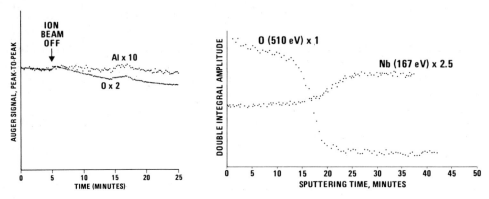

Fig. 22. Electron beam fragmentation of sapphire. E_P = 5 kV.

Fig. 23. Double integral amplitude profile of anodized niobium.

electron beam sources [49]. As an example, aluminum and oxygen peak-to-peak amplitudes from Al_2O_3 were measured as a function of time with and without ion beam sputtering. As indicated in Fig. 22, the oxygen signal decreased relative to the aluminum when the ion beam was turned off. The data is interpreted as dissociation and release of the oxygen from the sapphire surface. Subsequent studies showed that beam damage is negligible for electron beam current densities of 10^{-4} A cm^{-2} or less.

It has been pointed out [50] that in those cases where the shape of the peak is altered by the chemical matrix, measuring the total Auger current could be an improved method of measurement. In a recent study [51], the differentiated Auger signals of metal oxides were doubly integrated electronically and compared with pure metals. The electronic scheme accounted for a linear slope in the background through the Auger peak in the $N(E)$ curve. In Fig. 23, the double integral amplitudes were monitored as a function of time while sputtering a 1000 Å anodized niobium film. Assuming that the anodized region is predominantly Nb_2O_5, the ratio of the niobium atomic density in pure niobium to that in Nb_2O_5 is 2.8. In Fig. 23, the ratio of the Nb signals in pure Nb to those from Nb_2O_5 is 1.5. Since the weight densities of the two matrices are similar, the backscattering factors should be nearly equal and, therefore, cannot account for this large discrepancy. The discrepancy must be related to differences in the electron mean escape depths between the metal and metal oxide. These results suggest that absolute mean escape depth measurements may be required for quantitative analysis.

The results presented here demonstrate that quantitative analysis with an accuracy of a few percent is possible with elemental sensitivity factors when the matrix of the test specimen is similar to that of the specimens used to derive the sensitivity factors. In more routine analysis where matrix variations are large, the accuracy is reduced to 30—50%. Future developments in quantitative analysis will require precise measurement of backscattering factors and escape depths. In these experiments, techniques such as in situ fracture or cleaving provide a convenient means to prepare surfaces with known homogeneous concentration. Ion sputtering or scribing can also be used, but with some caution due to preferential sputtering. The validity of any of these techniques can be easily tested by examining several compositions of the alloy system of interest [52,53].

V. Applications

Numerous applications of AES have appeared in the literature during the past few years. While the initial applications were mostly fundamental in nature, recent trends and developments in high sensitivity and versatile instrumentation have led to many practical applications. In the following sections, numerous examples of typical AES applications in several fields are presented.

A. FUNDAMENTAL SURFACE SCIENCE

Initial applications of AES were primarily related to fundamental surface studies because of the simplicity of converting the LEED systems to AES instruments. In most of these studies, LEED was used to determine the atomic structure and AES for chemical analysis. In more general applications, AES is used to monitor the elemental composition of surfaces during physical property measurements. Several phenomena such as adsorption, desorption, surface segregation from the bulk, measurement of diffusion coefficients, and catalytic activity of surfaces have been investigated. AES has also been used to study the surface compositional changes in alloys during ion sputtering which is covered in greater detail elsewhere [53].

1. Identification of anomalous structures

LEED provides direct information of the two-dimensional translational symmetry of surface atomic arrangements, but does not identify the chemical species giving rise to the structure. In combination with LEED, AES has often been utilized to provide chemical identification of surface species. An example of this application is the identification of "ring" structures on the Pt(100) surface [1(a)]. The structure forms irreversibly when the specimens are heated in the 600—800°C range. The Auger spectrum [Fig. 24(a)] obtained from a Pt(100) surface before any heating shows primarily carbon. Little or no Pt is detected indicating that the carbon layer is sufficiently

188

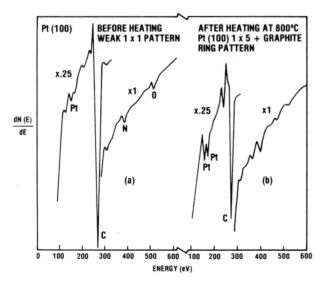

Fig. 24. Auger spectra from Pt(100): (a) Immediately after insertion into vacuum, and (b) after heating at 800°C. After Palmberg [1(a)].

thick to prevent the escape of Pt Auger electrons. A very weak 1 × 1 LEED pattern indicated a highly disordered structure. After briefly heating the specimen at 800°C, the LEED pattern showed a ring structure superimposed on the LEED pattern corresponding to the substrate. The Auger spectrum obtained is shown in Fig. 24(b). Auger peaks from the substrate as well as a change in the fine structure of the C Auger peak are observed. These results coupled with observations of the ring pattern suggests that heating causes the carbon to aggregate into graphite islands. Several other super structures on the single crystal surfaces have been related to the presence of impurities [1(a)].

2. Adsorption—desorption studies

In many adsorption and desorption studies, AES has been used to monitor surface coverage. The coverage is usually calibrated from supplementary techniques such as LEED, interferometry, quartz microbalance measurements, or ellipsometry. Palmberg [54] used the peak-to-peak amplitude in the dN/dE curve as a measure of surface coverage in a study of Xe adsorption on a Pd(100) surface. The variation of Xe peak height as a function of surface coverage at 77°K (Fig. 25) indicates that the sticking coefficient of Xe on Pd(100) remains near unity up to approximately a full monolayer coverage and then drops abruptly to zero. Also shown in the same plot is the variation of work function, $\Delta\phi$, with surface coverage. Many other studies of

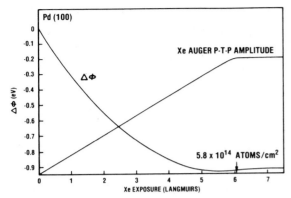

Fig. 25. Change in work function ($\Delta\phi$) and Xe Auger peak-to-peak amplitudes as a function of Xe exposure on a Pd(100) surface. After Palmberg [54].

adsorption and desorption of gases [55—60], as well as other elements [61—63] on solid surfaces, have appeared in the literature.

Adsorption—desorption studies are sometimes complicated by dissociation or desorption caused by electron beam irradiation. For example, the decomposition of CO into C and O makes the adsorption kinetics of CO impossible to study by using electron beams [64]. In most cases, however, the electron beam effects can easily be recognized in the experiment. Utilization of low primary beam energies and current densities and pulsed current techniques [65] often reduces electron beam effects sufficiently to allow examination by AES.

3. Surface segregation

Harris [4,66] was the first to report observations of surface segregation using AES. The specimen was positioned in front of the analyzer and heated from the back by electron bombardment. Harris observed that S segregates at the surface when Ni containing small amounts of S is heated between 600°C and 900°C. Since the S was readily removed by a brief exposure to air at room temperature and because its presence did not obscure the spectral lines of Ni, it was concluded that the extent of surface segregation was only one or two atomic layers. Harris has also observed that S segregates at the free surfaces of Ti, V, Fe, Cu, Nb, Mo, Ta, and W and that Cr and Sb segregate at the free surfaces of steel. Jenkins and Chung [67] observed C and S segregation to Cu(111) surfaces at 750°C. In the latter study it was also shown that segregated S occupies preferred sites on the surface, thereby giving a $\sqrt{7} \times \sqrt{7}$ surface structure. Taylor [68] reported S and P segregation at the surface of 304 stainless steel.

Results of a recent scanning Auger study of S and P segregation in 304 stainless steel are shown in Figs. 26 and 27. It is clear from Fig. 26 that S

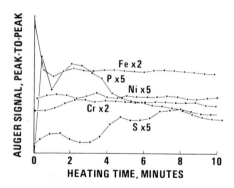

Fig. 26. Variation of surface composition during heating a 304 stainless steel foil specimen at 750°C.

diffuses to the surface more slowly than P (the bulk concentrations are known to be approximately 0.07 atomic percent each).

After 10 min at 650°C, the specimen was cooled to room temperature. The surface distribution of S and P (Fig. 27) indicates a nearly uniform distribution of P on the surface, while S is highly localized. Ion sputtering

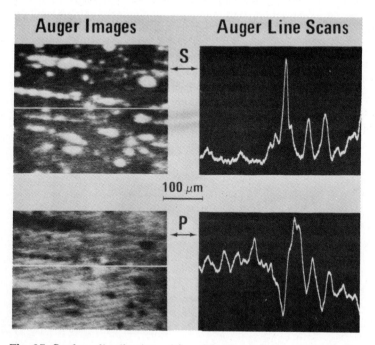

Fig. 27. Surface distribution of S and P after heating: Auger images and line scans.

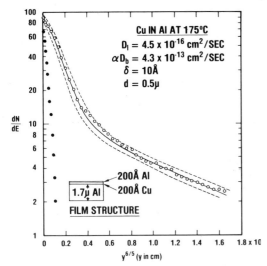

Fig. 28. Diffusion profile for Cu in Al thin film obtained after annealing for 85 min at 175°C. The solid circles show the unannealed profile. After Wildman et al. [73].

experiments in conjunction with AES have clearly demonstrated that the extent of segregation from the surface for both S and P is very small.

Diffusion parameters have been determined in some systems by studying the desorption and dissolution [69] or the segregation [70] kinetics of solute at the external surfaces. AES measurements of diffusion and surface segregation of deposited species from one side of a thin film to the other have been reported by several workers [71,72]. From such measurements the diffusion of Si through W [71] and of Cr through Pt [72] in the temperature ranges examined were determined to occur primarily via grain boundaries. Concentration variation of the diffusing species within the thin film has been measured by utilizing ion sputtering and AES [72,73]. Results of a diffusion profile for Cu in Al at 175°C after 85 min of annealing are shown in Fig. 28, where Y is the penetration distance measured from the center of the oxygen signal obtained at the Al/Cu interface. Based on the Whipple [74] analysis for grain boundary diffusion, these measurements enabled Wildman et al. [73] to separate the grain boundary contribution from lattice diffusion of Cu in Al.

Equilibrium segregation, which refers to the increase in surface concentration of one or more components as thermal equilibrium is established under conditions that correspond to a one-phase field in the phase diagram, has been studied in many systems [57,58,75—81] by monitoring the surface composition as a function of time at different temperatures. The thermodynamics of such segregation can be expressed [82] as

$$\theta_2/(1-\theta_2) = [X_2/(1-X_2)] \exp(Q/RT)$$

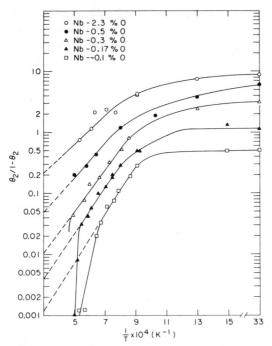

Fig. 29. Variation of oxygen surface coverage as a function of temperature in Nb—O alloys. θ_2 is the fraction of occupied sites to those actually available. After Joshi and Strongin [58].

where θ_2 is the ratio of the number of surface sites occupied by component 2 to the total number of available surface sites, X_2 is the ratio of the number of lattice sites, and Q is the energy change involved in transferring one mole of solute from the interior to the free surface of the sample.

Equilibrium surface segregation of O in Nb—O and Ta—O alloys has been observed in the 800—1700°C temperature range. The temperature dependence of oxygen segregation to surfaces in a series of alloys is shown in Fig. 29. In the low temperature region, surface saturation and kinetically limited diffusion occur, while in the high temperature region, loss of solute due to evaporation of Nb oxides is evident. In the intermediate temperature range, the surface reaches a steady equilibrium state from which the distribution coefficient, $K = K_0 \exp(\Delta H/RT)$, could be obtained. In this temperature range (825—1200°C), ΔH for oxygen segregation varied from 17 to 11 kcal/(g atom) in Nb and from 19 to 6.6 kcal/(g atom) in Ta as the bulk oxygen varied from 0.1 to 2.2 at.%, an actual decrease with increasing bulk concentration of O. The AES studies performed in conjunction with ion sputtering [58] clearly indicated that the segregation is confined to the one to five atomic layer region near the surface.

Segregation of carbon at Ni surfaces has been studied extensively [78, 79,83—85]. Segregation studies at single crystal surfaces [78] and stepped surfaces [79] indicate that the C segregation exhibits three distinct equilibrium states at the surface in different temperature regions. From $\theta/(1 - \theta)$ vs. $1/T$ plots, Isett and Blakely [79] obtained three different heats of segregation corresponding to different amounts of surface coverage, suggesting changes in surface chemical bonding.

Ferrante [77] reported equilibrium segregation of Al in Cu—Al alloys (discussed in Section V.B.1.c) and obtained [76] Q values of 1.38 kcal/mole for Al segregation in the Cu—10 at.% Al alloy. Bonzel and Aaron [75] studied equilibrium surface segregation of Ag on a Cu—14.3 wt. % Al alloy. The Cu—Al alloy was plated with silver on one face and then equilibrated at various temperatures in UHV, while the composition of the other face was monitored with AES. From the temperature dependence of the equilibrium surface composition, they arrived at an enthalpy of adsorption for Ag on the Cu—Al alloy of 14.6—14.9 kcal/mole.

4. Surface reactions

Several examples of "gas—metal" as well as "gas—surface impurity" reactions have appeared in the literature. Reaction of oxygen with S segregated on Ni(111) surfaces was studied by AES in conjunction with mass spectrometry [86]. By studying the rate of growth of the oxygen Auger peak and the disappearance of the S peak at various temperatures and partial pressures of oxygen, Holloway and Hudson [86] have observed that the surface reaction consists of three steps, viz.

O_2(gas) $\rightarrow$ O_2 (ads.)

O_2(ads.) + S (ads.) $\rightarrow$ SO_2(ads.) + n σ

and

SO_2(ads.) $\rightarrow$ SO_2(gas)

where σ represents an adsorption site for oxygen. Kinetic studies of the growth of surface oxygen and the decrease in S led them to conclude that the formation of SO_2 (ads.) is the rate-controlling step, and that the reaction occurs by nucleation at a limited number of sulfur surface sites and outward spreading of the reaction zone.

AES in conjunction with LEED has been utilized for studying the interaction of O with C on a Mo(111) surface [60] and with CO on Pd(111) surfaces [59]. Adsorption and interaction of O and N on Nb surfaces have been studied by Farrell et al. [57].

B. METALLURGY AND MATERIALS SCIENCE

Chemical properties such as corrosion, stress corrosion, oxidation, and catalytic activity and mechanical properties such as fatigue, wear, adhesion, resistance to deformation processes, and surface cracking depend on surface properties. Similarly, grain boundary "chemistry" influences mechanical properties, such as low and high temperature ductility and fatigue, chemical properties, such as intergranular corrosion and stress corrosion cracking, and electrical properties. The near surface or grain boundary chemistry can be affected by penetration of external contamination or segregation from the bulk.

It is well known that free surfaces and interfaces are characterized by atomic structure and energies which are different from the bulk. The free energy of these surfaces is often reduced by adsorption or segregation of impurity or alloying elements from the bulk. When the segregation is in thermodynamic equilibrium with the bulk (i.e. not diffusion limited), it is viewed as equilibrium segregation. Since surface forces govern such segregation, the extent of segregation is expected to be very limited, at most a few atomic layers. Non-equilibrium segregation at surfaces and interfaces results when thermodynamic equilibrium is not established and occurs, for example, during freezing and solid state phase transformations, segregation resulting from drag of alloying elements by the vacancies during heat treatments, and concentration build-ups that occur during mechanical treatments. The non-equilibrium segregation is thus characterized by diffusion limitations, and its occurrence can either be short or large in extent from the surface or grain boundary. In several studies involving segregation, AES has provided non-destructive surface chemical analysis and proved to be extremely valuable compared to most other techniques which are limited by either large sampling depth and/or poor sensitivity. AES experiments in conjunction with ion sputtering have, in some cases, helped to distinguish equilibrium from non-equilibrium segregation.

In the following sections, several examples are presented in which AES has been used to relate surface and grain boundary chemistry to properties of materials.

1. Surface properties

a. Mechanical properties inferred from surface segregation studies
Boron segregation to grain boundaries or to carbide inclusions at the boundaries was thought to be at least partially responsible for high temperature fracture in precipitation-annealed (650—750°C range) steels containing boron. On the presumption that surface segregation would be similar to that at the grain boundary, Bishop and Riviere [87] conducted experiments to study the surface segregation of boron-doped Fe. They observed maximum

segregation rates for B in the 700—800°C range, where precipitation treatment is carried out in steels, and confirms the likelihood that B is one of the elements involved in embrittlement of steels. In addition to B, they also observed segregation of N and S on the surface of Fe.

The temperature dependence of surface segregation has been examined by AES in zircaloy specimens [88]. Surface segregation of carbon and subsequent metal carbide formation above 300°C was inferred from the occurrence of a C Auger peak and the detailed shape of the C peak. The formation of a brittle carbide surface layer is believed to contribute to the ductility minimum observed at certain strain rates in the 300—350°C range.

Surface segregation and formation of surface phases in alloys within limited temperature ranges strongly influence their behavior during mechanical processing. AES studies performed [89] on Al and Cu base alloys have substantially aided in understanding many processing problems, such as resistance to deformation in rolling and drawing, surface defects in finished products, and poor adhesion properties during final plating operations.

Thompson et al. utilized AES to study the role of lead [90] in the machining of steel. By measuring the composition, thickness, and coverage of the lead and oxide films on the chip and tool surfaces, they concluded that a discontinuous lead film 20—40 Å thick forms on both the tool and chip surfaces during the machining of a leaded steel and is responsible for reducing the metal—metal contact between the work material and the tool. The presence of appreciable quantities of oxygen on the chip surface is believed to result from surface oxidation caused at the machining temperatures.

b. Corrosion and oxidation

The addition of small amounts of Zr substantially improves corrosion properties of some systems. For example, as little as 100 ppm Zr added to liquid Bi prevents liquid metal corrosion of low alloy steel containers. It has also been found that preferential corrosion of Nb grain boundaries by liquid Li can be retarded by adding 1% Zr. The effect of Zr in inhibiting corrosion is believed to arise from the formation of stable compounds with C, N, and O.

AES studies were performed [91] on a commercial Nb—1% Zr alloy to examine the formation and stability of these surface compounds during heat treatment. Some of the results of this study are shown in Fig. 30. Prior to AES analysis, the specimen was cooled to room temperature after heating for a few minutes at the temperature designated on the abscissa. As the specimen temperature was raised progressively to higher values, observable surface segregation of oxygen first occurred in specimens cooled from 500°C and reached a maximum at about 1300°C. When specimens were held at higher temperatures, some oxygen diffused back into the bulk, while the remaining fraction as well as the segregated Zr appeared to be stable on the

196

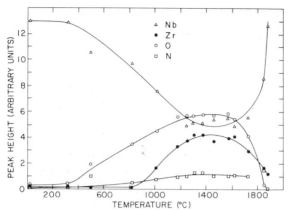

Fig. 30. Peak heights of Nb, Zr, O, and N as a function of the temperature from which the sample was cooled to room temperature. After Joshi et al. [91].

surface. The surface concentration of the irreversible and segregated Zr was estimated at about 30 at.%. Segregation of N was observed after heating to temperatures beyond 1000°C. At temperatures beyond 1500°C, Zr was desorbed from the surface, probably in the form of oxides. Sputter profiling measurements indicated that segregation of O, Zr, and N was limited to a few atomic layers. These measurements clearly indicate the formation of stable surface compounds at various annealing temperatures and thus explain the beneficial effect of Zr additions on the corrosion properties.

Wehner and co-workers [92] have studied surfaces of 304 stainless steel, Inconel, and CoCrAlY alloys corroded by fused salt eutectic mixtures of (a) $NaCl + NaNO_3$ and (b) $KF + NaF + KNO_3$ solutions. AES analysis and in-depth profiling gave compositional information of corrosion products on the surface and in the near surface region of the specimens. In some specimens which were corroded for a long time in mixture B, the corrosion products did not adhere well and could be peeled off easily. Scanning Auger analysis on these specimens provided chemical information near the reaction zones. Pitting and the formation of mesas and holes with a distinctive surrounding halo were studied in microscopic detail. An example of the type of information obtained in this study on 304 stainless steel specimens is given in Fig. 31. The Auger images were taken after sputter removing approximately 1500 Å. In addition, information on the compositional variation as a function of depth in various areas on the surface has led to a better understanding of the corrosion process.

Several studies on the initial stages of oxidation using AES have appeared in the literature. Ferrante [93] studied the oxidation of a Cu—19.6% Al alloy in 5×10^{-4} torr oxygen at 700°C and observed complete coverage of the surface by aluminum oxide. Studies of oxide growth at various tempera-

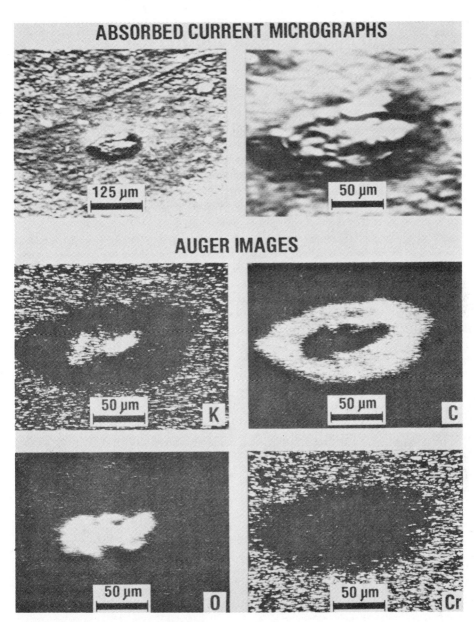

Fig. 31. Pit in 304 stainless steel after immersion in molten KF—NaF—NaNO$_3$ at 350°C for 7 days. After Wehner et al. [92].

tures indicated that a thermally activated diffusion process was important in oxide formation. Oxidation kinetics of Re in the 300—2200°K range were studied in the 10^{-9} to 10^{-6} torr range of oxygen pressure using a quadru-

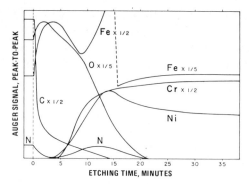

Fig. 32. Depth profile for a 304 stainless steel sample heated in air at 200°C for 2 h. Sputter rate = 5—10 Å/min. After Betz et al. [96].

pole mass spectrometer in conjunction with AES. In a different study, AES and ellipsometry were used to follow the growth of PbO on lead [94].

Studies on air oxidation of several commercial alloys [95,96] have been performed using AES. Betz et al. [96] have examined changes which occur at surfaces of 304 stainless steel, 321 stainless steel, and Inconel when exposed to air at various temperatures. Composition—depth profiles were obtained by AES and simultaneous sputtering. Results from a 304 stainless steel specimen oxidized at 200°C for two hours are shown in Fig. 32. Oxidation at temperatures below approximately 200°C caused formation of iron oxides within a narrow region near the surface, while oxidation at higher temperatures resulted in chromium-rich oxides. These results were explained on the basis that the chromium-rich oxide is the thermodynamically stable phase in the oxidation products of these alloys. The formation of an iron-rich oxide at low temperature occurs, because the kinetics of Cr migration to the surface is limited.

c. Adhesion, friction, and wear

Buckley [97] used a LEED—AES apparatus to study surface contamination and its influence on the adhesive properties of Fe single crystals. He showed that the effects of methane or ethane [98] on the Fe—Fe junction strength are minimal, while O [97] and H_2S [99] reduced the junction strength to values approaching zero. Similar studies were also performed in alloy systems. Ferrante and Buckley [100] found that surface segregation of Al occurs in Al—Cu alloys containing 1, 5, and 10 at. % Al. The segregated Al amounts to one atomic layer of ~100% Al. The enhanced adhesion observed for these alloy surfaces relative to pure copper was correlated [77,101] with the segregation of Al at free surfaces.

Adhesive properties of paint on galvanized steel surfaces has been studied as a function of pretreatment [102]. Auger analysis of steel showed that surfaces cleaned in a kerosene—acetone—alcohol mixture were nearly free of surface carbon' and exhibited excellent adhesion qualities. By performing AES analysis on several surfaces, Gjostein and Chavka [102] arrived at an empirical expression for an Adhesion Index (A.I.)

A.I. = (Zn) (O)/(C)

where the quantities in parenthesis are the Auger peak heights of the indicated surface species.

Auger measurements [103] from ground alumina substrates, which had been annealed in the $1400-1500°C$ range, demonstrated that Ca and Si segregated to the surface. Surfaces containing Ca and Si segregation exhibited good adhesion to Ta_2N—TiPdAu films, while etched specimens or annealed sapphire exhibited poor adhesion.

AES has succeeded in identifying contaminants in the plating industry which cause peeling, bulging, poor adhesion, or discoloration of platings. Electrodeposited Ni laminate material containing undesirably weak and fracture-sensitive Ni—Ni bonds was examined using an AES—SEM system [104,105]. Auger analysis of the fracture surfaces showed that Ag was a major constituent and definitely responsible for observed failures in structures fabricated by this technique.

AES data have led to a better understanding of wear properties [102] in friction materials. In a sintered Fe base brake material consisting of a wear additive MX, AES showed that the wear surface contained M in a concentration seven times larger than the bulk. The extent of segregation was estimated at 35 Å from an ion sputtering profile. These results demonstrate that the additive segregates to the wear interface in sufficient concentration to provide an effective wear interface.

In an effort to understand surface properties, such as corrosion, adhesion, wear, and wetting, Watterson and Simons [106] conducted AES investigations on the surface segregation of S, P, N, C, Sn, Sb, and Ca in heat-treated steels. Sulfur segregation was the most dominant feature, and the argon ion sputtering showed that the extent of segregation from the surface is very small. However, no direct correlations with any of the properties were reported in this study.

2. Grain boundary phenomena

Numerous mechanical properties and corrosion phenomena are related to grain boundary chemistry. Grain boundary embrittlement, intergranular corrosion, intergranular stress corrosion cracking, and grain growth are strongly influenced by impurity segregation to grain boundaries and have been successfully studied utilizing AES. The boundary chemistry is made amenable to study by obtaining intergranular fractures in these materials. Since the external surfaces so formed are envelopes containing grain boundary facets, analysis of these fracture surfaces by AES is representative of grain boundary or near boundary chemistry. The fractures are usually carried out in a UHV and analyzed immediately to minimize contamination from the ambient. Pressures in the range of 10^{-9} torr or better are essential to prevent excessive contamination on the active surfaces. In most cases, fracture and analysis are preceded by a bakeout of the entire system. In this section, AES studies of fracture surface chemistry are reviewed and related to various properties. Also included are studies on some powder metallurgy materials in which the surface preparation of particles plays a dominant role in the properties of final sintered compacts.

a. Embrittlement in ferrous alloys

AES has been highly successful in identifying grain boundary segregation of impurities in many steels and other Fe-base alloys. The first experiments [107,108], on low alloy steels with a nominal composition of 3.5% Ni, 1.6% Cr, 0.4% C, and 300 ppm Sb by weight, clearly demonstrated that Sb segregates to grain boundaries during embrittling treatment. Enrichment of Cr and Ni at the grain boundaries was also detected. Segregation was not detected in steels fractured in the intergranular mode after a non-embrittling or de-embrittling treatment. The extent of segregation [109] was estimated by ion sputtering in conjunction with AES; it was shown that all impurity elements, including Sb [107—111], P [109,111,112], and Sn [111,113] were concentrated within a very narrow region (1—4 atomic layers) near the grain boundary (Fig. 33). Segregation of Ni near the grain boundaries extended [110,112]

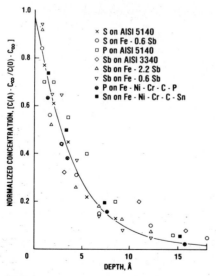

Fig. 33. Normalized concentration of segregated solute on fracture surface vs. average amount removed by ion sputtering. After Marcus et al. [111]. Reproduced by permission of the ASTM.

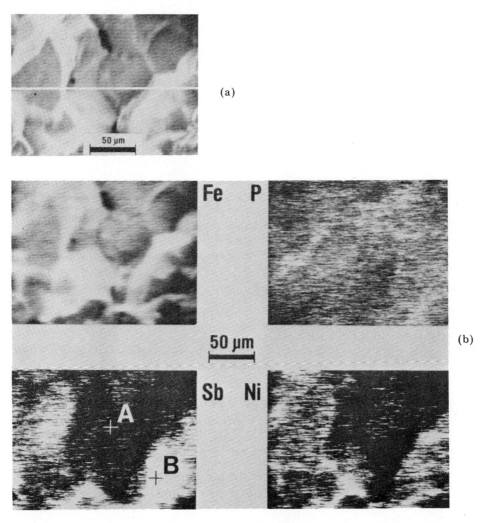

Fig. 34. (a) Specimen current micrograph showing the fracture surface of an embrittled 3340 steel, and (b) Auger elemental images of Fe, P, Sb and Ni obtained from the surface shown in (a).

over a larger range (as much as several hundred angstroms) and did not decrease exponentially with sputtering time; this result suggests [110] a segregation mechanism based on non-equilibrium processes.

The lateral distribution of segregated species on fracture surfaces from a 3340 steel has been examined with scanning AES [114]. These results, illustrated in Fig. 34, show that P segregates uniformly on all grain boundaries, and that Ni and Sb segregate preferentially on specific grain facets. Spectra obtained from two such facets are shown in Fig. 35. These observa-

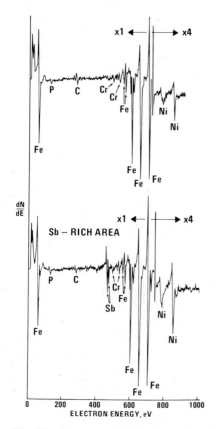

Fig. 35. Auger spectra obtained from the low Sb and Sb-rich grain boundary facets.

tions suggest a very strong dependence of segregation on grain boundary structure and energy. In most cases, the degree of embrittlement could be directly related to the observed segregation at grain boundaries; the variation in grain boundary composition as a function of the ductile to brittle transition temperature is shown in Fig. 36. Similar studies made in Ni—Cr—Mo—V steels [115] also indicate a direct correlation of segregation to the embrittlement.

The combined presence of Ni and Cr in low alloy steels leads to more segregation of Sb, P, or Sn to grain boundaries than when either is present alone [110,111]. The addition of Mo [111] to a Sb bearing steel only changes the kinetics of embrittlement, while extended aging at 480°C enhances segregation and consequent embrittlement. In a Sb steel containing additions of 0.1% Ti, however, Sb segregation was found to be diminished upon prolonged aging [116].

AES examination of grain boundary embrittlement in Fe—P and Fe—P—S alloys showed that P and S irreversibly segregated to grain boundaries in the

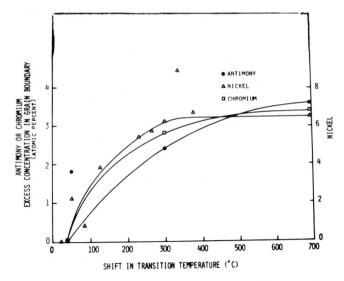

Fig. 36. Relation of severity of temper embrittlement to grain boundary concentrations of Sb, Ni, and Cr in 3340 steels. After Joshi and Stein [110]. Reproduced by permission of the ASTM.

entire ferrite range, while no segregation occurred in the austenite. It was also shown that, in the absence of C, the elements Ni and Cr are not required for the embrittlement and that the presence of P, S, or possibly Sb at the grain boundaries is sufficient to cause extreme brittleness in Fe. Irreversible embrittlement in Fe—Te alloys [118] was shown to be due to Te segregation to grain boundaries.

Segregation of Si, Sn, and S to grain boundaries of Fe was studied by Hondros and Seah [119] in an effort to measure their grain boundary activity directly. It was shown that enrichment ratios of 3, 462, and 10^4, respectively, are inversely related to maximum solid solubility, a phenomenon that was found to be consistent with other measurements of interfacial activity.

b. Embrittlement in non-ferrous alloys

Tungsten made by powder metallurgical methods is often extremely brittle with ductile—brittle transition temperatures as high as 825°C. It was believed that the embrittlement was associated with the segregation of oxygen to grain boundaries, but AES studies [120] did not indicate any oxygen enrichment in the brittle materials. Instead, AES experiments have shown that the phosphorus concentration at grain boundaries is a strong function of the transition temperature as indicated in Fig. 37 [120]. The P concentration at grain boundaries was found to be a function of grain size of the material examined and in all cases the P enrichment was localized to <20 Å near the grain boundary.

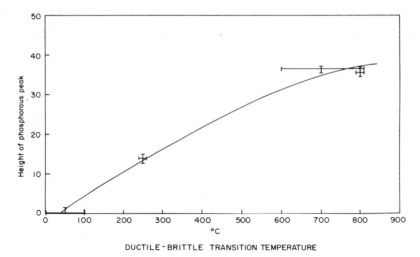

DUCTILE-BRITTLE TRANSITION TEMPERATURE

Fig. 37. Effect of P segregation to the grain boundaries on the ductile—brittle transition temperature of tungsten. After Joshi and Stein [120].

AES [121,122] was used to establish a correlation of embrittlement in copper, containing various amounts of Bi, with Bi segregation at grain boundaries. The concentration of Bi in the grain boundary region was estimated at 30 at.% and extended over a very limited region. In a different study, embrittlement in Cu containing 25—100 ppm Te was studied using AES [123]. A solution-treated material aged at $540°C$ for one day resulted in grain boundary segregation of Te and consequent embrittlement. Further aging resulted in a second phase formation in both the matrix and grain boundary region. This second phase formation resulted in de-embrittlement, which was believed to result from desegregation of Te from grain boundaries.

Polycrystalline Ni charged with hydrogen results in brittle, intergranular fractures when tested in slow tension. From AES studies, the embrittlement of Ni by hydrogen was correlated with segregation of hydrogen recombination poisons at the grain boundaries [124]. It was suggested that the entry of hydrogen into the Ni specimens occurs preferentially in the proximity of grain boundary intersections with the free surface. The Sb and Sn present at grain boundaries act as hydrogen recombination poisons and stimulate absorption of hydrogen by the metal.

In another AES study [125] on S-doped Ni base alloys, it was shown that embrittlement results from S segregation at grain boundaries in concentrations up to 12 at.%. It was also shown that ductility enhancement induced by doping the alloy with small amounts of elements such as Hf, Zr, and La was associated with a reduction in grain boundary sulfur concentration.

The role of TiC precipitation in the thermal embrittlement of 250 series maraging steels has been elucidated by AES [126]. It was shown that the segregated TiC is highly localized near the prior austenite grain boundaries. Further loss in fracture toughness with aging was shown to result from a loss in cohesive strength at the TiC/matrix interface due to segregation of B.

c. Embrittlement in powder metallurgy materials

Under identical conditions of pressing and sintering, compacts of iron powders prepared by various methods differ in density and tensile strength. Differences in mechanical properties were believed to arise from variations in initial surface preparation techniques or from changes that occurred during sintering. The interface chemistry between particles can be examined with AES by in situ fracture which occurs primarily along interparticle interfaces. In a recent AES study [127], the density and tensile strength of various specimens were related to the fracture surface concentration of impurity elements; the results are presented in Fig. 38. The relationship between the concentration of impurity elements and the density of powder-metallurgy compacts was similar; only oxygen appears to be detrimental. Low density (also high strength) irons exhibited a shallow in-depth profile, while the lowest density iron exhibited a region of segregation extending well over 300 Å. In a scanning AES study [70] of a low-density Fe, it was shown that O, C, and Ca are localized in the same regions on the fracture surface. These observations suggest that the contamination in the low-density iron consists of O, C, and Ca in a compound form. The distribution of O and C were not highly correlated on the fracture surface of a high-density Fe specimen.

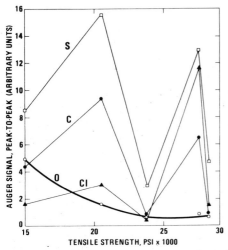

Fig. 38. Variation of S, Cl, C, and O at the fracture surface of powder metallurgy iron vs. tensile strength.

d. Intergranular corrosion

Austenitic stainless steels that have been solution-treated are often susceptible to grain boundary corrosion after heating in the 450—900°C temperature range. This phenomenon, known as sensitization, was strongly believed to result from precipitation of chromium-rich carbides at the grain boundaries. Since the carbides were chromium-rich, it was expected that the region adjacent to the grain boundaries would be depleted of Cr; impoverishment in Cr was hypothesized to be the reason for poor corrosion resistance. This explanation appeared satisfactory for explaining the corrosion behaviors in weakly oxidizing solutions, but does not account for observed intergranular corrosion of solution-treated steels which show no detectable carbide precipitation. Chaudron [128] and Armijo [129] have shown that high-purity alloys are immune to attack. These and other experiments indicate that Cr depletion and the strain theories [130] cannot account for all the observed properties. In order to understand the corrosion behaviors in austenitic stainless steels better, Joshi and Stein [131] conducted an AES study on freshly fractured surfaces of selected alloys and related the grain boundary chemistry to the corrosion behavior in strongly and weakly oxidizing solutions. The results of the study are briefly summarized below and in Table 2.

Sulfur, a common impurity element and sometimes an additive in stainless steels, segregates to grain boundaries during the high-temperature ($\sim 1050°C$) non-sensitization treatment. When the steel is subsequently treated in the 600—850°C range, the grain boundaries undergo depletion of Cr and enrichment of Ni. The corrosion properties, measured as weight loss/cm^2/h, in nitric—dichromate (strongly oxidizing) solutions relate very well with S concentration at grain boundaries and apparently have no relation to Cr depletion. The corrosion properties in H_2SO_4—$CuSO_4$ solutions, on the other hand, strongly depended upon Cr depletion, but not on S or Si segregation. The minimum Cr content at grain boundaries, required to provide good resistance to intergranular corrosion, was estimated at 13 wt.%. In other steels examined the segregation of P and N was observed and believed to cause poor corrosion resistance in nitric—dichromate solutions.

There is considerable evidence that intergranular corrosion under stress is influenced by impurities in many materials. AES studies on these materials in conjunction with stress corrosion studies is needed to assess the impurity effects fully. Another interesting area of application for AES studies is in materials where the penetration of liquid metals along grain boundaries seems to depend on the composition of the boundary.

e. Stability of grain structures

Powder metallurgy tungsten wire produced from WO_3, which is doped with potassium silicate and aluminum chloride prior to hydrogen reduction, is used extensively for filaments in incandescent lamps because of its resis-

TABLE 2

Auger spectroscopic analysis of 304 stainless steel fracture surfaces [131]

Treatment	C/Fe	Si/Fe	S/Fe	Cr/Fe	Ni/Fe	Corrosion rate (mg/cm²/h)	
						Boiling nitric—dichromate solution[a]	Modified Strauss test[b]
(1) 2 h 1050°C, W.Q.[d]	0.0128	N.D.[c]	1.230	0.664	0.102	2.25	Negligible
(2) 2 h 1050°C, W.Q. +2 h 850°C, W.Q.	0.177	0.066	0.588	0.515	0.111	0.84	0.016
(3) 2 h 1050°C, W.Q. +2 h 650°C, W.Q.	0.011	0.111	0.920	0.525	0.109	2.93	Negligible
(4) 2 h 1050°C, W.Q. +2 h 600°C, W.Q.	0.0645	N.D.[c]	0.881	0.548	0.103	2.64	Negligible
(5) 2 h 1050°C, W.Q. +3 days 650°C, W.Q.	0.283	N.D.[c]	0.850	0.459	0.119	0.87	0.447

[a]Weight loss rate determined at the end of 14 h corrosion test in nitric—dichromate solutions.
[b]Weight loss rate determined from a 60 h corrosion test.
[c]None detected.
[d]W.Q. = water quenched.

tance to creep at high temperatures. The cause of this phenomenon has been attributed to an interlocked grain structure which is formed as a consequence of either mullite particles, as suggested by Walter [132], or bubbles, as suggested by Moon and Koo [133]. An AES study performed on the intergranular fracture surfaces of tungsten rods doped with Al, K, and Si showed that K was segregated at the grain boundaries. No Al or Si was detected, but a strong correlation between bubble density and K content was established [134]. The results demonstrate that the bubbles are formed by elemental K and that K, rather than mullite, is responsible for the interlocked grain structures. Additions of ~1000 ppm of MgO to Al_2O_3 are known to increase the sintering rates and substantially increase the density of sintered Al_2O_3. The role of MgO was thought to result from reduced grain boundary motion caused by solute segregation at grain boundaries [135]. An AES study of intergranular fracture surfaces showed no large increase in Mg over bulk levels [136,137]; the grain boundary concentration was estimated at 0.4 at.% Mg^{2+} in a specimen containing 0.2 at.% Mg^{2+} in the bulk [138]. The only other element present in significant amounts at fracture surfaces was Ca. A closer examination [89] of the fresh fracture surfaces in a scanning Auger system indicated that the Mg signal arises from localized precipitates present at the grain corners where the Mg^{2+} content was estimated at 4 at.%. In Fig. 39(a), an optical micrograph of a relief polished section of slow-cooled Al_2O_3 sintered with MgO is shown, and in Fig. 39(b) the Mg Auger image of the fracture surface of a similar specimen demonstrates that the precipitates are indeed Mg-rich.

AES studies [137] in NiO-doped Al_2O_3 showed no significant segregation of the additive, but did exhibit heavy Ca segregation, which was observed in

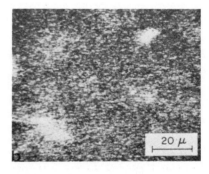

Fig. 39. (a) Micrograph of relief polished alumina, and (b) Mg Auger image obtained from a fracture surface of alumina. (Courtesy of W.C. Johnson, Ford Scientific Laboratory.)

CaO-doped Al_2O_3, as well as commercially pure Al_2O_3. Sputtering studies indicate that the segregated species were localized in the first 5—10 atomic layers near the boundaries.

Fracture studies of hot-pressed MgO doped with NaF and LiF have also been studied utilizing AES [137]. In the hot-pressed condition, the additives were present at the grain boundaries; after annealing only Ca and Ti were observed. Grain boundary chemistry changes affected by annealing were shown to be consistent with a liquid phase-assisted hot pressing mechanism and responsible for changes in physical properties.

C. CATALYTIC ACTIVITY

Catalytic activity of many materials is a surface phenomenon and hence a strong function of surface structure and chemistry. It is likely that surface impurities influence catalytic activity through induced changes in surface structure [139—141] as well as by more direct means. The surface chemistry in catalytic materials may be altered by adsorption from the environment, surface segregation from bulk, and contamination during usage.

Several Pt and Pt alloy catalyst surfaces have been studied using AES. After normal polishing and etching, Pt surfaces consist almost entirely of C and S to a depth of several atomic layers [142,143]. These are believed to be products of decomposition and can only be removed by excessive heating, ion bombardment, or heating in low oxygen pressures. A heat treatment of contaminated Pt [144] at 400°C in O_2 at 1 atm for 4 h removed the C and S from the surfaces. The Pt surface thus formed contained no more than a few monolayers of oxygen and was passive toward contamination in atmosphere that normally occurs on untreated Pt surfaces.

The kinetics of CO oxidation on a Pt(110) surface [145] was investigated using AES in conjunction with LEED and residual gas analysis. The experiments demonstrate that two different catalytic reaction mechanisms prevail in the oxidation process. At low temperatures, $100°C < T < 200°C$, CO remains adsorbed on the Pt surface resulting in a competitive adsorption by oxygen and an induction period for the onset of CO_2 formation which is a strong function of temperature. At high temperatures or in cases where oxygen was preadsorbed on Pt surfaces, the reaction with CO occurred immediately and was temperature-independent.

Catalytic platinum gauzes are often used in the oxidation of ammonia and in the synthesis of HCN. A 4% increase in HCN production and 20°C rise in temperature results when the catalyst, a Pt—10% Rh alloy, is treated with ~100 ppm H_2S for 110 h during continued HCN production. An examination of such a catalyst using AES [139] indicated that the surface consists of large amounts of C along with Rh, Ag, S, and N. Sputtering studies indicated that C extends over a great depth, while S and N are concentrated in the first few layers. This shows that treatment with H_2S does not produce bulk contamination with S. Silver, the origin of which is unknown, was also present to depths on the order of several thousand angstroms. It was also noticed that Rh was present in excessive amounts near the surface compared with the bulk.

Copper catalysts that are used in a variety of oxidation and dehydrogenation reactions were examined [146] by AES to determine the cause of differences in their performance. The poorly performing catalyst was found to contain 4.9 at.% Pb near the surface, which was almost three times that of a good catalyst surface. Lead is a known poison for many copper catalysts and apparently migrated to the surface during the manufacturing operation. In a similar study [147] on a Pd—alumina catalyst used for selective hydrogenation of diolefins in an olefin—aromatic—paraffin stream, it was found that small amounts of Fe acted as a poison by preferentially masking the active Pd surface.

Surface composition, which can differ significantly from bulk composition with little or no external influence, plays an even more important role in alloy catalysis. AES analysis of Ni—Au alloy foil surfaces [148] indicated that Au enrichment occurs at the alloy surfaces upon thermal equilibration

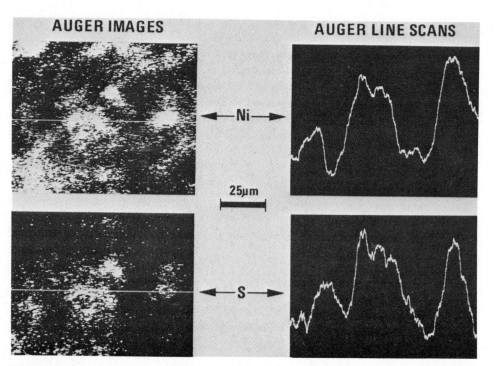

Fig. 40. Auger images and corresponding Auger line scans for Ni and S from spent methanation catalyst.

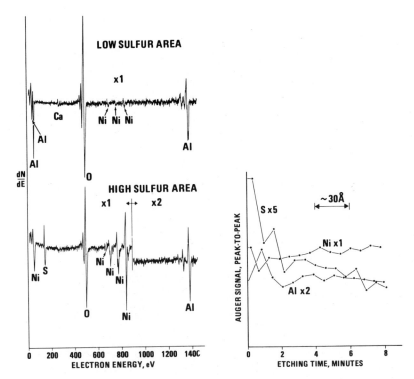

Fig. 41. Auger spectra from areas (5 μm diam.) of high and low sulfur (concentration on the spent catalyst.

Fig. 42. Depth—composition profile of the spent catalyst.

at high temperature. Chemisorption of hydrogen or oxygen at room temperature caused the surface concentration of Ni to increase. Since the Ni—O and Ni—H bonds are stronger than Au—O and Au—H bonds, preferential migration of Ni to the surface from the near surface region is expected.

In a 50% Ni—50% Al alloy catalyst used in methanation of natural gas, AES experiments have shown [89] that deactivation resulted from coverage of the surface by a small amount of S. A scanning Auger system was used to study the actual sites of S enrichment at the surface. The Auger images (Fig. 40) of Ni, S, Al, and O clearly show that Ni and S are present in the same localized regions. The spectra of Fig. 41 taken from areas of high and low sulfur content also support this observation. These data suggest that S was either preferentially adsorbed or segregated to Ni-rich surfaces. Depth composition profiles obtained from the surface (Fig. 42) of the spent catalyst demonstrate that the S is localized within a depth of about 30 Å on the Ni surfaces.

D. SEMICONDUCTOR TECHNOLOGY

Semiconductor technology was one of the first areas for application of AES to failure analysis and is now one in which AES plays a critical role in process development. In many respects, semiconductor devices are ideal specimens for the AES technique. Thin film structures on these devices are well-defined in thickness and lateral extent. The depth profiling technique can be used to elucidate surface and interfacial contamination, as well as inter-diffusion with excellent in-depth resolution. Scanning Auger microscopy can be used to localize the area of examination on a bonding pad, an oxide window, a resistor, or any other feature of interest.

1. Failure analysis

Failure of a device can often be traced to impurity poisoning of a surface during a particular process step. For example, impurities such as hydrocarbons and alkalis left on the surface after cleaning can lead to poor thin film adherence or poisoning of the device. Incomplete removal of a photoresist also can cause poor adherence between metallic thin films. Similarly, incomplete removal of the passivation layer over a metallic contact pad can produce a bond with poor mechanical strength and/or a high resistance contact.

Thermal compression bond failure on gold-coated lead frames or bonding pads is most often caused by the formation of impurity metal oxides at the gold surface. These metal oxides usually form in the thermal annealing process step necessary for good thin film adherence. All active metallic elements (such as Fe, Ni, Cr, Ag, Cu, etc.) within the gold layer are potential bond inhibitors. Similarly, metallic components may diffuse to the surface from the underlying layers and form oxides at the gold surface. Usually, these oxide layers are less than a few hundred angstroms in thickness and thus a surface-sensitive technique is required for positive identification.

The role of AES in detecting and identifying the impurities detrimental to bonding is exemplified in a recent publication by Wagner et al. [149]. A thick gold film on an alumina substrate was examined and found to have a high surface concentration of copper oxide, even though the manufacturers had specified 99% purity. Wagner also showed that ultrasonic bonding of aluminum wires to electroplated gold failed "pull" tests because of impurities (C, Ca, Ag, Cd, and O) at the surface following thermal treatment of the device. It is important to point out that these impurities would not affect bond strength if they were uniformly distributed throughout the gold layer.

Figure 43 shows Auger spectra of gold-plated lead frames which had demonstrated both good and poor mechanical bonding strengths [150]. As indicated, the poor bonding characteristics were associated with the presence of thallium at the surface. Depth profile analysis demonstrated that the thallium was segregated within a few atomic layers of the surface. A com-

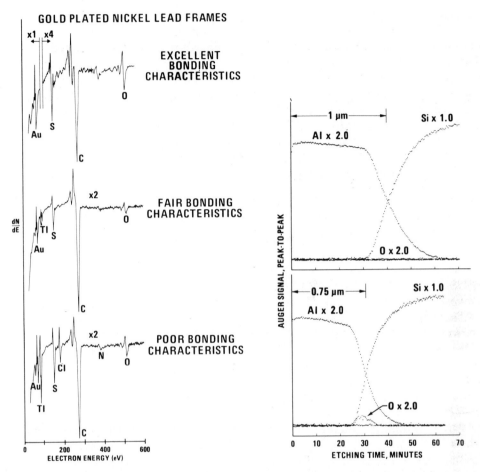

Fig. 43. Correlation of thallium Auger signal amplitude with bonding characteristics on gold-plated lead frames.

Fig. 44. Depth—composition profiles of Al—Si contacts; low resistance contact (top), and high resistance contact (bottom).

plete study showed that the mechanical bonding strength could be predicted by the amount (relative peak-to-peak height) of thallium present on the surface. After the cause of failure was determined, it was then possible to trace the source of thallium to the gold plating solution.

The formation of oxides at the surface or an interface can cause electrical as well as mechanical failure. Using AES, Morabito [151] has shown that zinc oxide at the surface of a gold-plated spring contact was the direct cause of high resistivity in the contact area. Further, the oxide layer was at most a few hundred angstroms thick and localized in the contact area.

Depth—composition profiles of aluminum/silicon-alloyed contacts with good and poor electrical resistance characteristics are shown in Fig. 44. The depth profile demonstrates that at least a partial oxide layer existed at the aluminum/silicon interface in the poor contact area and very likely increased

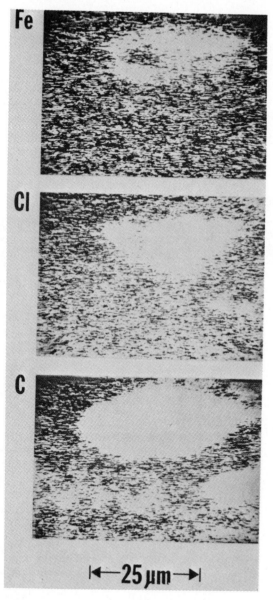

Fig. 45. Elemental analysis of particles on gold-plated lead frame showing Auger images for Fe, Cl, and C. After MacDonald [43].

the contact resistance. The source of contamination was traced to the accumulation of impurities from the vacuum environment on the silicon surface prior to aluminum evaporation.

An example of a scanning AES application in evaluating a bonding failure

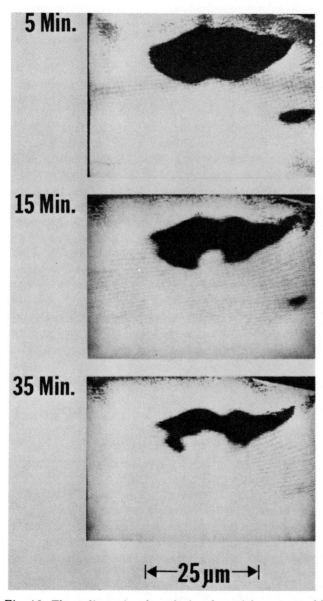

Fig. 46. Three-dimensional analysis of particles on a gold-plated lead frame. Au Auger images as a function of sputter etch time. After MacDonald [43].

References pp. 218—222

is given in Fig. 45 which is a two-dimensional compositional map of a gold-plated beam lead frame [43]. As indicated, AES analysis revealed localized iron, chlorine, and carbon impurities at the surface. Three-dimensional analysis of the impurity particle is demonstrated in Fig. 46. Gold images taken after sputtering 5, 15, and 35 min (~ 700 Å) define the particle geometry as a function of depth. The ability to identify particulate contamination and measure the thickness of the particle led to the particular process step in which the contamination was introduced.

2. Process evaluation

AES surface analysis and depth composition profiling provide a useful means for evaluating and developing specific device processes. The surface or interface composition may be characterized before and after a cleaning process. By direct comparison, the inter-diffusion characteristics important in achieving specific mechanical, electrical, and optical thin film properties can be characterized.

It has been shown [149] that contaminants such as Cl, F, and S left after final cleaning of MOS/LSI circuits can affect the operation of the final device. AES characterization of the surface was instrumental in the development of improved cleaning processes. From a depth—composition profile it was possible to show that the discoloration of a thin film metallization system used for terminating a tantalum nitride resistor was caused by copper and sulfur contamination of a palladium overlayer [151]. Since the profile showed that the copper and sulfur were removed from the palladium at similar rates during sputtering, it was concluded that copper and sulfur were chemically associated.

An example of depth—composition profile analysis of a thin film metallization system is shown in Fig. 47. This particular composite is made up of three separate metallic layers bonded to alumina. (Thin layers of chromium and titanium typically are used to promote gold adherence to ceramic surfaces.) The profile itself suggests what parameters are important and can be used as a visual aid for discussing the different kinds of evaluation that may be performed. For example, the test may be focused on the inter-diffusion of the chromium and titanium layers as a function of annealing temperature. In this case, several test specimens or even the actual package devices could be treated at different temperatures. The depth—composition profiles could then be compared for the penetration of chromium throughout the gold film or into the ceramic substrate. In another approach, the profile could be used as a quality control monitor of chromium film thicknesses for different production runs. Combining the test to include both thickness and Cr—Ti inter-diffusion could reveal preferred film thicknesses as a function of annealing temperature, thereby empirically arriving at an improved metallization system. Finally, the test could also be aimed at determining the maxi-

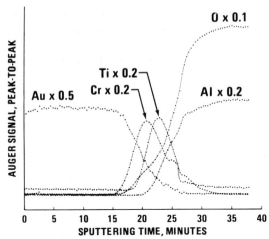

Fig. 47. Depth—composition profile of Au, Cr, and Ti metallization system on Al_2O_3.

mum allowable contamination at the titanium—alumina interface. Correlating adhesion tests and/or contact resistance measurements with depth composition profiles of carbon and oxygen, for example, could serve as a quality control of ceramic cleaning procedures and predeposition environment.

Finn et al. [152] have shown that AES/X-ray diffraction studies can be used to study alloying and chemical reaction occurring at Pt/GaAs interfaces after device thermal treatment in the temperature range of 300—400°C. Their study showed that the inter-diffusion of Pt and GaAs was directly observable from AES depth profiles by comparing treated and untreated specimens. A depth profile of a sample treated at 400°C for one hour showed that Ga and As were present on the Pt surface and that Pt had diffused into the GaAs substrate. X-ray diffraction data showed that compounds of $PtAs_2$, Pt_3Ga, and PtGa were formed after heat treatment.

In a depth—composition study of ohmic contacts, Robinson and Jarvis [153] examined nickel and gold—germanium thin films on GaAs substrates before and after alloying. Composition profiles from the surface showed that before annealing the 150 Å nickel and 340 Å gold—germanium films were well-defined on the GaAs substrate. Upon annealing (460°C for 40 sec), however, it was determined from the profiles that Ga, As, Au, and Ge had diffused to the nickel surface. The AES results, as well as scanning electron micrographs, indicated that the nickel overlayer was not a continuous and uniform coating after annealing. Additional process information about heated GaAs was gained from results which showed that differential out-diffusion of gallium depends on the crystalline orientation of the GaAs.

In a later study of ohmic contacts, Robinson [154] applied the same techniques and method to investigate the formation of palladium silicide on

References pp. 218—222

single crystal silicon. In a comparison of X-ray diffraction data of a thick Pd_2Si layer and the low-energy silicon Auger spectrum, it was possible to associate distinguishable electron peaks with free silicon and silicon in the compound form of Pd_2Si. In total, the results indicated silicon enrichment at the surface and diffusion-limited formation of the Pd_2Si layer.

Acknowledgments

The authors express their appreciation to Drs. G.E. Riach, R.E. Weber, G.K. Wehner, R.L. Gerlach, and N.C. MacDonald for critical reading of the manuscript.

References

1 (a) P.W. Palmberg, in G.A. Somorjai (Ed.), The Structure and Chemistry of Solid Surfaces, Wiley, New York, 1969, p. 29-1. (b) D.F. Stein, R.E. Weber and P.W. Palmberg, J. Metals, 23 (1971) 39. (c) C.C. Chang, Surface Sci., 25 (1971) 53. (d) P.W. Palmberg, in D.A. Shirley (Ed.), Electron Spectroscopy, North Holland, Amsterdam, 1972, p. 835. (e) R.E. Weber, J. Cryst. Growth, 17 (1972) 342. (f) T.E. Gallon and J.A.D. Matthew, Rev. Phys. Technol., III 1 (1972) 31. (g) J.C. Riviere, Contemp. Phys., 14 (1973) 513. (h) C.C. Chang, in P.F. Kane and G.B. Larrabee (Eds.), Characterization of Solid Surfaces, Plenum Press, New York, 1974, p. 509.
2 P. Auger, J. Phys. Radium, 6 (1925) 205.
3 J.J. Lander, Phys. Rev., 91 (1953) 1382.
4 L.A. Harris, J. Appl. Phys., 39 (1968) 1419.
5 R.E. Weber and W.T. Peria, J. Appl. Phys., 38 (1967) 4355.
6 H. Hafner, J. Arol Simpson and C.E. Kuyatt, Rev. Sci. Instrum., 39 (1968) 33; V.V. Zashkvara, M.I. Korsuwskii and O.S. Kosmackei, Zh. Tekh. Fiz., 36 (1966) 132; H.Z. Sar-el, Rev. Sci. Instrum., 38 (1967) 1210; E. Blauth, Z. Phys., 160 (1957) 228.
7 P.W. Palmberg, G.K. Bohn and J.C. Tracy, Appl. Phys. Lett. 15 (1969) 254.
8 P.W. Palmberg, J. Vac. Sci. Technol., 9 (1972) 160.
9 N.C. MacDonald, Appl. Phys. Lett., 16 (1970) 76.
10 H.E. Bishop and J.C. Riviere, J. Appl. Phys., 40 (1969) 1740.
11 J.F. Hannequin and P. Viaris de Lesegno, Surface Sci., 42 (1974) 50; T.W. Haas, R.W. Springer, M.P. Hooker and J.T. Grant, Phys. Lett. A, 47 (1974) 317; R.G. Musket and W. Bauer, Thin Solid Films, 19 (1973) 69.
12 M.F. Chung and L.H. Jenkins, Surface Sci., 22 (1970) 479.
13 W.N. Assad and E.H.S. Burhop, Proc. Phys. Soc., 71 (1958) 369.
14 W.N. Assad, Proc. Roy. Soc. Ser. A, 249 (1959) 559; W.N. Assad, Nucl. Phys., 44 (1963) 399, 415; 63 (1965) 337; 66 (1965) 494.
15 I. Bergstrom and C. Nordling, in K. Siegbahn (Ed.), Alpha-, Beta-, and Gamma-Ray Spectroscopy, Vol. II, North Holland, Amsterdam. 1965.
16 D.A. Shirley, Phys. Rev. A, 7 (1973) 1520.
17 P.W. Palmberg, G.E. Riach, R.E. Weber and N.C. MacDonald, Handbook of Auger Electron Spectroscopy, Physical Electronics Industries, Edina, 1972.
18 J.J. Vrakking and F. Meyer, Surface Sci., 47 (1975) 50.

19 A.R. DuCharme and R.L. Gerlach, J. Vac. Sci. Technol., 10 (1973) 188.
20 E.H.S. Burhop, Proc. Cambridge Phil. Soc., 36 (1940) 43; A.M. Arthurs and B.L. Moiseiwitsch, Proc. Roy. Soc. Ser. A, 247 (1958) 550.
21 M.R.H. Rudge and S.B. Schwartz, Proc. Phys. Soc., 88 (1966) 563.
22 R.L. Gerlach and A.R. DuCharme, Surface Sci., 32 (1972) 329.
23 H.E. Bishop, Brit. J. Appl. Phys., 18 (1967) 703.
24 K. Gato, K. Ishikawa, T. Koshikawa and R. Shimizu, Appl. Phys. Lett., 24 (1974) 358.
25 T.E. Gallon, J. Phys., 5 (1972) 822.
26 F.J. Szalkowski and G.A. Somorjai, J. Chem. Phys., 56 (1972) 6097.
27 C.D. Wagner and P. Biloen, Surface Sci., 35 (1972) 82.
28 C.D. Wagner, American Chemical Society Meeting, Houston, Texas, Nov. 1974.
29 S.P. Kowalczyk, R.A. Pollak, L. Ley and D.A. Shirley, Phys. Rev. B, 8 (1973) 2387.
30 T.W. Haas, J.T. Grant and G.J. Dooley, in F. Ricca (Ed.), Adsorption—Desorption Phenomena, Academic Press, New York, 1972, p. 359.
31 J.P. Coad, Phys. Lett. A, 35 (1971) 185.
32 R.J. Fortner and R.G. Musket, Surface Sci., 28 (1971) 339.
33 T.W. Haas and J.T. Grant, Phys. Lett. A, 30 (1969) 272.
34 G.G. Amelio, Surface Sci., 22 (1970) 301.
35 H.D. Hagstrum, Phys. Rev., 150 (1966) 495.
36 G.E. Becker and H.S. Hagstrum, Surface Sci., 30 (1972) 125.
37 E.N. Sickafus, J. Vac. Sci. Technol., 11 (1974) 308.
38 D. Pines, Elementary Excitations in Solids, Benjamin, New York, 1964.
39 W.M. Mularie and W.T. Peria, Surface Sci., 26 (1971) 125.
40 K.K. Spangenberg, Vacuum Tubes, McGraw-Hill, New York, 1948.
41 N.C. MacDonald and J.R. Waldrop, Appl. Phys. Lett., 19 (1971) 315.
42 D.J. Pocker and T.W. Haas, J. Vac. Sci. Technol., 12 (1975) 370.
43 N.C. MacDonald, in D. Beaman and B. Siegel (Eds.), Electron Microscopy: Physical Aspects, Wiley, New York, 1974.
44 P.W. Palmberg, Anal. Chem., 45 (1973) 549A.
45 P.H. Holloway, to be published.
46 R.E. Weber and A.L. Johnson, J. Appl. Phys., 40 (1969) 314.
47 M.L. Tarng and G.K. Wehner, J. Appl. Phys., 44 (1973) 1534.
48 L.E. Davis and A. Joshi, to be published.
49 S. Thomas, J. Appl. Phys., 45 (1974) 161; P.W. Palmberg and T.N. Rhodin, J. Phys. Chem. Solids, 29 (1968) 1917.
50 J.T. Grant, T.W. Haas and J.E. Houston, J. Vac. Sci. Technol., 11 (1974) 227; J.E. Houston, Rev. Sci. Instrum., 45 (1974) 897; J.T. Grant, M.P. Hooker and T.W. Haas, Surface Sci., 46 (1974) 672.
51 R.L. Gerlach and R. Hedtke, Physical Electronics Industries, Inc., Eden Prairie, Minnesota, private communication.
52 W. Farber and P. Braun, Vak. Tech., to be published.
53 G.K. Wehner, this volume, Chap. 1.
54 P.W. Palmberg, Surface Sci., 25 (1971) 598.
55 J.V. Florio and W.D. Robertson, Surface Sci., 18 (1969) 398.
56 R.G. Musket and J. Ferrante, J. Vac. Sci. Technol., 7 (1970) 14.
57 H.H. Farrell, H.S. Isaacs and M. Strongin, Surface Sci., 38 (1973) 31.
58 A. Joshi and M. Strongin, Scr. Met., 8 (1974) 413.
59 G. Ertl and J. Koch, in F. Ricca (Ed.), Adsorption—Desorption Phenomena, Academic Press, New York, 1972, p. 345.
60 R.M. Lambert, J.W. Linnett and J.A. Schwartz, in F. Ricca (Ed.), Adsorption—Desorption Phenomena, Academic Press, New York, 1972, p. 381.

61 S. Thomas and T.W. Haas, J. Vac. Sci. Technol., 9 (1972) 840.
62 E. Bauer, H. Poppa, G. Todd and F. Bonczek, J. Appl. Phys., 45 (1974) 5164.
63 L.E. Davis, L.L. Levenson and J.J. Melles, J. Cryst. Growth, 17 (1972) 354.
64 J.C. Tracy and P.W. Palmberg, Surface Sci., 14 (1969) 274.
65 B.W. Byrum, Rev. Sci. Instrum., 45 (1974) 707.
66 L.A. Harris, J. Appl. Phys., 39 (1968) 1428.
67 L.H. Jenkins and M.F. Chung, Surface Sci., 24 (1971) 125.
68 N.J. Taylor, J. Vac. Sci. Technol., 6 (1969) 241.
69 M. LaGues and J.L. Domange, Surface Sci., 47 (1975) 77.
70 A. Joshi, to be published.
71 C.C. Chang and G. Quintana, J. Electron Spectrosc. Relat. Phenom., 2 (1973) 363.
72 S. Danyluk, G.E. McGuire, K.M. Koliwad and M.G. Yang, Thin Solid Films, 25 (1975) 483.
73 H.S. Wildman, J.K. Howard and P.S. Ho, J. Vac. Sci. Technol., 12 (1975) 75.
74 R.T. Whipple, Phil. Mag., 45 (1954) 1225.
75 H.P. Bonzel and H.B. Aaron, Scr. Met., 5 (1971) 1057.
76 J. Ferrante, Scr. Met., 5 (1971) 1129.
77 J. Ferrante, Acta Met., 19 (1971) 743.
78 J.C. Shelton, H.R. Patil and J.M. Blakely, Surface Sci., 43 (1974) 493.
79 L.C. Isett and M. Blakely, J. Vac. Sci. Technol., 12 (1975) 237.
80 K. Kunimori, T. Kawai, T. Kondow, T. Onishi and K. Tamaru, Surface Sci., 46 (1974) 567.
81 J.J. Bellina and B.B. Rath, AIME Meeting, Pittsburgh, Pennsylvania, May 20, 1974.
82 D. McLean, Grain Boundaries in Metals, Oxford University Press, London, 1957.
83 J.M. Blakely, J.S. Kim and H.C. Potter, J. Appl. Phys., 41 (1970) 2693.
84 J.P. Coad and J.C. Riviere, Surface Sci., 25 (1971) 609.
85 E.N. Sickafus, Surface Sci., 19 (1970) 181.
86 P.H. Holloway and J.B. Hudson, Surface Sci., 33 (1972) 56.
87 H.E. Bishop and J.C. Riviere, Acta Met., 18 (1970) 813.
88 G.J. Dooley, III, J. Vac. Sci. Technol., 9 (1972) 145.
89 Analytical Services Laboratory, Physical Electronics Industries, Inc., Eden Prairie, Minnesota.
90 R.W. Thompson, D.T. Quinto and B.S. Levy, Inland Steel Co., East Chicago, private communication.
91 A. Joshi, M.N. Varma and M. Strongin, Met. Trans., 5 (1974) 861.
92 E.M. Sparrow, J.W. Ramsey, G.K. Wehner, and co-workers, Rep. No. NSF/RANN/SE/GI-34871/PR/74/2, 1974.
93 J. Ferrante, NASA Tech. Note, TND 7479, 1973.
94 N.J. Chow, J.M. Eldridge, R. Hammer and D. Dong, IBM, T.J. Watson Research Center, private communication.
95 C.T.H. Stoddart and E.D. Hondros, Nat. Phys. Sci., 237 (1972) 90.
96 G. Betz, G.K. Wehner, L.E. Toth and A. Joshi, J. Appl. Phys., 45 (1974) 5312.
97 D.H. Buckley, NASA Tech. Note, TND 5756, 1970.
98 D.H. Buckley, NASA Tech. Note, TND 5322, 1970.
99 D.H. Buckley, Int. J. Nondestr. Test., 2 (1970) 171.
100 J. Ferrante and D.H. Buckley, NASA Tech. Note, TND 6095, 1970.
101 D.H. Buckley, J. Adhesion, 1 (1969) 264.
102 N.A. Gjosten and N.G. Chavka, J. Test. Eval., 1 (1973) 183.
103 R.C. Sundahl, J. Vac. Sci. Technol., 9 (1972) 181.
104 J.R. Waldrop and H.L. Marcus, J. Test. Eval., 3 (1973) 194.
105 H.L. Marcus, J.R. Waldrop, F.T. Schuler and E.F.C. Cain, J. Electrochem. Soc., 119 (1972) 1348.

106 K.F. Watterson and G.W. Simmons, Paper presented before the Gl Committee of ASTM, Williamsburg, Virginia, December 6—8, 1972.
107 H.L. Marcus and P.W. Palmberg, Trans. AIME, 245 (1969) 1164.
108 D.F. Stein, A. Joshi and R.P. LaForce, ASM Trans. Quart., 62 (1969) 776.
109 P.W. Palmberg and H.L. Marcus, ASM Trans. Quart., 62 (1969) 1016.
110 A. Joshi and D.F. Stein, ASTM-STP, 499, 1972, p. 59.
111 H.L. Marcus, L.H. Hackett and P.W. Palmberg, ASTM-STP, 499, 1972, p. 90.
112 R. Viswanathan, Met. Trans., 2 (1971) 809.
113 H.G. Suzuki and M. Ono, Presented at the Japan Institute of Metals, Kanazawa, October 1971.
114 A. Joshi, Scr. Met., 9 (1975) 251.
115 R. Viswanathan and T.P. Sherlock, Met. Trans., 3 (1972) 459.
116 B.J. Schulz and C.J. McMahon, Jr., ASTM-STP, 499, 1972, p. 104.
117 P.V. Ramasubramanian and D.F. Stein, Met. Trans., 3 (1972) 2939.
118 J.R. Rellick, C.J. McMahon, Jr., H.L. Marcus and P.W. Palmberg, Met. Trans., 2 (1971) 1492.
119 E.D. Hondros and M.P. Seah, Scr. Met., 6 (1972) 1007.
120 A. Joshi and D.F. Stein, Met. Trans., 1 (1970) 2543.
121 A. Joshi and D.F. Stein, J. Inst. Metals, 99 (1971) 178.
122 B.D. Powell and H. Mykura, Acta Met., 21 (1973) 1151.
123 H.L. Marcus and N.E. Paton, Met. Trans., 5 (1974) 2135.
124 R.M. Latinision and H. Opperhauser, Met. Trans., 5 (1974) 483.
125 W.C. Johnson, J.E. Doherty, B.H. Kear and A.F. Giamei, Scr. Met., 8 (1974) 971.
126 W.C. Johnson and D.F. Stein, Met. Trans., 5 (1974) 549.
127 A. Joshi, J. Wildermuth and D.F. Stein, Int. J. Powder Met., 11 (1975) 137.
128 G. Chaudron, EURAEC-976, Quart. Rept. 6, October—December, 1963.
129 J.S. Armijo, Corrosion, 24 (1968) 24.
130 M.A. Streicher, J. Electrochem. Soc., 106 (1959) 161; H. Coriou, J. Hurl and G. Plante, Electrochim. Acta, 5 (1961) 105.
131 A. Joshi and D.F. Stein, Corrosion, 28 (1972) 321.
132 J.L. Walter, Trans. AIME, 239 (1967) 272.
133 D.M. Moon and R.C. Koo, Met. Trans., 2 (1971) 2115.
134 H.G. Sell, D.F. Stein, R. Stickler, A. Joshi and E. Berkey, J. Inst. Metals, 100 (1972) 275.
135 P.J. Jorgenson and J.H. Westbrook, J. Amer. Ceram. Soc., 47 (1964) 332.
136 H.L. Marcus and M.E. Fine, J. Amer. Ceram. Soc., 55 (1972) 568.
137 W.C. Johnson, D.F. Stein and R.W. Rice, Proc. Bolton Landing Conf., August, 1974.
138 D.F. Stein, J. Vac. Sci. Technol., 12 (1975) 268.
139 L.D. Schmidt and D. Luss, J. Catal., 22 (1971) 269.
140 J.J. McCarroll, T. Edmonds and R.C. Pikethly, Nature (London), 223 (1969) 1260.
141 G.A. Somorjai, J. Catal., 23 (1972) 453.
142 P.W. Palmberg, Proc. Int. Mater. Conf., 4th, Univ. Calif., Berkeley, June, 1968, 1969, p. 29-1.
143 G.A. Somorjai, Catal. Rev., 7 (1972) 87.
144 T.P. Pignet, L.D. Schmidt and N.L. Jarvis, J. Catal., 31 (1973) 145.
145 H.P. Bonzel and R. Ku, J. Vac. Sci. Technol., 9 (1972) 663.
146 M.M. Bhasin, J. Catal., 34 (1974) 356.
147 M.M. Bhasin, J. Catal., (1975) to be published.
148 F.L. Williams and M. Boudart, J. Catal., 30 (1973) 438.
149 N.K. Wagner, A.R. Hart and D.W. McQuitty, Government Microelectronics Conf., Boulder, Colorado, June, 1974.
150 N.C. MacDonald and G.E. Riach, Electron. Packaging Prod., 13 (1973) 50.

151 J.M. Morabito, Thin Solid Films, 19 (1973) 21.
152 M.C. Finn, H.Y.P. Hong, W.T. Lindley, R.A. Murphy, E.B. Owens and A.J. Strauss, presented at the 15th Electronics Materials Conference, Las Vegas, August 1973.
153 G.Y. Robinson and N.L. Jarvis, Appl. Phys. Lett., 21 (1972) 507.
154 G.Y. Robinson, Appl. Phys. Lett., 25 (1974) 158.

Chapter 6

SECONDARY ION MASS SPECTROMETRY

J.A. McHUGH

Nomenclature

$S_A^{\pm}$ Positive or negative secondary ion yield for element A (ions produced per incident ion).

$\gamma_A^{\pm}$ Ratio of the number of secondary ions produced to the number of particles (neutral and charged) of element A sputtered.

C_A Atomic concentration of element A in the sample matrix.

S Sputter atom yield (atoms per incident ion).

$i_A^{\pm}$ Secondary ion current measured in the instrument (ions $\sec^{-1}$).

f_a Isotopic abundance for isotope a of element A.

η_A The isotope ion collection efficiency of the SIMS instrument.

I_p Total primary or incident ion current (ions $\sec^{-1}$).

D_p Primary ion current density (ions $\mathrm{cm}^{-2}\ \sec^{-1}$).

d Diameter of the primary ion beam.

R_p Mean projection of the incident ion range on the initial beam direction; the mean projected range.

ΔR_p Standard deviation of the mean projected range, R_p.

δ_A Relative elemental sensitivity factor for element A (measured relative to a given reference element).

ϵ_s Parameter that characterizes the electronic properties of the secondary ion-emitting surface.

I. Introduction

The ability to obtain compositional information from the outermost atomic layer of a solid has progressed considerably during the past 10 years with the development and perfection of secondary ion mass spectrometry (SIMS) and other techniques (see the preceding and following chapters). The majority of these methods approach a true surface analysis capability, since most of the compositional information originates from the outer 10 Å of the

References pp. 273—278

surface and since the sensitivity of each of these methods is sufficient to detect a small fraction of a monolayer for most elements. The application of ion sputtering mass spectrometric methods to surface and solids analysis, with the major emphasis on SIMS methods, will be discussed in this chapter.

The interaction of energetic ions with a solid results in the ejection of substrate atoms and molecules in both neutral and charged states. This moderately efficient production of charged particles (secondary ions) coupled with high sensitivity mass spectrometric techniques forms the basis of the SIMS method. Although SIMS possesses certain limitations, as does any technique, it nevertheless provides in a single instrument the broadest capability for both surface and bulk analysis of solids. The significant characteristics that have generated the considerable interest in SIMS methods are: high detection sensitivity for the majority of elements ($<10^{-4}$ of a monolayer), depth concentration profiling of trace constituents with depth resolutions $\leqslant 50$ Å, lateral characterization of the surface on a micrometer (μm) scale, isotopic analysis, and analysis of low atomic number elements (H, Li, Be, etc.).

The first experiments dealing with SIMS were performed in the late 1930's by Arnot and coworkers [1,2] and Sloane and Press [3] as part of a general study of negative ion formation resulting from ion bombardment of metal surfaces. This early beginning did not produce any immediate interest or any profound appreciation for the potential usefulness of the method. In 1949, Herzog and Viehböck [4] described a sputtering ion source for mass spectrometers, but it was almost 10 years later before interest on the part of a number of researchers was evident.

The investigations of Veksler and Ben'iaminovich [5] and Honig [6] signaled the beginning of a widening interest in the process of secondary ion emission and the technique of SIMS. Through the middle 1960's a number of other workers [7—25] added to the expanding bank of knowledge on the general features of secondary ion emission. During the same period, specialized SIMS instruments for microarea analysis were developed [26—31], and investigations illustrating particular analytical applications were reported [28,32—41].

The interest in SIMS as a tool for surface and bulk solid analysis has grown steadily from the mid-1960's to the present time as is evident from the number of publications related to SIMS that have appeared in the past 7 years. Good reviews of early literature on secondary ion emission and associated phenomena can be found in the publications of Carter and Colligon [42] and Kaminsky [43]. The techniques of SIMS as applied to surface and solids analysis have been the subject of general reviews by Honig [44], Benninghoven [45], Werner and De Grefte [46,47] Fogel [48,49], Evans [50], and others [51—53].

The important physical and instrumental parameters that are the basis of SIMS are best expressed in eqns. (1)—(3). The positive or negative secondary ion yield $S_A^\pm$ (ions per incident ion) for element A in the sample matrix is given by

$$S_A^\pm = \gamma_A^\pm C_A S \qquad (1)$$

The term $\gamma_A^\pm$ is the ratio of secondary ions (positive or negative) of element A to the total number of neutral plus charged particles of element A that are sputtered and C_A is the atomic concentration of element A in the sample. The term S is the total sputter atom yield (atoms per incident ion) of the substrate. It includes all particles leaving the surface, neutrals and ions. The quantities $\gamma_A^\pm$ and S are strongly dependent on the composition of the sample matrix because $\gamma_A^\pm$ is sensitive to the electronic properties of the surface and S is controlled to a great extent by the elemental binding energies or the heat of atomization of the solid. Any theoretical quantitation procedure used to reduce measured secondary ion yields to atomic concentration must predict the absolute $\gamma_A^\pm$ or a normalized set of $\gamma_A^\pm$ for any sample matrix.

The secondary ion current $i_A^\pm$ (ions sec^{-1}) measured in a SIMS instrument is given by

$$i_A^\pm = \eta_A S_A^\pm I_p \qquad (2)$$

The quantity $i_A^\pm$ is the ion current for a monoisotopic element, and the ion current for a given isotopic component of a multi-isotope element is $f_a\, i_A^\pm$, where f_a is the isotopic abundance for isotope a of element A. The term η_A is the ion collection efficiency of the particular SIMS instrument for a given isotope. It is the product of the secondary ion mass analyzer transmission efficiency and the ion detector efficiency. Generally, η_A can be treated as a constant independent of element and isotopic mass if the secondary ion energy distributions are similar, peak at a few electron volts and the mass-dependent bias of the ion detector is small. Finally, I_p is the total primary ion current (ions sec^{-1}) delivered to the sample.

Of course, I_p is related to the primary ion current density D_p (ions cm^{-2} sec^{-1}) and the diameter d (cm) of the beam. For simplicity, let us assume a circular spot of uniform D_p, then, we obtain

$$I_p = (0.25\,\pi)D_p d^2 \qquad (3)$$

The brightness limits of the primary ion sources used in SIMS instruments restrict current densities at the sample to values generally <100 mA cm^{-2} (1 mA of singly-charged ions equals 6.2×10^{15} ions sec^{-1}). The values that the various terms in eqns. (1)—(3) assumed for a variety of elements, samples and instruments are given in Table 1.

The most important concept necessary for a basic understanding of the surface analysis capabilities of SIMS pertains to the interdependence of incident ion beam parameters, surface removal rate, and elemental detection

TABLE 1

Typical values for parameters of eqns. (1)—(3)

$\gamma_A^{\pm}$	$10^{-5} - 10^{-1}$
S	$1 - 10$
η_A	$10^{-5} - 10^{-2}$
D_p	$10^{-6} - 10^{-2}$ mA cm^{-2}
d	$10^{-4} - 10^{-1}$ cm

sensitivity. Many times, through a lack of knowledge regarding the inter-dependence of these parameters, one generates misconceptions of SIMS capabilities for surface analysis. The relationships between incident ion current, density and beam diameter, surface removal rate, and SIMS detection sensitivity for a typical situation are given in Fig. 1. The atomic layer removal rate is proportional to primary ion current density D_p at a given primary ion energy, and the SIMS detection sensitivity (minimum detectable level of an element, ignoring spectra interferences) is inversely proportional to the total incident ion current I_p. The proportionality constant that relates I_p and the SIMS detection sensitivity was derived from SIMS analyses of a number of elements in a variety of matrices [54]. It represents a realistic estimate based on practical experience for a typical element—matrix situa-

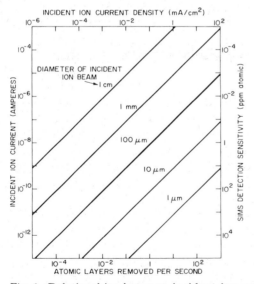

Fig. 1. Relationships between incident ion current, diameter and current density, atomic layer removal rate and SIMS detection sensitivity. The atomic layer removal rate is based on a typical matrix sputter atom yield and the SIMS detection sensitivity is based on experimentally derived sensitivities for a typical element—matrix situation.

tion. Also with regard to Fig. 1, one must assume that the secondary ion analyzer acceptance area is equal to or greater than the primary ion beam diameter. Generally, this begins to be a problem in mass spectrometry when the ion source area is > 1 mm diameter.

Ion beam sputtering is a destructive process. However, when there is a requirement that the surface remain virtually undisturbed, the SIMS analysis can be accomplished at very low surface removal rates ($< 10^{-4}$ atomic layers sec^{-1}). To meet this condition and to provide a detection sensitivity of 100 ppm atomic for surface atoms, we see from Fig. 1 that a 10^{-10} A primary ion beam ~ 1 mm in diameter is required. At this low primary ion current density ($\sim 10^{-5}$ mA cm^{-2}), the arrival rate at the sample surface of gaseous species from the vacuum environment can exceed the arrival rate of incoming ions. SIMS experiments operating under these conditions are usually performed in a clean or ultrahigh vacuum (UHV) environment to avoid these complications.

The instrumental conditions for the above situation are not universally applicable to all analysis requirements. For example, an elemental depth concentration profile of a trace constituent in a surface film over 500 Å thick is conveniently performed if one employs a 100 μm diameter beam and a surface removal rate $> 10^{-1}$ atom layer sec^{-1}. Even higher primary ion current densities are required to provide statistically significant quantities of secondary ions per unit surface area; this is necessary to obtain lateral elemental distributions of trace constituents with the ion microprobe [31] or direct imaging instruments [27]. From Fig. 1 and this brief discussion, it should be evident that one cannot achieve a lateral resolution of a few micrometers for a constituent at the 1 ppm level and at the same time experience a surface removal rate of $< 10^{-3}$ atomic layers per second. These conditions are mutually exclusive.

Surface analysis by SIMS falls into two categories: low current density sputtering and high current density sputtering. The categories are determined by the characteristics of the primary ion beam. A low current density sputtering analysis results in a very small fraction of the surface being disturbed, a result that approaches a basic requirement of a true surface analysis method. High current density sputtering with its attendant high surface removal rates, on the other hand, is required for obtaining elemental depth profiles, microarea analysis, and trace element analysis (< 1 ppm). Using various primary ion generation and focusing concepts, and secondary ion analyzers, a number of SIMS instruments have been designed to provide one or more of these capabilities (see Section III). The intent of this chapter is to provide the reader with sufficient information of a relevant and practical nature so that he might develop a basic awareness of the advantages and limitations of SIMS methods. Today, the analyst needs to possess the capability of critically evaluating his surface analysis problem in light of the various tools available to him.

To present the subject matter in a logical and organized manner, the chapter is divided into five sections. Section II deals with the phenomena of secondary ion emission (i.e. mechanisms, efficiency of ion production, species generated, and incident ion effects). Basic instrument concepts including glow discharge or ionized neutral mass spectrometry and the advantages and limitations of each are discussed in Section III. Quantitation of SIMS data, depth profiling with special attention to the problem of obtaining a meaningful elemental profile, and the applications of SIMS to various surface analysis problems are covered in Sections IV, V, and VI, respectively.

II. Secondary ion emission

An energetic ion impinging on a solid target is either back-scattered from a surface atom (a low probability event) or it enters the solid and dissipates its energy to lattice atoms through a number of elastic and inelastic biparticle collisions. This collision cascade process is depicted in Fig. 2. The phenomenon of sputtering takes place when the recoil atoms produced at or near the surface have the necessary energy and direction to escape the solid. The sputtered atoms depart the surface in a neutral state, excited state, or an ionic state (either positive or negative). The mean kinetic energy of sputtered particles typically is of the order of 10 eV with the tails of the distribution extending to hundreds of electron volts [25].

The escape depth for sputtered particles ranges from the surface to values greater than 20 Å and is strongly dependent on the characteristics of the collision cascade (recoil atom angular and energy distributions) produced in the solid by the incident ion (Fig. 2). The characteristics of the collision cascade are controlled by many parameters, principally, the energy of the primary ion and the atomic numbers and masses of the primary ion and atoms of the target. The distribution of sputtered particles as a function of

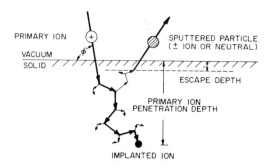

Fig. 2. Schematic representation of an energetic ion—solid interaction and the sputtering process.

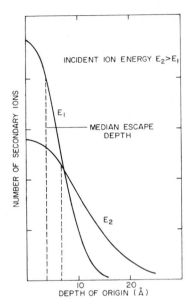

Fig. 3. Hypothetical secondary ion depth-of-origin distributions for incident ion energies E_1 and E_2.

depth of origin below the surface also depends on the atomic binding energies in the solid. Materials with low surface binding energies give sputter atom depth-of-origin distributions with a lower mean escape depth. The postulated distributions in Fig. 3 illustrate the effect of incident ion energy on the secondary ion depths of origin. Even though the median escape depth may be <6 Å, a significant fraction of secondary ions originate from much deeper in the solid. For example, if we consider the distribution of Fig. 3 for primary ion energy E_2, approximately 15% of the ions originate at a depth >15 Å. Reducing the primary ion energy to E_1 produces a distribution where the contribution from depths >15 Å is less than 1%. Experimental information on this subject is difficult to obtain; however, some indirect estimates have been made by Honig [44].

In most SIMS analyses, sample surfaces are covered by at least a monolayer of material foreign to the sample matrix. Yet in these analyses, underlying substrate atoms are detected in the secondary ion mass spectra. The reason for this is the nature of the sputtered ion depth-of-origin distributions mentioned previously and the high ion detection sensitivity of the SIMS method. To reduce the number of secondary ions that originate from layers deep in the solid, one can reduce the primary ion energy to <1 keV. In this situation, the majority of atomic displacements occur at or near the surface, and not deep (i.e. >50 Å) in the solid. All surface analysis methods that utilize sputtered particles generated by moderately energetic primary ions for compositional information will encounter the problem illustrated in Fig. 3. Specifically, SIMS will show a greater variability in the information depth zone than other methods, i.e. ion scattering spectrometry, X-ray photoelectron spectroscopy, and Auger spectrometry, described in Chapters 3—5.

A. MECHANISM

To determine the probability that a sputtered atom or molecule departs the surface as an ion is a complex quantum mechanical problem involving ground- and excited-state interactions of the atom or molecule with the solid and surface electronic states. A number of processes contribute to the secondary ion yield: ionization as a de-excitation path for excited atoms ejected from the substrate, electron redistribution and ionization from molecular dissociation of surface compounds, and surface or resonance ionization processes. The contribution that each process makes to the total ion yield is dependent on the primary ion beam conditions, the element, the matrix and the surface conditions. Joyes [55] has reviewed the various theoretical mechanisms of secondary ion emission.

For moderate energy sputter ion beams (>5 keV), a major contributor to the secondary positive ion yield is resonance and autoionization of *excited* species that emerge from the solid phase to the vacuum. A second and, normally, less important contribution to the ion yield is resonance ionization of *ground-state* species. The work of Hagstrum and Becker [56] provides a summary of resonance, Auger, and autoionization processes. The degree of ionization, $\gamma_A^{\pm}$, for this mechanism will depend heavily on the survival of excited species of element A from the point of origin to and out of the surface. The major factors in the survival of an excited state is the velocity and mean lifetime (or relaxation time) of the excited species in the surface layers and the characteristics of the surface electronic structure through or from which it must emerge.

The production of species initially excited in the collision cascade should, to a first approximation, be similar for a variety of matrices; however, the conduction band structure of metals would promote fast de-excitation because of a greater probability of tunneling transitions, and the band structure of insulators with their large forbidden energy gaps would favor a much slower de-excitation of excited species.

A number of experimental observations on insulators and metals lend support to such processes. For example, the secondary ion yields of insulator constituents are typically 10^2-10^3 times the yield of the corresponding metals [27,57,58], and the photon excitation efficiencies are similarly 10^2-10^3 times greater than the efficiencies of typical metals [59]. The mean velocity of sputtered species is greater in materials with high binding energies [60] (e.g. insulator compounds), and the higher velocity components of the sputtered particle distribution possesses a greater ionization efficiency [61].

The mechanism underlying the production of negative secondary ions follows almost the reverse requirements of positive secondary ion emission. The fate of the initial excitation in the sputtered species is not a major factor. The probability of negative ion formation is dependent for the most part on low energy resonance processes controlled by the electron affinity of the ground-state configuration and the surface-state electron binding energies.

Many models have been proposed to account for the absolute and relative secondary ionization probabilities for elements in a variety of matrices. The early attempts to explain positive secondary ion emission probability on the basis of the Saha—Langmuir and Dobrestov surface ionization equations were unsuccessful [5]. Veksler and Tsipinyuk [62,63] have considered a nonadiabatic electron exchange mechanism to explain deviations in the ion yields predicted by the Saha—Langmuir equation. Jurela [64,65] has recently found that a thermodynamic nonequilibrium surface ionization model for both positive and negative ions gives reasonably good results for a number of conducting, semiconducting, and nonconducting samples. The reader is referred to Jurela's work [65] for a review of the surface ionization approach to the interpretation of secondary ion yields.

Joyes and Castaing [66] proposed that secondary positive ions are produced near the surface as a result of Auger de-excitation or autoionization of excited species ejected from the sample. Joyes and Hennequin [67] and Joyes [68—70] have discussed this mechanism in considerable detail. Blaise [71] suggests that autoionizing states are formed as a result of the perturbation acting on the valence electrons of the atom while passing through the metal—vacuum interface. His model explains the general trends in the probability for secondary ion emission observed for transition elements in dilute alloys.

The treatment by Schroeer et al. [72—74] of positive ion emission from metals assumes that the sputtered particle leaves the surface as a neutral atom in the ground state and is ionized via transitions of the atom valence electrons to the top of the conduction band in the metal. The quantum mechanical transition probabilities are computed using the adiabatic approximation and simplified matrix elements. The expression he derived for the probability of ionization bears a functional relationship to cross-section formulas for atomic charge transfer collisions [75]. Joyes and Toulouse [76] proposed that any time-varying perturbation excites conduction electrons near the Fermi level and this nonadiabatic effect results in excitation of the emitted particle. A mechanism for secondary ion formation involving Auger and resonance neutralization of ejected ions has been discussed by Benninghoven [77].

Andersen [78,79] and Andersen and Hinthorne [52,80] have had some success in predicting the relative behavior of positive secondary ion yields from an oxide or oxygen-rich matrix. They make the assumption that the population of atomic states in the sputtered particles is determined by the equilibrium statistical thermodynamics of an electron, ion, and a neutral atom ensemble. Under this condition, one can employ the Saha—Eggert relationships to derive the ratio of ion to neutral states for a given elemental species.

The mechanisms described above, although involving quite different concepts, have had varying degrees of success in predicting the observed general trends of secondary ion emission. It is not unusual that a number of models could satisfy the general trends, since the relative differences in atomic structure and the population of excited atomic states control to a first approximation the relative ionization probabilities of sputtered particles. The dominant trend would be established in all models by the same basic parameters or by parameters that relate indirectly to the same entity. At the present time, however, no single theoretical treatment has emerged to predict accurately from first principles the ionization probability of an element in any matrix.

B. SECONDARY ION YIELDS

The major factors influencing secondary ion yield are the electronic and chemical properties of the surface brought about by the basic characteristics of the matrix and the equilibrium concentration of adsorbed active species. The absolute yields S_A^+ for Al, Cr and V, which are dramatic examples of this effect, increase by as much as a factor of 10^3 from a clean metal surface condition to a fully oxidized surface [81]. In addition, the relative ion yields for different elements in the same matrix can exceed 10^4. The wide spread in relative yield, however, decreases as the base matrix changes from a metallic character to an oxide or insulator. The relative positive secondary ion yields of a number of elements in the insulator matrix, Al_2O_3 are given in Fig. 4. The experimental measurement of absolute secondary ion yields by mass spectrometric methods alone is a difficult task because of the uncertainty in ion collection efficiency. The collection efficiency is a strong function of initial secondary ion energy and direction of ejection. The relative differences between elements given in Fig. 4 are derived from a single SIMS instrument with no attempt to correct to a total ion yield [54]. This figure provides an overview of the relative differences that can be expected between elements.

The observed variation of secondary ion yield with crystal orientation [82—88] is another matrix effect. This effect is not surprising considering that the sputtering yield [89] and the electronic properties of surfaces are dependent on which crystal planes are exposed [43], e.g. the densely packed planes of metals have a greater work function. Furthermore, the deposition

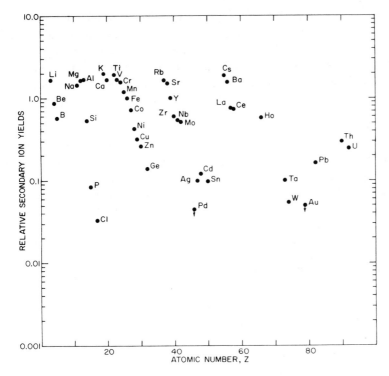

Fig. 4. Relative secondary ion yields for a number of elements present in a refractory oxide matrix. The yields are relative values derived from a particular SIMS instrument and do not represent absolute relative secondary ion yields. (From McHugh [54].)

of primary ion energy in atom layers near the surface is affected by channeling phenomena. The temperature of the sample has an effect on the secondary ion emission from single crystals [85]; however, no important influence on sputter yields [90,91] or secondary ion yields [24,85] of clean or oxidized polycrystalline metals has been observed. The temperature of the sample can affect the concentration of active species adsorbed on the surface and, therefore, affect S_A^+. However, the effect is due to the surface concentration of active species and not directly to temperature. For some oxides or insulators, temperature does affect the total sputter yield [92]; therefore, one would expect that the secondary ion yields should exhibit an effect in these particular cases.

C. SECONDARY ION SPECIES

The predominant ion species observed in the secondary ion mass spectrum are singly charged atomic and molecular ions. Molecular ions take the form of dimers, trimers, etc. of the matrix constituents (M and N) and if the constituents involved form oxides, one observes in the spectrum ions of the form $M_x^\pm$, $M_x N_y^\pm$, $M_x O_y^\pm$, $M_x N_y O_z^\pm$. The yield of secondary molecular ions depends on the electronic properties of the molecular ion, particularly the dissociation energy of the complex. Joyes [93] and Leleyter and Joyes [94,95] have used semi-empirical quantum chemical calculations to explain a number of experimental observations relating to relative molecular ion yields.

Elements with many isotopes that combine to form molecular ions yield complex secondary ion spectra. Fortunately, from an analytical standpoint, the abundance of molecular ion species falls off rapidly as the number of particles in combination increases. If this were not so, the secondary ion spectra and the resultant interferences would be unmanageable. The ratio of charge states, M^{2+}/M^+, and the molecular ion fraction, MN^+/M^+, depend on the initial energy distribution of sputtered ions [25,96—99]. Examples of secondary ion energy distributions are given in Fig. 5 for atomic, molecular, and multiply-charged species [25]. High sensitivity SIMS utilizes the low energy portion of the distribution for two reasons. First, the abundance of

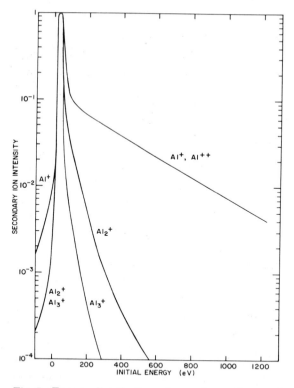

Fig. 5. Energy distributions for various Al secondary ion species sputtered from an Al—Mg alloy by 12 keV Ar^+. (From Herzog et al. [25].)

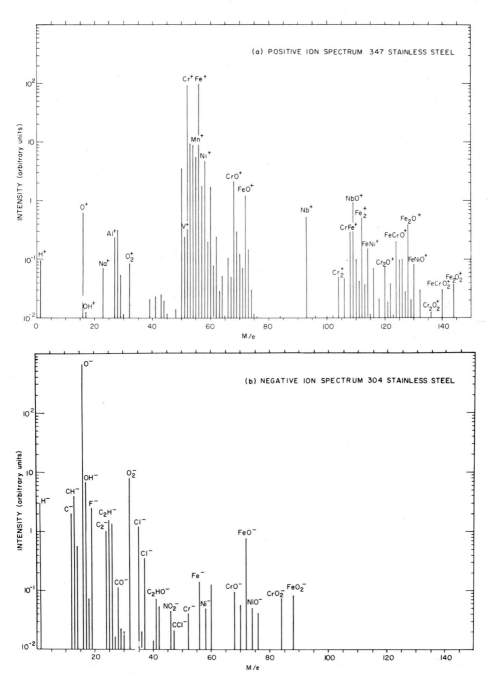

Fig. 6. Mass spectra for (a) positive secondary ions from a 347 stainless steel sample and (b) negative secondary ions from a 304 stainless steel sample. The incident $^{16}O^-$ beam energy and current were 15 keV and $\sim 10^{-9}$ A respectively. (From McHugh [54].)

secondary ions is greater and second, the ion collection efficiency of all SIMS instruments is at a maximum for low energy ions. Under these conditions, M^{2+}/M^+ is typically $<10^{-2}$, and the dimer molecular ion fraction typically falls in the range of 10^{-3} to 1. Figures 6(a) and 6(b) show typical positive and negative secondary ion spectra from two stainless steel samples. If one limits the ion collection to secondary ions with high initial energy (i.e. >100 eV), a significant change occurs in the mass spectra. The spectrum under these conditions is dominated by atomic ions [25].

The atomic and molecular secondary ions ejected from the sample provide a means for obtaining information on the chemical composition of atomic layers at or very near the surface. The secondary ion intensities of minor matrix constituents are proportional to the atomic concentration of the constituents in the particular matrix. It is possible to obtain some molecular structure and atomic bonding information from secondary ion molecular spectra of surface films provided radiation damage to the structure of interest is not significant. This can be achieved by keeping the incident ion dose low, $\lesssim 10^{14}$ ions cm^{-2}, or by continual reestablishment of the molecular film on the substrate surface by adsorption from the gas phase. Benninghoven [45] has shown that secondary ion fragmentation patterns for hydrocarbon films aid in identification of the parent molecule.

D. INCIDENT ION EFFECTS

Consideration must be given to certain characteristics of the incident or primary beam that could influence secondary ion emission either directly or indirectly. These characteristics are energy, mass and angle of incidence of the primary ion beam; primary ion current density; and the chemical nature of the bombarding ion.

The energy, mass, and angle of incidence of the primary ion beam control the characteristics of the collision cascade in surface layers of a particular solid. These parameters have an influence on the sputter yield S and likewise influence $S_A^\pm$, since, to a first approximation, $S_A^\pm$ is directly proportional to S. For example, consider the observations of McHugh and Sheffield [24] (Fig. 7) regarding the effect of primary ion energy on S_A^+. The yield S_A^+ rises sharply with energy for the first few keV and then reaches a plateau at about 10 keV. The similar trends in sputter atom yield (dashed curves) and S_{Ta}^+ (solid curve) suggest that $S_A^\pm$ is directly proportional to S. The energy dependence noted in Fig. 7 is not limited to the Hg^+–Ta system but is a general trend common to all incident ion–target combinations. Most SIMS instruments employ a primary ion beam with energies > 4 keV.

The principal effect of primary ion energy is manifested in $S_A^\pm$. However, secondary effects of energy, which may be of major importance from an analytical standpoint, relate to the mean escape depth of secondary ions, mixing in subsurface layers, and the incident ion implant zone. (These topics

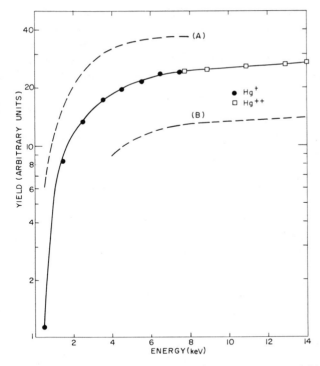

Fig. 7. The dependence of the Ta$^+$ secondary ion yield on the energy of the incident ion ^{202}Hg$^+$. The broken curves are examples of the energy dependence of the total sputtering yield for typical elements. The relative position of the curves on the ordinate has no significance. (From McHugh and Sheffield [24].)

are discussed in appropriate sections of this chapter.)

The primary ion current density D_p is observed to affect $S_A^\pm$ in certain situations; however, this effect is not a direct effect dependent on the simultaneous occurrence of overlapping collision cascades. Collision cascades can be treated as isolated events even at a D_p equal to 100 mA cm^{-2}. For example, if it is assumed that the collision cascade relaxation time is $< 10^{-12}$ sec (time to stop an ion in a solid is $\sim 10^{-13}$ sec), then the mean distance between cascade centers is > 2500 Å for D_p of 100 mA cm^{-2}. This distance is large compared with the diameter of a typical collision cascade volume (typical diameter < 200 Å).

An effect of current density that does have a potential influence on $S_A^\pm$ is temperature. If the thermal time constants of the solid are significant and the primary ion beam diameter is large, then the total primary ion beam energy dissipation is slowed, and the temperature rise experienced in the surface layers of the solid could influence $S_A^\pm$. The magnitude of the effect depends on the temperature rise and the characteristics of the matrix.

The major effect of D_p on $S_A^\pm$ is related to the influence of D_p on the equilibrium concentration of active species adsorbed on the surface. D_p controls the surface layer removal rate and, therefore, the equilibrium concentration of adsorbed species; a change in the equilibrium concentration of active species adsorbed on the surface produces a corresponding change in $S_A^{\pm}$. This is especially evident with inert gas ion sputtering of a reactive metal in the presence of an active gas such as O_2. High current density primary ion beams are commonly employed in SIMS bulk solids analyses for the distinct purpose of maximizing the ion signals of bulk constituents relative to ion signals produced from adsorbed contaminants that originate from the sample environment.

The chemical character of the incident ion does not influence $S_A^{\pm}$ in analyses that are directed at the outermost layers of a solid where, through necessity, the primary ion density and dose to the sample must be kept low ($<10^{14}$ ions cm^{-2}). The chemical character of the incident ion does, however, become important at high ion doses ($>10^{15}$ ions cm^{-2}) where sufficient implantation has occurred to effect a change in the character of the matrix and in the nature of the exposed surface. Under these conditions, the nature of the incident ion becomes an important parameter because $S_A^{\pm}$ is very sensitive to the properties of the matrix. Sputtering with electronegative species (e.g. oxygen) has been used extensively by Andersen [78,79] to enhance and stabilize positive secondary ion yields from a variety of matrices. On the other hand, sputtering with electropositive ions (e.g. Cs^+) was shown by Krohn [16] and Andersen [78,79] to enhance negative secondary ion yields.

The effect of the incident ion on $S_A^{\pm}$ and a number of other important effects observed in a SIMS analysis are emphasized in Fig. 8, where the secondary $^{56}Fe^+$ intensity is plotted as a function of sputtering time (or depth) for normal incident ions of $^{16}O^-$ and $^{40}Ar^+$. The initial rise in the leading portion of the $^{56}Fe^+$ curve is the result of the sputter removal of foreign surface contaminants, thus exposing a pre-existing surface oxide film. The $^{56}Fe^+$ intensity reaches a peak and begins to decrease when the oxide—metal interface region is reached, since S_{Fe}^+ decreases on moving from an oxide matrix to a metal matrix. The effect that the incident ion has on S_{Fe}^+ becomes apparent when a depth equal to the mean penetration depth of the ion in the solid is approached. For ions at normal incidence, the mean penetration depth equals the mean projected range, R_p. At this point, the implanted oxygen concentration is great enough to produce a noticeable effect on S_{Fe}^+.

The energy and the angle of incidence control the mean penetration depth of a particular ion in the solid and, therefore, control the location of the implantation zone where the character of the matrix is quite different from the surface or normal matrix. If the energy of the $^{16}O^-$ is reduced to <1 keV, the mean penetration depth is ~ 50 Å, and the implant zone is near enough to the surface to eliminate the decrease in the $^{56}Fe^+$ intensity that

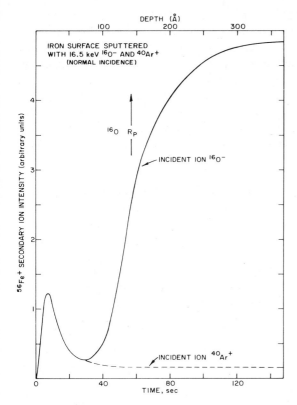

Fig. 8. The time or depth dependence of the $^{56}Fe^+$ secondary ion intensity from an iron surface for normal incidence $^{16}O^-$ and $^{40}Ar^+$ ion bombardment. The $^{56}Fe^+$ peak heights and positions corresponding to the surface oxide film were normalized in the $^{40}Ar^+$ and $^{16}O^-$ cases for ease of comparison. $R_p \cong 140$ Å for 16.5 keV ^{16}O ions incident on an Fe target. (From McHugh [54].)

occurs as the surface oxide film is sputtered away. There is no significant difference in the profiles produced by $^{40}Ar^+$ and $^{16}O^-$ within 40 Å of the surface when the chemically active ion has a mean penetration depth $\geqslant 40$ Å. Since Ar is inert, no enhancement occurs, and the $^{56}Fe^+$ signal approaches a base level determined by the partial pressure of active gas species in the vicinity of the sample and the primary ion current density. For oxygen bombardment, the $^{56}Fe^+$ signal continues to increase beyond the point equal to the mean penetration depth of oxygen in the Fe matrix until the rate of oxygen injection equals the rate of oxygen sputtered (i.e. the steady-state concentration of implanted oxygen in the solid is attained). This point occurs at a depth of the order of $R_p + 2\Delta R_p$, where ΔR_p is the standard deviation of the mean projected range, R_p [100]. Lewis et al. [101] have studied some of the effects described above.

III. SIMS instrumentation

The four basic components of a SIMS instrument are: a primary ion source and beam conditioning system, a sample positioner and secondary ion extraction lens, a mass spectrometer for mass/charge analysis, and a high sensitivity ion detection system. The primary ion sources used in the majority of instruments are of the gas discharge or plasma variety. Such ion sources coupled with the proper beam conditioning system, make available a wide range of surface removal rates, from 10^{-5} to 10^3 Å sec^{-1}. The mass/charge separation is accomplished with either a magnetic or quadrupole analyzer. Double-focusing magnetic instruments (energy and momentum analysis) are the more popular mass/charge analyzer in SIMS, both in number of instruments in use and from an analytical and trace analysis standpoint because of their inherent high abundance sensitivity. In other words, for multistage double-focusing magnetic instruments the background signal that results from tailing of major matrix peaks (wall scattering, gas scattering, etc.) can be $<10^{-9}$ for general background and as low as 10^{-6} for a mass position adjacent to a major peak [35]. Depending on the particular application, however, the less costly quadrupole analyzer approach may be the most practical. Rudenauer [102] has compared the qualities of the quadrupole and magnetic instruments for secondary ion mass analysis.

Recent reviews by Evans [50,103] and Liebl [104] provide information on the current state of SIMS instrumentation, and the reader is referred to these reviews and the original literature on a given instrument for details. This section does not discuss specific instruments in detail but, rather emphasizes basic instrument concepts and the important advantages and limitations that the different instrument designs bring to a SIMS analysis.

A. INSTRUMENT CONCEPTS

There are two basic concepts utilized in secondary ion mass analysis, the conventional mass spectrometric method and the direct imaging method [27]. The two methods are schematically compared in Fig. 9. In the conventional method, the object is to provide a moderate resolution analyzer that transmits to a high sensitivity ion detector a significant fraction of the energetic secondary ions from a large area of the sample (~ 1 mm^2). The mass-analyzed ion species are brought to a single-point focus at the entrance slit to the detector. For this static situation, there is no readily retrievable information regarding the exact point (i.e. within 1 μm) at which a particular secondary ion was created within the secondary ion acceptance area on the sample surface. The direct imaging method, on the other hand, produces a stigmatic ion image in the focal plane of the analyzer, and with appropriate aperturing (or image conversion through ion or electron sensitive emulsions)

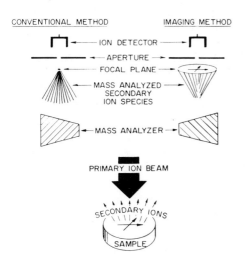

Fig. 9. Schematic representations of the conventional and direct imaging approaches to the mass spectrometric analysis of secondary ions.

information regarding the point of origin can readily be derived for a particular m/e secondary ion species.

The basic instrument concept of Castaing and Slodzian [27] is used in all direct-imaging instruments; all other instruments are variations of the conventional mass spectrometric approach. To obtain secondary ion images of a surface with the conventional approach, it is necessary to perform a sequential analysis by scanning the surface with a small-diameter primary ion beam. Electrical scanning of the primary ion beam, as opposed to translating the sample, is the preferred method for producing secondary ion images on a cathode ray tube (CRT); the primary ion beam and the electron beam in the CRT are synchronized, and the amplified signal from the secondary ion detector modulates the intensity of the CRT electron beam. The image magnification in this method is the ratio of the CRT horizontal trace dimension to the horizontal dimension of the area on the sample swept by the primary ion beam.

All SIMS instruments possess a capability for surface and elemental depth concentration analyses. The important differences that set instruments apart from one another are: detection sensitivity, mass resolution, primary ion current density, vacuum characteristics of the region surrounding the sample, and the ability to provide lateral, or xy, surface characterization through probe scanning or direct imaging. By definition, an instrument provides lateral, or xy, information if it can produce elemental images or point-to-point analyses with a spatial resolution of <10 μm. It is possible to place each SIMS instrument into one of three categories determined by its design concept and its capability for microarea analysis — instruments that provide

1) no lateral, or xy, surface characterization, 2) lateral information through probe imaging, or 3) lateral information through direct imaging.

1. Instruments providing no lateral surface characterization

A number of sputter ion-source mass spectrometers have been constructed for investigating particular analytical problems or various features of secondary ion emission. The single focusing analyzers (single magnetic sectors) used in early studies of secondary ion emission phenomena were severely limited in mass resolution and abundance sensitivity because of the large spread in secondary ion initial energies. Liebl and Herzog [29,30] designed a double-focusing sputter-source mass spectrometer for solids analysis, which later was marketed by GCA Corp. [105]. It operates at moderate vacuum (10^{-6} to 10^{-7} torr) with a focused primary ion beam diameter of ~ 0.5 mm and a maximum current density > 50 mA cm^{-2}. Nishimura and Okano [106] described an instrument with capabilities similar to the GCA instrument.

Benninghoven's [107,108] approach to SIMS surface analyses has been to use low current density primary ion beams and UHV techniques to define the surface better. He has employed magnetic and quadrupole analyzers in UHV SIMS. The most recent UHV instrument of Benninghoven and Loebach [109,110] employs a mass-analyzed primary ion beam, a UHV target chamber, and a quadrupole mass/charge analyzer. At the present time, considerable interest is being directed towards the quadrupole approach [109—113] because of its simplicity and low cost and because it can accomplish most surface and depth profile analyses where information on trace constituents and xy surface characterization is not of prime importance. Wittmaack et al. [113] have achieved low background with the quadrupole mass filter by pre-selecting the secondary ions with an aperture-limited parallel-plate electrostatic analyzer and by employing an ion detector that is located off-axis.

2. Instruments that provide lateral information through probe imaging

SIMS instruments that fall in this category are generally called ion microprobes. Ion microprobes with a focused primary ion beam of a few micrometers diameter have been described by Liebl [31], and Long and Drummond [114,115]. The first commercial instrument of this type was built by Applied Research Laboratories [116] and is based on the design of Liebl [31]. In this instrument, the primary ion beam is mass-analyzed and can be focused to a probe diameter from < 2 to $300\,\mu$m. The mass spectrometer is a high transmission stigmatic-imaging double-focusing instrument of moderate mass resolution. A schematic representation of this instrument is given in Fig. 10. Liebl [117] has recently designed and built a UHV surface analysis instrument that permits either independent or simultaneous ion or electron microprobe analyses of a sample. Hitachi [118] is marketing an instrument

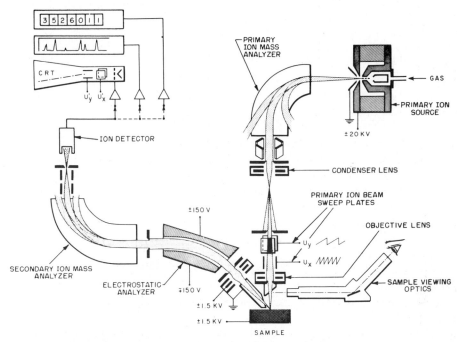

Fig. 10. Schematic drawing of the Applied Research Laboratories ion microprobe mass analyzer. (From McHugh and Stevens [134].)

that has, in many respects, capabilities similar to the ARL microprobe; however, it lacks a number of the features possessed by the ARL instrument. The Hitachi design has been discussed in the literature by Tamura et al. [119–121]. AEI Scientific Apparatus [122] offers a microprobe attachment to convert one of its Mattauch—Herzog double-focusing spark-source mass spectrographs into a scanning ion microprobe [50]. The unique capability of the AEI instrument compared with other instruments is its high mass resolution ($M/\Delta M$ = 10,000 with a 50% valley definition).

3. Direct imaging instrument

The first SIMS instrument to produce m/e separated ion images and xy surface characterization was the direct-imaging instrument of Castaing and Slodzian [27]. An instrument patterned after the Castaing and Slodzian concept is commercially available from CAMECA Instruments [123,124]. A schematic layout of this instrument is shown in Fig. 11. The unique capability of the direct-imaging instrument is its ability to monitor the intensity of a secondary ion species from a specifically defined microarea of the sample independent of the size and location of the primary ion beam as long as part of the beam illuminates the region of interest. This capability is advanta-

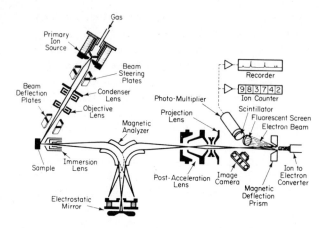

Fig. 11. Schematic drawing of the CAMECA Ion Microanalyzer. (From Evans [50].)

geous in certain analytical situations relating to xy surface characterization and bulk solids analyses by SIMS. In the material that follows, some of these important advantages and limitations possessed by various instrument concepts will be considered.

B. DETECTION SENSITIVITY

The minimum detectable level of an element in a given sample depends on the characteristics of the element itself, the chemical composition of the matrix in which the element is present, the primary ion species and total primary current delivered to the sample, the solid acceptance angle and the secondary ion transmission efficiency of the analyzer, the general background produced in the analyzer, and the secondary ion detector background and efficiency. All but the first two of the above parameters are controllable through instrument design and, therefore, in theory can be optimized for high sensitivity. Since sputtering is a destructive process, a high-efficiency mass analyzer and high sensitivity are desirable to minimize the volume of sample consumed. Because various SIMS instrument designs exist that either emphasize some desirable feature or convenience, there exist wide differences in sensitivities. A convenient measure of instrument sensitivity is the number of ions detected per incident primary ion for a standard set of conditions: sample, primary ion species, and some minimum mass resolution. A SIMS instrument that detects $\sim 10^6$ ions sec^{-1} of a typical element from an oxide matrix (e.g. Fe^+ from Fe_2O_3 sample) per 10^{-9} A of primary ion beam is classified as having a sensitivity suitable for trace and microarea analysis.

The chemical composition of the sample matrix has a direct effect on the elemental detection sensitivity and is the greatest source of uncontrollable variability in detection sensitivity. The matrix influences the detection sensitivity in two ways: it influences the absolute secondary ion yield $S_A^\pm$ through differences in the electronic properties of materials, and the matrix can be the source of unwanted molecular and multiply-charged ion species that appear at mass positions of interest (see Section II.C). It is fortunate that the abundance of molecular ion species falls off rapidly as the number of particles in combination increases, and for most analyses where the element concentration is >100 ppm atomic, no problem is encountered from molecular ion interferences.

Molecular ion interferences can be controlled by two methods: analyze secondary ions with high initial energy (see Section II.C and Fig. 5) and/or employ a mass/charge analyzer with a resolution $M/\Delta M > 3000$. In using the first method, the ion yields are typically reduced by approximately the same order of magnitude as the reduction in molecular ion yields relative to atomic ion yields. In some cases, this technique is acceptable; however, for most trace and microarea analysis problems, the large reduction in detection sensitivity is unacceptable. The second method is the most direct and is desirable from an analytical standpoint. A number of investigators [125–128] have employed high mass resolution in SIMS to show the complex nature of a single m/e peak in the spectrum. In Fig. 12, taken from the work of Bakale et al. [128] the mass 43 peak scanned under different resolution conditions is shown. To reduce or eliminate uncertainties in all m/e peak assignments, high mass resolution is essential, especially if one is concerned

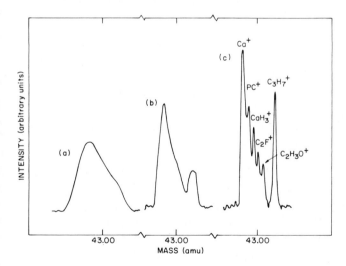

Fig. 12. The mass 43 position of the secondary ion mass spectrum of fluorapatite at an approximate mass resolution of (a) 300, (b) 1000, and (c) 3000. (From Bakale et al. [128].)

routinely with general trace analyses at concentration levels <10 ppm atomic.

The SIMS elemental detection sensitivity has been discussed in terms of theoretical expectations and experimental measurements by a number of authors [46,47,108,129—131]. Projected detection sensitivities of $<10^{-7}$ of a monolayer, 1 ppb atomic, and $<10^{-18}$ g of an element can be derived from theoretical and experimental information and verified in specific well-defined situations. However, these are specific instances and are not the norm for the practical situation. In the practical situation, we are confronted with complex spectra with many intereferences that result in the detection sensitivity depending heavily on the particular sample matrix involved. It is therefore important that one specify additional factors, in particular the matrix, when stating a given detection sensitivity and not make blanket statements regarding a particular element.

The minimum detectable level of an element in a matrix (ignoring inter-ferences) is inversely proportional to I_p, the primary ion current delivered to the sample. Figures 1 and 13 show the variation of detection sensitivity with the various parameters that define the total primary ion current incident on the sample. The detection sensitivities in Figs. 1 and 13 are based on experi-

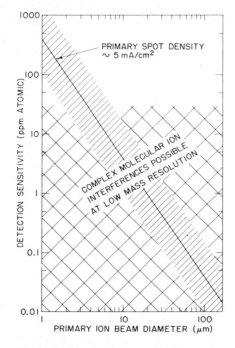

Fig. 13. The detection sensitivity of a typical SIMS instrument as a function of primary ion beam diameter is given by the single solid line (the line corresponds to a constant current density of 5 mA cm^{-2}). The spread about the line represents the range of typical primary ion current densities employed in microprobe or direct-imaging instruments. (From McHugh [131].)

mental data [54] for a typical analytical situation where O_2^+ is the primary ion. The shaded region about the 5 mA cm^{-2} line in Fig. 13 represents the range of common primary ion current densities employed in microprobe or direct-imaging instruments. The cross-hatched region shows the typical level at which complex molecular ion interferences become important and appropriate concern for these interferences should be exercised in m/e peak identification. The actual location or height of the cross-hatched region depends on the sample matrix and the mass resolution and abundance sensitivity of the secondary ion analyzer. In microarea analyses (areas $< 3 \mu$m in diameter) with microprobe or direct-imaging techniques, the minimum detectable level is above the level where molecular ion interferences are of major importance (see Fig. 13); therefore, it is not necessary to have high mass resolution if the only concern is microarea solids analysis. On the other hand, if one's interest is trace element analysis in complex matrices, then a high mass resolution capability is essential.

C. TRACE ANALYSIS

The projected SIMS detection sensitivity approaches 1 ppb for many elements. However, to realize a general instrument detection sensitivity in the ppb range, it is essential to employ (as seen in Section III.B) a mass analyzer with high resolution and abundance sensitivity and to have control over the effects discussed below.

Most of the secondary ions originate from the first few atomic layers of the solid; therefore, species that adsorb on the surface become an apparent major constituent of the solid or solid surface. Hydrocarbons, H_2, N_2, O_2, H_2O, CO_2, and CO, are the normally more abundant species found in the environs of the sample. Therefore, trace elemental analyses for C, H, N, and O are difficult unless special precautions are taken to minimize their effect in analyses. These methods usually involve one or all of the following: hydrocarbon free UHV methods, cryogenic and getter pumping in the vicinity of sample, and high primary ion current densities such that the surface layer removal rate is much greater than the arrival rate of contaminant species. The arrival rate per unit area for a gaseous species at 10^{-8} torr is approximately equal to the arrival rate of ions at a primary ion current density of 10^{-3} mA cm^{-2}.

Other sources of contaminants are the surfaces close to the sample that collect appreciable quantities of sputtered material. Some of this material can return to the sample by evaporation and/or by sputtering caused by secondary and back-scattered ions. These are memory effects, and the degree to which memory is a problem in a particular analysis depends on the previous sample history. The effect is more pronounced in instruments that employ large primary ion currents and have the extraction lens close to the sample surface.

The chemical purity of the primary ion beam is important to avoid implanting the sample with an element that one is interested in determining. For a typical set of conditions (sputter rate and primary ion range and straggling), together with the assumption that sputtering has proceeded for a time sufficient to expose the implant zone and resputtering of implanted species is the major source of the impurity ion, a 1 ppm impurity in the primary ion beam would appear as ~0.1 ppm atomic concentration in the solid. To ensure purity of the primary ion beam and to eliminate the possibility of complications at trace levels, it is desirable to mass-separate the primary ion beam.

High-current density primary ion beams are used in analyses of bulk solids to minimize the effect of surface contamination by residual gases. In this situation, the low current density region or periphery of the primary ion beam becomes a major contributor to the secondary ion signal of an element that is present in both the residual gas and the solid as a trace constituent.

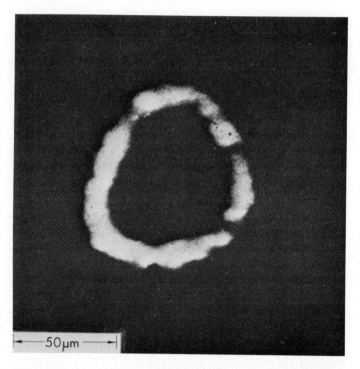

Fig. 14. Al^+ secondary ion micrograph of an Al surface bombarded with Ar^+ ions. The micrograph was produced with the direct-imaging CAMECA instrument and illustrates the Al^+ enhancement at the periphery of the primary ion beam because of adsorption of O_2 from the residual gas atmosphere. An O^+ or AlO^+ micrograph would be similar to the above Al^+ micrograph. (From Evans [103].)

This effect is evident in the Al⁺ secondary ion micrograph shown in Fig. 14. The micrograph was produced with the direct-imaging CAMECA instrument and shows a region on the Al surface that is bombarded with Ar⁺. The "halo" in the micrograph is the result of enhanced Al⁺ emission from the periphery of the primary ion beam caused by oxygen adsorption. At the beam periphery, the rate of residual gas contamination competes favorably with the rate of surface removal. These effects are not limited to residual gas contamination (usually the most important) but apply to any source of contamination that is active during an analysis. SIMS methods employing the conventional mass spectrometric approach are affected to a greater extent by the above problem than is the direct-imaging approach (see Section III.A). In the direct-imaging method, an aperture can be placed in the image plane to detect only those ions from the central high-sputter-rate region. The equilibrium surface concentration of adsorbed impurities is lowest in this region.

Another situation, paralleling the above beam periphery effect and schematically represented in Fig. 15, concerns the energetic neutral component produced by charge-exchange collisions, and the scattered ion component, that accompanies the focused primary ion beam and illuminates a large area of the sample. The dimension of the illuminated area is set by the limiting apertures in the primary beam column and is usually $\geqslant 250\,\mu$m in diameter. The magnitude of this problem is determined by residual gas pressure, lens

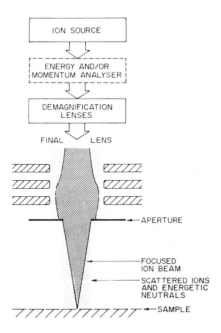

Fig. 15. Schematic representation of the scattered ion and neutral component of a focused ion beam.

References pp. 273—278

design, location and size of the apertures, and the geometrical arrangement of the column elements. The nonfocused component does not depend directly on the focused ion current incident on the sample, but is controlled, for the most part, by the total ion current present in the beam column.

These effects are a great deal more serious in SIMS analyses employing microprobe methods than with direct-imaging methods. However, the only time this becomes an important factor in a microprobe analysis is when the large area illuminated by the nonfocused component is quite different in composition compared with the analysis point. Nonfocused components illuminating a large area of the sample and exceeding 1% of the focused primary ion current can be encountered with ion beams of a few micrometers in diameter. The most unfavorable condition exists when the components of the primary beam column (ion source, lenses, deflector, and apertures) are arranged on a common axis intersecting the secondary ion analyzer axis. This problem can be reduced considerably by deflecting the primary ion beam off axis and aperturing the beam at a point close to the sample.

A second technique [54], useful in bulk analyses where there exists severe heterogeneities in surface composition or surface contamination with elements that are of particular interest in the bulk, involves overcoating the surface of the sample with a 200—500 Å thick layer of highly purified carbon or any element not of interest in the analysis. The layer can be easily penetrated by the high-current density primary beam at the point of interest. The very low-current density tail and the nonfocused components of the primary ion beam impact on the clean carbon surface and, therefore, contribute nothing from points other than the point of interest.

D. ION IMAGING

Secondary ion images that provide a two-dimensional elemental characterization of a surface can be produced either by the direct-imaging or by the scanning microprobe method. Slodzian [27,28,132] and Liebl [104,117, 133] have discussed the different parameters that control the lateral resolution in these two approaches.

In the direct-imaging method, lateral resolution is independent of primary ion-beam size but depends on the aberrations in the secondary ion-analyzer optics and the chromatic aberrations resulting from the energy spread of secondary ions. Also, instrument transmission (secondary ions detected per incident ion) decreases when the resolution is optimized by narrow-band energy filtering. Lateral resolutions of approximately $1 \mu m$ have been achieved with direct-imaging instruments [50,123,130].

The lateral resolution of scanning microprobes is limited by the primary beam diameter and is only as good as the primary ion focusing column. At high demagnifications ($< 1 \mu m$ beams) the nonfocused component discussed in Section III.C may be a significant fraction of the total particle current incident on the sample and, therefore, must be eliminated from the beam. The instrument transmission in the microprobe method remains constant for any lateral resolution because transmission is a characteristic of the secondary ion analyzer and not the dimensions of the primary ion beam. Lateral resolutions of 1—2 μm have been achieved with microprobe instruments [31,104]. The ultimate lateral resolution that one can hope to attain with secondary ion emission is of the order of 100 Å. This projected resolution is the physical limit resulting from the characteristics of the collision cascade, the primary ion-beam-induced mixing in subsurface layers, and the mean escape depth of secondary ions. However, with present methods, the practical lower limit to beam size is about 1000 Å because of limited primary ion source brightness and the poor optical quality of electrostatic lenses.

When operating at similar lateral resolutions and primary ion current den-

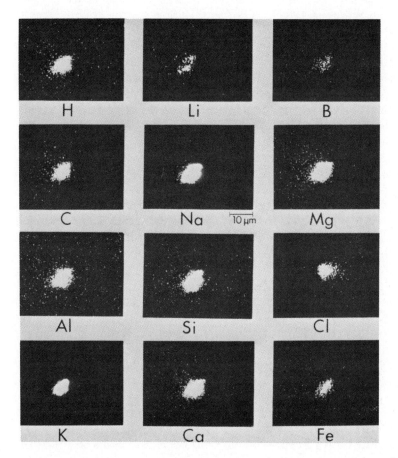

Fig. 16. Secondary ion micrographs taken with the ion microprobe for a number of elements present in a 6 μm dust particle (From McHugh and Stevens [134].)

sities, the direct-imaging method utilizes less time to produce an ion image from large surface areas (i.e. 200×200 μm) compared with the scanning microprobe method. The reason is that direct-imaging information is accumulated simultaneously from all points and not sequentially. However, for areas of interest that are $< 20 \times 20$ μm, the image formation time for the microprobe method is comparable to or shorter than that of the direct-imaging method. The reason lies in the inherent greater gain of the scanning microprobe image display system (electron multiplier and modulated CRT) compared with the sensitivity of the ion-electron sensitive emulsions employed for direct-image conversion.

Two additional advantages of the scanning microprobe method in small-area analyses are: less surface area is disturbed, and image information is available on the CRT screen for immediate analysis. An example of a small-area scan is shown in Fig. 16. The ion images of the 6 μm dust particle [134] were taken with an ARL ion microprobe employing a low noise ion detector sensitive to single ions. The exposure time required to generate an ion micrograph (Polaroid film picture of the CRT display) is from 1 sec to 3 min depending on the concentration and the secondary ion emission characteristics of the element in the sample. In Fig. 16, the concentrations of H, Li, and B are approximately 0.01, 0.002, and 0.001 wt.%, respectively.

E. PRIMARY ION BEAM CONSIDERATIONS

The primary ion beam is an important component of a SIMS instrument; therefore, it is worthwhile to consider some desirable characteristics of a primary ion beam and a beam-forming system that have not already been mentioned. The primary system should be capable of producing focused and stable ion beams of inert gas ions (e.g. Ar^+) and positive and negative ions of reactive gases (e.g. O_2^+ and O^-). The reactive gas ions are useful in solids analyses [52,79] and negative ion bombardment reduces surface charging effects [135].

The degree of surface charge buildup depends on the primary ion species, charge, energy and current density, the dimensions of the bombarded area and the conductivity and thickness of the insulating layer. A localized change in surface potential can result in a number of adverse effects: displacement of the primary ion beam, element migration in the implant zone (see Sections IV and V.B), and a change in secondary ion energy and reduced secondary ion collection efficiency resulting from the perturbation of the ion extraction field in the immediate vicinity of the sample surface. Richards et al. [136] have calculated the influence of surface charging on the trajectories of secondary ions sputtered from an insulating surface.

A number of techniques have been used in SIMS to reduce the positive charge buildup on surfaces, electron flooding by an adjacent hot filament [25], negative primary ion beams [135], and high pressures of O_2 (10^{-4}

torr) in the vicinity of the sample [137]. The analysis of bulk insulators (as opposed to thin insulating films on conductors) requires that the proper electric field gradient be established between the sample and secondary ion extractor electrode for efficient extraction of ions into the mass analyzer. This is usually accomplished by overcoating the insulator with a conducting film or by placing a conducting grid over the insulator surface.

The gas discharge sources commonly employed as sources of primary ions contribute significantly to the background gas of the system; therefore, it is highly desirable to pump the primary column differentially. Mass separation of the primary ion beam is important in establishing beam purity, not only the elemental purity as mentioned previously but also molecular purity needed to control the characteristics of the collision cascade and its effect on depth resolution and atomic mixing in subsurface layers (see Section V). The primary column should have a beam raster capability for producing areas of uniform current density necessary for depth profiling. Finally, in microarea analyses it is desirable to have the ability to view directly with good quality high-magnification optics the area of the sample about the point of primary beam impact.

F. MASS SPECTROMETRIC ANALYSIS OF THE SPUTTERED NEUTRAL COMPONENT

In sputtering, the magnitude of the neutral atom component is significantly greater than the magnitude of the secondary ion component for many matrices. Therefore, a logical alternative or complement to secondary ion analysis is a method of ionizing and analyzing the neutral species. This method has been termed *ionized neutral mass spectrometry* [44]. One attractive feature of the method is that the neutral component can be ionized by a process that is not matrix- or surface-dependent. The major drawbacks to ionized neutral mass spectrometry are that all gas-phase species are ionized in the process, and the extraction efficiency of ionized neutrals into the mass analyzer is low compared with SIMS. If the secondary ion fraction $\gamma_A^{\pm} \geqslant 10^{-4}$ (which is the case for most elements and matrices), ionized neutral mass spectrometry cannot compete favorably with SIMS in absolute sensitivity, $S_A^{\pm}$. Coburn and coworkers [138—141] with their approach to the analysis of neutrals have demonstrated some important and attractive characteristics of the method. The subject of ionized neutral mass spectrometry has been reviewed by Honig [44].

The neutrals sputtered from the target by gas ions from a discharge are subsequently ionized by an electron impact, charge-exchange reactions, or a Penning ionization process in the discharge plasma. The sensitivity and usefulness of the method depend on the ion extraction efficiency and the optimum discharge conditions for maximizing the ionized neutral component relative to the gas discharge species. Honig [8] employed a low-pressure d.c. discharge in early studies of ionized neutral mass spectrometry. Under these experimental conditions, the discharge gas ion peak was sufficiently intense to cause severe spectral interferences and, therefore, limited the usefulness of the method.

Coburn and Kay [138—140] and Oechsner and Gerhard [142] have investigated rare gas, radiofrequency discharge sputtering, and neutral ionization. These workers obtained results that indicated the method has practical applications. The energy spread of ions sampled from the rf glow discharge sputtering source is <1 eV and a quadrupole mass analyzer without prior energy filtering has been used effectively [141]. In the discharge conditions employed in their experiments, Coburn and Kay have established that neutral ionization occurs principally by the Penning mechanism [139]. This being the case, the ionization efficiency of neutral species is only slightly dependent on the element and is totally independent of the matrix or surface condition of the sample (as contrasted to SIMS). Therefore, the relative ion signals are, to a good approximation, equal to the relative atomic concentrations in the matrix. Using large area ($\sim$10 cm^2) bulk samples, a 1 ppm concentration is detectable with this instrument [141]. Typical surface removal rates range between 10^{-2} and 10 monolayer sec^{-1}, and insulators and conductors can be analyzed with equal ease. Provided one is not seeking lateral characterization of the surface, the method of glow discharge mass spectrometry provides chemical characterization of the surface and solid, and provides a depth profiling capability. The technique offers a number of desirable characteristics that should stimulate future development of the method.

IV. Quantitation

The secondary ion yield $S_A^\pm$ is sensitive to the state of the surface, the matrix, and the effects induced by the primary ion beam (see Section II.B). Therefore, a comparison of secondary ion intensities of an element from various points on a surface is not always a measure of relative element concentration. Caution must be exercised, especially when interpreting ion images of the surface. Such a case is illustrated in the ion micrograph of an inconel sample [143] shown in Fig. 17. The Ni, Fe, and Cr content of each grain does not vary, yet the absolute ion yield of an element shows considerable variation (as much as a factor of 3). The variation is brought about by effects dependent on the chemical nature and the crystal structure of the material, and the relative orientation of the grains in the plane of the surface; such effects are primary ion channeling, radiation-induced recrystallization [144], differences in the concentration of implanted oxygen and differences in the angular distribution of secondary ions ejected from the various grain orientations. The relative changes in ion intensity between grains is approxi-

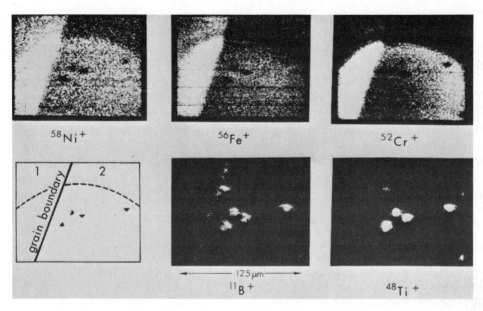

Fig. 17. Ion micrographs from a grain boundary region of an inconel sample that previously had been bombarded with a large diameter 18.5 keV $^{16}O^-$ beam. The periphery of the bombarded region is shown by the dotted line, the position of the grain boundary by the solid line, and the Ti-rich inclusions by the small triangles. Enhanced secondary ion emission is noted from grain 1 compared to grain 2. (From McHugh [143].)

mately equal for all elements, and if the intensities are referenced to the ion intensity of the major constituent at each point, the differences between grains are reduced or eliminated. One concludes from observations of this type that any quantitation procedure is better based on relative ion yields from a point of analysis rather than on an absolute yield, $S_A^\pm$.

The precision or reproducibility of relative ion yield measurements by SIMS is quite satisfactory ($\leqslant 5\%$, under controlled conditions). It would not be worthwhile to attempt quantitation of SIMS data if reasonable precision did not exist. The models of secondary ion emission discussed in Section II.A, except for that of Andersen and Hinthorne [80] have seen limited application in analytical SIMS. Andersen and Hinthorne have used an internal standard approach (i.e. two or more known concentrations of elements in the sample) to fix the parameters of their model that are necessary to compute the relative atomic concentration from secondary ion intensities. This model has been applied to many materials ranging from minerals to metal alloys with reasonable success. The general conclusion of Andersen and Hinthorne [80] is that their model will yield, as a minimum, an analysis accurate to within a factor of two of the true concentration for almost any trace element in any solid matrix.

Considering the complexity of secondary ion emission and the basic differences in SIMS instruments, it is difficult to envision a purely theoretical model applicable to all SIMS instruments, samples, and analysis conditions. For example, SIMS instruments do not collect all the secondary ions produced at the sample, nor do they transmit the same representative fraction of ions. This results from inherent differences in transmission efficiency as a function of secondary ion initial energy. Too many variables exist to control the situation effectively on a purely theoretical basis. Therefore, any practical quantitation approach must reduce variables to a minimum and be adaptable to any SIMS instrument. By necessity, such an approach will be empirical or semiempirical in nature and employ standards for determining relative elemental sensitivities.

To achieve success with a quantitation method based on calibration with standards (or any method, for that matter), it is important to standardize instrument operating parameters: primary ion species, energy and current density, sample environment, detector efficiency, and the energy-band pass of the secondary ion analyzer. Once one fixes these conditions, meaningful sample analyses are possible using relative elemental sensitivity factors derived from standards of the same or similar composition to the sample. If eqns. (1) and (2) (see Section I) are combined and ratioed to a reference element to eliminate constants, one obtains

$$\frac{i_A^\pm C_A^{-1}}{i_{ref}^\pm C_{ref}^{-1}} = \frac{\gamma_A^\pm}{\gamma_{ref}^\pm} = \delta_A \; ; \delta_{ref} = 1.00 \tag{4}$$

where $i_A^\pm$ and C_A, and $i_{ref}^\pm$ and C_{ref} are the secondary ion signals and atomic concentration of an element and reference element for the matrix, respectively. The set of relative elemental sensitivity factors so produced δ_A can be employed to reduce the measured secondary ion signals of the unknown sample to relative atomic concentration, i.e. $(i_A^\pm / i_{ref}^\pm) \delta_A^{-1} = C_A / C_{ref}$. The set of relative atomic concentrations can be normalized to 100%, and the result will be the composition of the matrix in atomic percent provided the SIMS instrument is operating at a sensitivity sufficient to detect all major constituents of the matrix. The approach outlined above or a variation of it has been used for bulk solids analysis of a variety of matrices [54,134,143]. In general, quantitative accuracies of 10% can be expected. However, such an approach places great demands on the standards and on sample homogeneity. The results generated by this method are, therefore, no better than the standard or the sample homogeneity.

A standard for every matrix that one could encounter is an unreasonable requirement. Therefore, empirical methods are needed to extend and adjust sensitivity factors from a few known standards to any desired matrix. The sensitivity factor δ_A will show a dependence on ϵ_s, a parameter that characterizes the electronic properties of the secondary ion-emitting surface. The

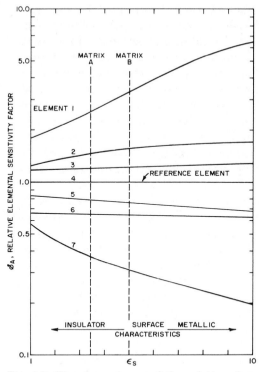

Fig. 18. The dependence of the relative elemental sensitivity factor on the character of the sample surface. Information of the type illustrated in this figure can be derived for any SIMS instrument from a set of standard samples. Establishing ϵ_s for a given matrix determines the $\mathcal{S}_A$ values required for quantitating relative secondary ion yield data, e.g. compare the set of $\mathcal{S}_A$ values for elements 1 through 7 in matrices A and B. (From McHugh [54].)

general behavior of $\mathcal{S}_A(\epsilon_s)$ will follow the pattern shown in Fig. 18. The type of information given in Fig. 18 is derived from standards that differ widely in matrix composition (but possess a common reference element). The simplest approach for defining ϵ_s is $\epsilon_s = k(\mathcal{S}_1/\mathcal{S}_2)$ where k is an arbitrary constant and $\mathcal{S}_1$ and $\mathcal{S}_2$ are relative sensitivity factors for two elements that yield a matrix-sensitive $\mathcal{S}_1/\mathcal{S}_2$ ratio. Other types of information present in the secondary ion spectra that could characterize ϵ_s are the secondary ion ratios M^{2+}/M^+, MO^+/M^+ and MN^+/M^+. Once one standardizes the instrument operating conditions and generates the empirical information of Fig. 18 from standard samples, one has the ability to quantitate with reasonable accuracy secondary ion signals from most any matrix provided that ϵ_s can be ascertained.

The preferable means for establishing a value for ϵ_s should involve methods based solely on information present in the secondary ion spectra or relative secondary ion signals from the unknown sample, information such as the M^{2+}/M^+, MO^+/M^+, MN^+/M^+ or any other secondary ion ratio sensitive to ϵ_s and independent of element concentration. Also, ϵ_s can be derived from relative ion signals and the concentrations of two or more constituents that possess a matrix-sensitive $\delta_A/\delta_{A'}$ ratio. This method is useful for trace element analyses in a well-characterized matrix. The last alternative is to choose ϵ_s solely from analytical experience and judgement. Such an approach is not as undesirable as one might expect. At least the extremes are eliminated (see Fig. 18).

The analysis of solids with an oxygen primary ion beam and the methods outlined above have given favorable results [54,143]. In fact, a single set of sensitivity factors gives accuracies for elemental concentrations much better than a factor of 2 for most elements in a number of quite different matrices (Al_2O_3 and Zr, Fe and Ni base alloys). Therefore, any model that applies a matrix-sensitive correction to relative sensitivity factors could only improve the results. The attractive features of the approach described above are: the model is simple, should be applicable to any instrument, does not depend on theoretical or physical constants, and is based solely on standards and measurements within the instrument.

The major emphasis in the foregoing discussions has been on analyses of bulk solids and not on thin surface films. Since bulk solids can be well-characterized, they become standards for testing quantitative models. Standards for surface films <50 Å thick do not exist or are extremely difficult to produce. Therefore, quantitation of SIMS data gathered from the outermost 50 Å will, by necessity, follow the same procedures as those for the bulk sample where no standard exists.

The accuracy of any quantitation procedure, not involving a direct sample—standard comparison, depends on the susceptibility of the matrix to surface and bulk element fractionation caused by sputtering and ion implantation effects. The enrichment or depletion of certain species at the surface has been the subject of studies by Tarng and Wehner [145] and Dahlgren and McClanahan [146]. Feng and Chen [147] have observed that low energy Ar^+ bombardment of the layer compound MoS_2 causes extensive preferential sputtering and the formation of islands of metallic Mo on the surface. This differential sputtering effect and its importance in surface analytical methods have been discussed by Honig [44]. The important conclusion is that the sputtered particles carry information that reflects the composition of the underlying layers and is not adversely affected by the fractionated outermost layer. The other effect, element fractionation in the implant zone, depends on the characteristics of the matrix, the particular constituent and its mobility, the chemical character of the implant ion, and electric field gradients established in the sample from surface charging (see Section V.B). SIMS results for many different samples suggest that this problem is not a major threat to analytical accuracy in most situations; nevertheless, there are situations where anomalous effects are observed and fractionation in the implant zone is a reasonable explanation for the effect.

The topography of the surface and surface charging are two matrix-independent effects that could adversely influence an analysis. A topographical feature of micrometer dimensions (present initially or induced by the primary ion beam) could change the relative secondary ion collection efficiency of the analyzer for ions of different initial energy; the analyzer collection efficiency is very sensitive to the initial energy and ejection angle of the ion. Likewise, surface charging induced by the primary ion beam could also change the relative secondary ion collection efficiency by disturbing the ion extraction field in the immediate vicinity of the sample surface. These effects would be minimal or nonexistent in electrically conducting specimens that are metallographically polished; however, if one departs from these conditions, the variables and sources for error mentioned above become factors for consideration.

V. Elemental depth concentration profiling

To obtain depth concentration profiles of elements distributed in layers parallel to the surface, ion sputtering (not an exclusive tool of SIMS) is used in most surface analysis methods as an in situ tool for exposing the underlying layers of the solid. Thus, the depth resolution inherent in the individual surface analysis method becomes less important, and subsurface layer mixing and other problems associated with ion etching of surfaces become the major factors in the depth resolution of the technique.

The profile broadening produced by the measurement process in a thin subsurface layer or in a sharp boundary between dissimilar materials can be used to characterize the depth resolution of the profiling method. If a layer or boundary exists at a depth greater than approximately $2 R_p$, the various contributors to profile broadening (ion-matrix and instrumental effects) give a distribution for a thin layer that resembles, to a first approximation, a normal distribution [54] with a standard deviation, σ_R. Depth resolution can be defined as σ_R of this distribution. If the layer has significant width, then the standard deviation of the experimentally observed or measured profile, σ_M, is related to σ_R and σ_T by $\sigma_M^2 = \sigma_R^2 + \sigma_T^2$, where σ_T is the standard deviation of the true layer distribution. In situations where $\sigma_M \gg \sigma_T$, e.g. a thin layer profile, the observed $\sigma_M \cong \sigma_R$, the depth resolution.

If it is assumed that the measured profile of a thin layer is a normal distribution, then one can consider the case of boundary-edge-broadening and its relationship to the depth resolution. The depth resolution can be derived from the profile of a well-defined step concentration gradient (length of the step $\gg \sigma_R$) when the shape of the true step edge resembles a cumulative normal with an associated standard deviation of σ_t. If the step concentration gradient is sharply defined ($\sigma_t \cong 0$), one half the distance between the depths corresponding to 84% and 16% of the experimentally measured step height represents $\sigma_m \cong \sigma_R$, the depth resolution. For a boundary region with significant initial broadening (i.e. a significant σ_t) then depth resolution is given by $\sigma_R = (\sigma_m^2 - \sigma_t^2)^{1/2}$, with due consideration for the errors in σ_m and σ_t. The case of a layer profile with a significant σ_T can be treated in a similar manner.

The material discussed in this section is related to the basic physical or instrumental effects associated with ion-beam sputter etching of surfaces and the problems of using ion etching to give an unaltered picture of an element concentration profile. Therefore, much of the information will apply to any surface analysis method using sputter etching.

The general procedure for producing a SIMS depth profile is to monitor the secondary ion signal of an element of interest as a function of sputter time. In a uniform matrix, time can be translated to depth through suitable calibration experiments (known film thickness, crater depth measurements, sputter yields, etc.). The variation of secondary ion intensity (as discussed in Sections II.B, II.D. and IV) is not always a measure of a relative difference in element concentration; therefore, care must be exercised in the interpretation of depth profiles, especially for depth profiles very near the surface, i.e. at depths $<R_p + 2\,\Delta R_p$ (see Section II.D), or of films composed of dissimilar layered phases or matrices with nonhomogeneous distributions of trace elements that have a major influence on the secondary ion emission characteristics of the matrix. To obtain meaningful results in this situation, procedures for quantitating the secondary ion intensities (see Section IV) must be applied to the profile data. If this is not possible, one must attempt, as a minimum, to reference the profiled element to the secondary ion intensity of one or more uniformly distributed constituents of the film. In general, the only instance where the absolute ion intensity gives depth concentration information directly is for a trace constituent in an amorphous or single-crystal matrix where the major constituents are homogeneously distributed and the information of interest is located >50 Å below the surface.

The adsorption of O_2 on surfaces generally enhances and stabilizes the secondary ion yields [28,148]. The result is that element sensitivity is improved and certain matrix effects are reduced or eliminated which gives more meaning to the secondary ion signal as a direct measure of element concentration. In a number of cases, the variation of secondary ion yield with crystal orientation is eliminated by O_2 adsorption on a surface [84,149].

Blanchard et al. [150,151] have demonstrated the usefulness of this technique in analytical applications, and their results [151] for a SiO_2—Si interface are given in Fig. 19. The sharp decrease in the $^{30}Si^+$ intensity in the SiO_2—Si interface region is eliminated at an O_2 partial pressure of 2×10^{-4} torr and a surface removal rate of <5 Å sec^{-1}. At this pressure, the Si^+ intensity is constant and equal for the two phases SiO_2 and Si. The broken

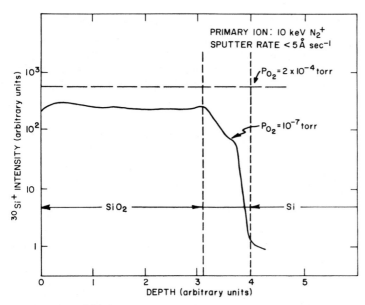

Fig. 19. The $^{30}Si^+$ secondary ion intensity as a function of depth for a SiO_2—Si sample. The broken and solid curves represent profiles performed with a partial pressure of O_2 equal to 2×10^{-4} and 10^{-7} torr, respectively, in the vicinity of the sample surface. Adsorption of O_2 on the surface eliminates the difference in Si^+ emission from the SiO_2 and Si phases. (From Blanchard et al. [151].)

line in Fig. 19 is a schematic representation of the observations of Blanchard et al. [151] at 2×10^{-4} torr.

The depth profile capability of SIMS coupled with its high detection sensitivity for a majority of the elements makes SIMS a very attractive tool for studying thin films, ion implantation, and diffusion. Hofker et al. [152] and Schulz et al. [153], in particular, have discussed factors that influence a SIMS depth analysis. The factors that affect the depth resolution of the method fall into two classifications: instrumental factors and ion-matrix effects.

A. INSTRUMENTAL FACTORS INFLUENCING PROFILE DEPTH RESOLUTION

It is virtually impossible to extract meaningful depth profile information from SIMS data unless one maintains constant primary ion intensity and uniform primary ion current density over the secondary ion extraction area of the sample. In a stationary focused ion beam, the ion current density incident on the sample is not constant over the beam diameter; therefore, surface layer removal cannot be uniform. If the information zone for secondary ions extends over the total area of the incident beam, the contribution of

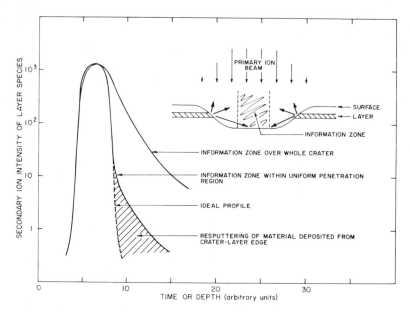

Fig. 20. Illustration of a concentration profile of a subsurface layer and various instrumental factors that produce a distorted profile of the true distribution.

ions from the crater edges will distort the profile of a subsurface layer as shown in Fig. 20.

This problem is handled a number of ways in SIMS depending on the particular instrument. The common methods used in SIMS instruments with no lateral resolution capability involve defocusing the primary beam over an area much larger than the secondary ion analyzer acceptance area or aperturing the defocused beam to produce a well-defined area of uniform current density. Another method is to place a thin mask (composed of a material that produces no interferences) over the sample to limit sample secondary ion emission to the uniform current density region of the primary ion beam. Guthrie and Blewer [154,155] have generated well-defined sharp-edge sputter craters of uniform depth using the dual-lens primary beam focusing column of an ARL ion microprobe. Although this method has attractive possibilities, it places too much demand on the focusing quality and stability of the primary beam column and never completely removes the edge contribution.

The most satisfactory approach to the problem is to electrically sweep the focused ion beam in a raster mode over an area of the surface large enough to provide a central region of uniform current density. To ensure that the sputtering is uniform over the central area of the rastered region, attention must be given to certain aspects of the beam sweep system: the xy sweep voltages must be linear with time, the beam retrace must be blanked or randomized, and the sweep rates must be consistent with the dimensions of the beam so that the individual traces overlap.

To take full advantage of the raster mode, the secondary ion information zone must be restricted to the uniform current density region. It is a simple matter to limit the information zone to this region with SIMS instruments that provide lateral characterization of the surface. In the direct-imaging method (CAMECA ion analyzer), an aperture can be placed in the image plane to pass only information from the desired area on the sample (see Section III.A). In scanning microprobe methods, a small diameter primary ion beam is rastered over the surface, and the ion detection system is gated to accept ions only when the primary ion beam is within a selected "window area" of uniform current density.

Even with the information zone restricted to the uniform current density region, a number of other instrumental effects come into play that distort the true depth profile. Croset [156] and Blanchard et al. [151] have discussed the redeposition phenomenon. Material from the crater edge is sputter-deposited over the surface and resputtered from the information zone of the surface (see Fig. 20). The effect is, generally, a problem only for the low concentration tail of a profile and can be reduced considerably by sharpening the sputter crater edges [151,156]. The adsorption of gas phase impurities, and memory effects also affect a profile in a similar manner. In addition, any factor that influences a SIMS trace analysis (molecular ion interferences, chemical purity of the primary ion beam, primary ion beam tail effects and the nonfocused component of the primary ion beam) can also distort a profile at low concentration levels (see Section III.C for a discussion of these effects).

B. ION—MATRIX EFFECTS INFLUENCING PROFILE DEPTH RESOLUTION

A number of effects that distort the true concentration profile are determined by the characteristics of the collision cascade that is induced in the solid by the incident ion. Two such effects are the mean escape depth of secondary ions (see Section II) and atomic mixing in subsurface layers. Most depth profiles extend to depths >100 Å and, in these situations, atomic mixing in subsurface layers, recoil implantation and matrix-associated effects are much greater contributors to profile distortion than the secondary ion escape depth.

McHugh [157] has used SIMS methods to study the effect of incident ion energy on atomic mixing in subsurface layers. A Ta_2O_5 film sample that contained a 50 Å ^{31}P-rich layer located 230 Å below the surface was employed in this work. The sample was profiled for ^{31}P using normal incidence ^{16}O primary particles at various energies from 1.75 to 18.5 keV. A few results from these experiments are shown in Fig. 21. A pronounced energy effect is observed in the widths of the ^{31}P profiles produced by >4 keV

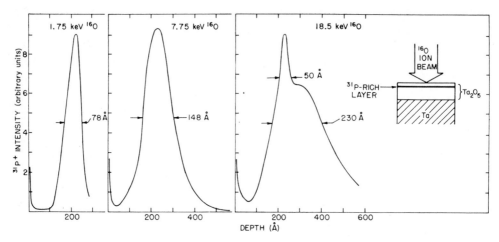

Fig. 21. Phosphorous depth profiles of a Ta_2O_5 film containing a ^{31}P-rich subsurface layer. The profiles were obtained using normal incident ^{16}O sputtering beams of 1.75, 7.75, and 18.5 keV. These results illustrate the effect of incident ion energy on a measured profile. (From McHugh [157].)

^{16}O. For 18.5 keV ^{16}O, the observed profile contains two distinct components, the ^{31}P-rich layer and the ^{31}P recoil distribution.

Knowing the true concentration gradient, the leading and trailing edges of a measured profile give information on the depth resolution of the method. Honig [44] has derived depth resolution values of 32 Å at 1.75 keV, 45 Å at 4.25 keV and 105 Å at 7.75 keV from McHugh's data. In the P—Ta_2O_5 system, all profile distortion effects except those related to recoil atom effects are minimal [157]. Therefore, the resolution values quoted above reflect the magnitude of the recoil or knock-on mixing in subsurface layers and would affect the profile of any surface analysis method that employs ion sputtering to remove surface layers. Using a 5—50 keV Ar^+ incident ion beam, Schulz et al. [153] have observed recoil broadening effects in SIMS depth profiles of B-implanted Si samples.

The important factors that determine the relative ranges of particles in a given film are the energy of the projectile particle, the atomic number Z, and the mass A of the projectiles and target atoms. The profile, then, should show a dependence on the Z and A of the projectile in a manner similar to the energy effect. Another factor that influences the profile is the angle of incidence that the primary ion beam makes with the surface (the angle with respect to the surface normal). Increasing the angle of incidence reduces the mean penetration depth of incoming particles, as measured with respect to the surface normal and should therefore give results equivalent to reducing the primary ion energy.

Other factors, the magnitudes of which are difficult to assess for all elements and matrices, concern the broadening or skewing of a profile because

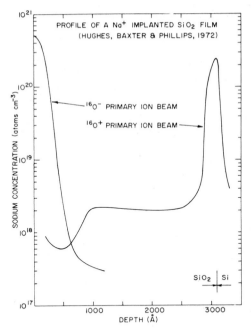

Fig. 22. Sodium profiles of a sodium-implanted SiO_2 film sample using $^{16}O^-$ and $^{16}O^+$ primary ion beams. The ion energy and fluence for the sodium implantation were 20 keV and 10^{15} ions cm^{-2}, respectively. Migration of sodium as a result of ion beam sputtering of the SiO_2 insulator film is apparent in these results. (From Hughes et al. [159].)

of the diffusion of elements brought on, or enhanced, by ion damage to the solid, an increase in the local temperature of the bombarded area, and high electric field gradients produced from primary ion beam charging of the surface. In surface charging, a 10^6 V cm^{-1} field can be produced across a 1000 Å film by a 10 V surface charge. Electric field-induced diffusion for a number of elements in Al_2O_3 has been observed [158]. Sodium migration resulting from ion beam surface charging has been observed in SiO_2 films by Hughes et al. [159]. Their results are shown in Fig. 22. They obtained the Na profile of a Na-implanted SiO_2 film by SIMS methods using O^+ and O^- primary ion beams. The positive charging of the surface from the O^+ sputtering caused the Na to migrate away from the surface to the SiO_2—Si interface. The opposite effect is apparent in the O^- bombardment.

A uniform primary ion current density does not in itself imply uniform etching of the surface. This characteristic is controlled by surface topography, cleanliness, and sample uniformity which depends on composition, physical state, and crystallographic orientation. Wilson [160] has employed scanning electron microscopy to study the topography of sputtered surfaces and has considered in detail these effects. Generally, one observes for controlled instrument and matrix conditions that the profile broadening due to the sputter etch process is equal to or less than 10% of the depth sampled [44,157,161].

The presence of the implanted primary ion and ion-induced lattice damage can influence a profile through its effect on atom yield and secondary ion emission probability. However, if the sputter etching is accomplished with a particle energy of ~100 eV/amu, the changing matrix condition caused by ion implantation (refer to Section II.D and Fig. 8) is confined within 100 Å of the surface; beyond 100 Å the implant ion concentration remains constant and a stable matrix condition is attained.

VI. Applications

SIMS applications fall into five broad, and sometimes overlapping, categories: surface studies, depth profiles, xy characterization, microanalysis, and bulk solids analysis. This section provides a brief and limited review of SIMS applications; it is intended to develop an appreciation for both established and potential applications.

A. SURFACE STUDIES

The capabilities of SIMS for surface analysis are well demonstrated in a study [143] of a vapor-deposited Au film surface. Figure 23 contains a number of ion micrographs taken with an ARL ion microprobe of the Au film surface. The dark area in the center of the micrograph is produced by raster-scanning a small diameter beam over the area for a time sufficient to remove the surface contaminants. At the completion of the cleaning step, the scan area is increased, and an ion micrograph of the surrounding region is produced (Polaroid film picture generated from the CRT display). The thickness of the contaminant layer is difficult to determine; however, a conservative estimate (based on primary ion current density, time, and reasonable sputter yields) places the thickness of the layer at $\leqslant 15$ Å. The concentration of Al in the layer is $\leqslant 0.1$ at.%, and the amount of surface removed to generate the Al micrograph is $\leqslant 5$ Å.

The $^1H^+$, $^7Li^+$, and $^{12}C^+$ micrographs demonstrate the light element capability of SIMS. The hydrocarbon contamination that gave the $^1H^+$ and $^{12}C^+$ micrographs is attributed to overall hydrocarbon contamination of the coating system caused by the mechanical vacuum pump used regularly in the preliminary evacuation step. After the Au charge was exhausted, the hot surfaces near the evaporator vaporized Al onto the fresh Au surface; aluminum had been previously evaporated in the system. The K contamination appears clumped, which indicates contamination by small particulates.

The analytical applications of SIMS in the field of surface analysis concern the identification of surface species (the foregoing example is a good practi-

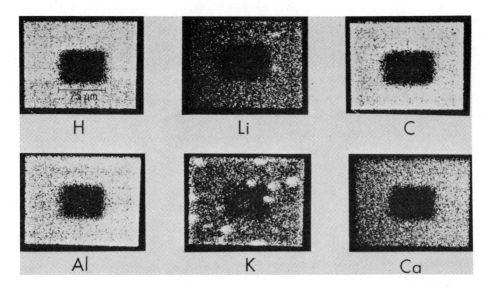

Fig. 23. Secondary ion micrographs of a number of elements present as surface contaminants on a Au film sample. The rectangular dark area in the center of the micrograph (approximately 50×75 μm area on the sample surface) is a region where the surface contaminant film has been removed by sputter etching. The thickness of the contaminant layer is $\leqslant 15$ Å. (From McHugh [143].)

cal application) and the study of dynamic surface processes. Benninghoven [107,108] has shown that SIMS sensitivities approach 10^{-6} of a monolayer for certain favorable situations. Also, it is possible to study dynamic surface processes without perturbing the process, since $<10^{-3}$ of the outermost surface layer is required for the complete analysis [48,162]. Some of the surface phenomena studied by SIMS are catalysis, corrosion, adsorption, and diffusion. Fogel [48,49] and Benninghoven [45] provide a rather comprehensive review on applications in these areas.

Surface processes can be studied in situ with SIMS instruments that employ gas introduction and heated substrates. The information present in the positive and negative secondary ion spectra provides an insight into surface bonding and the mechanisms of gas—surface interaction [45,48]. The catalytic decomposition and synthesis of NH_3 on iron [163,164] and the surface species present on a Ag catalyst used for the oxidation of ethylene [108] are two well-documented SIMS studies in the area of catalysis. Catalytic processes differ from metal oxidation or corrosion processes because the catalytic reaction zone is of monolayer dimensions. In metal oxidation, the reaction zone grows as a result of chemical driving forces developed across the adsorbed layer—metal interface. A number of SIMS studies have been directed at the initial phases of gaseous corrosion and the formation of surface compounds. A few of these studies concern the niobium—oxygen system [165], adsorption of oxygen on tungsten [166,167] and copper [168], and the oxidation of aluminum [45], chromium [169], and silicon [170].

The study of oxygen adsorption on tungsten by Rybalko et al. [166] is a good example of a dynamic surface process investigated by SIMS. These authors confirmed the two-stage interaction mechanism observed by other methods [171]. The secondary ion signals of O^+ and WO_2^+ were measured as a function of time for various surface temperatures. Their results are given in Fig. 24. In Stage I, O_2 dissociates into atomic oxygen when adsorbed on the surface; in Stage II, W compounds are in evidence by the appearance of the molecular ion species, WO_2^+. The saturation level attained in the lower-temperature ($900°$K) Stage II region corresponds to equilibrium conditions between gas phase, O_2, and the surface oxide layer.

The quantitation of secondary ion information from the outermost phase or reaction zone of the surface presents some difficulty (Section V). However, as shown above, surface processes can be studied without reducing the data to atomic concentration. The type of species and the variation of secondary ion signals with time, temperature, and gas pressure provide sufficient information to elucidate many processes occurring at surfaces. In many instances, the high sensitivity and broad capabilities of SIMS compensate for its lack of quantitative accuracy by providing qualitative and semiquantitative information that could not have been obtained by any other method, especially in the areas of general quality control of surface preparations and treatment processes [172,173].

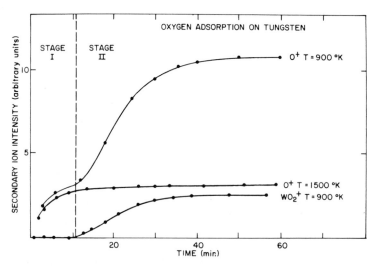

Fig. 24. The variation with time of the secondary ion intensities of O^+ and WO_2^+ from a W surface at 900 and $1500°$K. (At $t = 0$ the surface of the W sample was free of adsorbed gases.) The Stage II process or compound formation is evident in the lower temperature $900°$K O^+ and WO_2^+ curves. (From Rybalko [166].)

B. DEPTH PROFILES

SIMS is one of the most powerful of the surface analysis tools used for the measurement of depth concentration profiles. The depth resolution and sensitivity inherent in the method are <50 Å and $<10^{17}$ atom cm^{-3}, respectively (considering the factors mentioned in Sections V and III.C). A number of workers have applied SIMS methods to determine impurity and isotopic concentration gradients in studies related to the analysis of thin films, diffusion, and ion implantation.

Thin film studies cover many disciplines; a few of these studies where SIMS methods have been employed and depth information was of principal importance concern coatings and oxide layers on steel [174], erbium films [175], sputter-deposited Ta [176] and Cr films [51], CuO on Cu [177], Ag—Cu and Al—Ge—Nb films [178], Pt films on Si [179], ^{31}P impurity profiles in Ta$_2$O$_5$ [157,178,180], and oxygen isotopic intermixing in the anodic formation of Ta$_2$O$_5$ [178]. The ability to obtain isotopic concentration information adds another dimension to depth profiling. An example [54] of this is an ^{18}O profile taken of a Ta$_2$O$_5$ sample containing an enriched ^{18}O layer (Fig. 25). A Ta sample was anodized in three stages, each

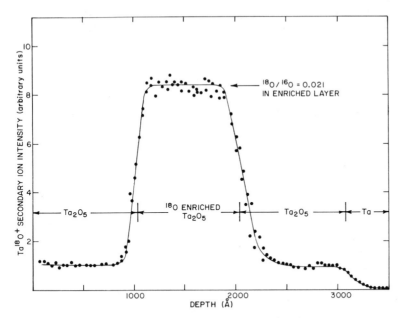

Fig. 25. The depth concentration profile of ^{18}O in a Ta$_2$O$_5$ sample containing a layer enriched in ^{18}O. The profile was performed with an ARL ion microprobe using 8.5 keV N$_2^+$ and the monitor of the ^{18}O concentration was Ta ^{18}O$^+$. The ^{18}O is ~2% of the oxygen in the enriched layer. The Ta ^{18}O$^+$ signal preceding and following the enriched layer peak is due to the ^{18}O level in normal oxygen. (From McHugh [54].)

References pp. 273—278

designed to produce a zone 1020 Å thick [181]. The middle stage employed an electrolyte enriched in ^{18}O. The ^{18}O gradient was established by monitoring the $Ta^{18}O^+$ secondary ion signal as a function of time [54]; time was converted to depth on the basis of the Ta_2O_5—Ta interface being 3060 Å from the surface.

SIMS methods have been used by Contamin and Slodzian [182] to study oxygen self-diffusion in stoichiometric UO_2. A thin layer of oxide enriched in ^{18}O was diffused into a large section of UO_2. The self-diffusion coefficients were derived from the SIMS depth profiles of the oxygen isotopic concentration gradients. The diffusion of In in Ge [46], oxygen in Zr [183,184], K in mica [185], Zn in InAs—P [51], Li in Nb, and B in Fe [186] represent a few other applications. Diffusion and impurity profiles in semiconductor devices have been studied by SIMS. The diffusion profile and impurities at low levels (<1 ppm) can have important effects on the electrical properties of a device. SIMS studies have been directed at B diffusion in Si and SiO_2 [150,151,156], problems involved in the manufacture of planar bipolar transistors [161], and oxide impurities in metal—oxide—silicon (MOS) structures [159].

Ion implantation is being widely employed to change the properties of

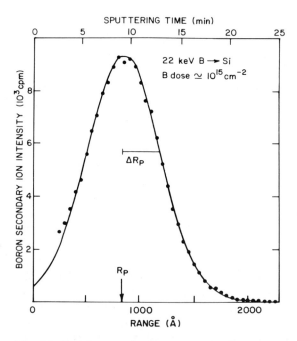

Fig. 26. Depth concentration profile of ^{11}B implanted in amorphous Si. Each experimental point represents removal of 50 Å of Si. The experimental points are fitted with a normal distribution where R_p and ΔR_p are 840 Å and 340 Å, respectively. (From Maul et al. [187].)

solids in layers near the surface. Many of the methods used to determine implant profiles (radiotracer, electrical activity, and X-ray generation by heavy ions) introduce complications and are not applicable to all ion—matrix situations. SIMS is free of many of these restrictions and, therefore, is a desirable tool for the measurement of implant profiles. One system receiving considerable attention by a number of workers is the implantation of B in Si [151,153,187—189]; other systems investigated are Ni^+ implanted in Cu [151], $^{14}N^+$ implanted in GaAs [190], B—As—B triple implant in Si [191] and various ions implanted in SiC [192].

The work of Maul et al. [187] demonstrates the capability of SIMS for measuring implant profiles. In their experiment (the results of which are shown in Fig. 26), they measured the profile of 22 keV ^{11}B implanted in amorphous Si. The boron secondary ion intensity was followed as a function of time and the sputter rate calculated from the experimentally measured crater depth. Crater edge effects were eliminated by maintaining constant primary ion current density over an area greater than the implanted area. A Gaussian range distribution was observed, and the measured projected range, R_p, and the standard deviation, ΔR_p, agreed quite well with theoretical values [187].

C. XY CHARACTERIZATION, MICRO AND BULK ANALYSIS

The electron microprobe, developed in the early 1950's by Castaing [193], opened the door to the microcharacterization of solids, and, today, it remains the principal tool in this area of solids analysis. The direct-imaging [27] and scanning microprobe [31] SIMS instruments have widened the capability for solids microcharacterization by providing greater sensitivity, an isotopic and surface analysis capability, and an ability to detect the presence of low Z elements. Ion-imaging methods provide information on the lateral distribution of elements in the surface and quantitative information under the proper conditions (see Sections IV and III.D). Ion imaging has been applied to studies of grain boundary precipitation [50,51,131,174,194, 195], metallurgical and single-crystal effects [28,57,84,149,174], diffusion (xy characterization of transverse sections) [182,196], mineral phase characterizations [52,132,195], and distribution of surface impurities [47,174].

The analysis of solids by SIMS is directed at two interests: the general characterization of the bulk solid and/or the characterization of specific points in the solid (i.e. microanalysis of areas ≤25 μm in diameter). In the domain of microanalysis, SIMS methods have been applied to trace element analysis in specific mineral grains [52,185,197], in situ Pb isotopic analysis of radiohalo inclusions (1—2 μm in diameter) [198], airborne particulate analysis [134], and the determination of crystallization ages of specific mineral phases by $^{207}Pb/^{206}Pb$ measurements and the rubidium—strontium method [52,199]. In the characterization of bulk solids, Andersen and Hin-

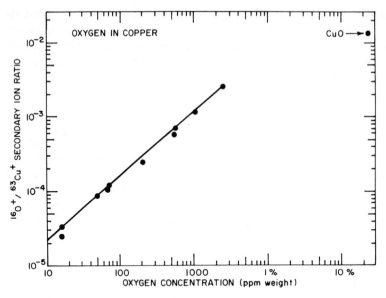

Fig. 27. Quantitative working curve for the determination of oxygen in Cu. The $^{16}O^+/^{63}Cu^+$ secondary ion ratio is used as the measure of oxygen in the Cu standards. (From Evans [205].)

thorne [80] have applied their quantitation procedures to a number of mineral and metal samples. SIMS methods have been used in the study of meteorites [106,200,201], binary alloys [186], low level B analyses in Si [156,161,202,203], trace element analysis of Si and GaAs [203], high energy nuclear reaction products [204], and gases in metals [205].

The analysis of active gases (e.g. H_2, N_2, and O_2) in metals presents some problems for a SIMS analysis (see Sections III.B and III.C). Evans [205], taking great care to minimize the oxygen background in his instrument, examined a number of Cu—O standards (20—2000 ppm by wt. oxygen). His quantitative working curve is shown in Fig. 27. The $^{16}O^+/^{63}Cu^+$ secondary ion ratio is plotted as a function of oxygen concentration in the Cu standard. This ratio provides a good quantitative relationship usable over a wide range of oxygen concentrations.

VII. Conclusions

The material presented in this chapter is intended to provide the reader with sufficient information for a basic understanding of SIMS and a balanced view of the capabilities and limitations of the method. At the present time, no surface analysis tool exists that satisfies all the needs of all surface analysts. SIMS is no exception; however, SIMS does occupy a unique position in the field of surface and bulk solids analysis because a number of its capabilities are unmatched by any other method. Its high detection sensitivity for the majority of elements, analysis of low atomic number elements, isotopic analysis, high resolution depth profiling and its capability for lateral characterization of surfaces, provide a tool for three-dimensional isotopic and trace elemental characterization of solids.

A great number of surface problems can be solved by qualitative or semi-quantitative surface analysis; therefore, the uncertainty in a SIMS quantitative analysis becomes a less important factor compared to the qualitative information provided by this high sensitivity technique. SIMS has already had a major impact on the microanalysis of solids in areas which are of fundamental and practical importance. Future developments in SIMS methods will center principally on areas relating to a better understanding of quantitation problems and the methods needed to improve quantitative accuracy.

References

1 F.L. Arnot and J.C. Milligan, Proc. Roy. Soc. Ser. A, 156 (1936) 538.
2 F.L. Arnot and C. Beckett, Proc. Roy. Soc. Ser. A, 168 (1938) 103.
3 R.H. Sloane and R. Press, Proc. Roy. Soc. Ser. A, 168 (1938) 284.
4 R.F.K. Herzog and F.P. Viehböck, Phys. Rev., 76 (1949) 855.
5 V.I. Veksler and M.B. Ben'iaminovich, Sov. Phys. Tech. Phys., 1 (1957) 1626.
6 R.E. Honig, J. Appl. Phys., 29 (1958) 549.
7 R.C. Bradley, J. Appl. Phys., 30 (1959) 1.
8 R.E. Honig, Advan. Mass Spectrom., 1 (1959) 162.
9 H.E. Stanton, J. Appl. Phys., 31 (1960) 678.
10 W.T. Leland and R. Olson, Los Alamos Rep., LA-2344, 1960.
11 V.I. Veksler, Sov. Phys. JETP, 11 (1960) 235.
12 R.C. Bradley, Phys. Rev., 117 (1960) 1204.
13 R.C. Bradley, A. Arking and D.S. Beers, J. Chem. Phys., 33 (1960) 764.
14 Ya.M. Fogel', R.P. Slabospitskiĭ and I.M. Karnaukhov, Sov. Phys. Tech. Phys., 5 (1961) 777.
15 R.E. Honig, Advan. Mass Spectrom., 2 (1961) 25.
16 V.E. Krohn, Jr., J. Appl. Phys., 33 (1962) 3523.
17 R.C. Bradley and E. Ruedl, J. Appl. Phys., 33 (1962) 880.
18 F.A. White, J.C. Sheffield and F.M. Rourke, J. Appl. Phys., 33 (1962) 2915.
19 H.E. Beske, Z. Angew. Phys., 14 (1962) 30.
20 E. Ruedl and R.C. Bradley, J. Phys. Chem. Solids, 23 (1962) 885.
21 V. Walther and H. Hintenberger, Z. Naturforsch. A, 18 (1963) 843.
22 A.J. Smith, L.A. Cambey and D.J. Marshall, J. Appl. Phys., 34 (1963) 2489.
23 A. Benninghoven, Ann. Phys., 15 (1965) 113.
24 J.A. McHugh and J.C. Sheffield, J. Appl. Phys., 35 (1964) 512.
25 R.F.K. Herzog, W.P. Poschenrieder and F.G. Satkiewicz, NASA Contract No. NAS5-9254, Final Report, GCA-TR-67-3N, 1967.
26 R. Castaing, B. Jouffrey and G. Slodzian, C.R. Acad. Sci., 251 (1960) 1010.
27 R. Castaing and G. Slodzian, J. Microsc., 1 (1962) 395.

28 G. Slodzian, Ann. Phys., 9 (1964) 591.
29 H.J. Liebl and R.F.K. Herzog, J. Appl. Phys., 34 (1963) 2893.
30 H.J. Liebl and R.F.K. Herzog, 12th Ann. Conf. Mass Spectrom. Allied Topics, Montreal, Canada, June 7—12, 1964, p. 393.
31 H. Liebl, J. Appl. Phys., 38 (1967) 5277.
32 J.A. McHugh and J.C. Sheffield, 12th Ann. Conf. Mass Spectrom. Allied Topics, Montreal, Canada, June 7—12, 1964, p. 388.
33 Ya.M. Fogel', B.T. Nadykto, V.F. Rybalko, R.P. Slabospitskiĭ, I.E. Korobchanskaya and V.I. Shvachko, Kinet. Katal., 5 (1964) 127.
34 T.L. Collins, Jr. and J.A. McHugh, Advan. Mass Spectrom., 3 (1965) 169.
35 J.A. McHugh and J.C. Sheffield, Anal. Chem., 37 (1965) 1099.
36 V.I. Shvachko, B.T. Nadykto, Ya.M. Fogel', B.M. Vasyutinskiĭ and G.N. Kartmazov, Sov. Phys. Solid State, 7 (1966) 1572.
37 P. Joyes and R. Castaing, C.R. Acad. Sci. Ser. B, 263 (1966) 384.
38 R. Castaing and J.F. Hennequin, C.R. Acad. Sci. Ser. B, 262 (1966) 1008.
39 J.A. McHugh and J.C. Sheffield, Anal. Chem., 39 (1967) 377.
40 H.E. Beske, Z. Naturforsch. A, 22 (1967) 459.
41 J.A. McHugh, 15th Ann. Conf. Mass Spectrom. Allied Topics, Denver, Colorado, May 14—19, 1967, p. 309.
42 G. Carter and J.S. Colligon, Ion Bombardment of Solids, Heinemann Educational Books, London, 1968.
43 M. Kaminsky, Atomic and Ionic Impact Phenomena on Metal Surfaces, Academic Press, New York, 1965.
44 R.E. Honig, Advan. Mass Spectrom., 6 (1974) 337.
45 A. Benninghoven, Surface Sci., 35 (1973) 427.
46 H.W. Werner, Develop. Appl. Spectrosc., 7A (1969) 239.
47 H.W. Werner and H.A.M. deGrefte, Surface Sci., 35 (1973) 458.
48 Ya.M. Fogel', Sov. Phys. Usp., 10 (1967) 17.
49 Ya.M. Fogel', Int. J. Mass Spectrom. Ion Phys., 9 (1972) 109.
50 C.A. Evans, Jr., Anal. Chem., 44 (1972) 67A.
51 A.J. Socha, Surface Sci., 25 (1971) 147.
52 C.A. Andersen and J.R. Hinthorne, Science, 175 (1972) 853.
53 R. Castaing and J.F. Hennequin, Advan. Mass Spectrom., 5 (1971) 419.
54 J.A. McHugh, unpublished results.
55 P. Joyes, Radiat. Eff., 19 (1973) 235.
56 H.D. Hagstrum and G.E. Becker, Phys. Rev. B, 8 (1973) 107.
57 R. Castaing and G. Slodzian, C.R. Acad. Sci., 255 (1962) 1893.
58 A. Benninghoven and A. Müller, Phys. Lett. A, 40 (1972) 169.
59 N.H. Tolk, D.L. Sims, E.B. Foley and C.W. White, Radiat. Eff., 18 (1973) 221.
60 R.V. Stuart, G.K. Wehner and G.S. Anderson, J. Appl. Phys., 40 (1969) 803.
61 V.I. Veksler, Sov. Phys. JETP, 11 (1960) 235.
62 V.I. Veksler, Izv. Akad. Nauk. Uzb. SSR, Ser. Fiz. Mat. Nauk, 4 (1959) 34.
63 V.I. Veksler and B.A. Tsipinyuk, Sov. Phys. JETP, 33 (1971) 753.
64 Z. Jurela, Radiat. Eff., 13 (1972) 167.
65 Z. Jurela, Int. J. Mass Spectrom. Ion Phys., 12 (1973) 33.
66 P. Joyes and R. Castaing, C.R. Acad. Sci. Ser. B, 263 (1966) 384.
67 P. Joyes and J.F. Hennequin, J. Phys., 29 (1968) 483.
68 P. Joyes, J. Phys., 29 (1968) 774.
69 P. Joyes, J. Phys., 30 (1969) 224.
70 P. Joyes, J. Phys., 30 (1969) 365.
71 G. Blaise, Radiat. Eff., 18 (1973) 235.

72 J.M. Schroeer, Bull. Amer. Phys. Soc., 10 (1965) 41.
73 J.M. Schroeer, Bull. Amer. Phys. Soc., 12 (1967) 137.
74 J.M. Schroeer, T.N. Rhodin and R.C. Bradley, Surface Sci., 34 (1973) 571.
75 J.B. Hasted, in D.R. Bates (Ed.), Atomic and Molecular Processes, Academic Press, London, 1962, p. 696.
76 P. Joyes and G. Toulouse, Phys. Lett. A, 39 (1972) 267.
77 A. Benninghoven, Z. Phys., 220 (1969) 159.
78 C.A. Andersen, J. Mass Spectrom. Ion Phys., 2 (1969) 61.
79 C.A. Andersen, J. Mass Spectrom. Ion Phys., 3 (1970) 413.
80 C.A. Andersen and J.R. Hinthorne, Anal. Chem., 45 (1973) 1421.
81 A. Benninghoven and A. Müller, Phys. Lett. A, 40 (1972) 169.
82 E. Dennis and R.J. MacDonald, Radiat. Eff., 13 (1972) 243.
83 G. Staudenmaier, Radiat. Eff., 18 (1973) 181.
84 M. Bernheim, Radiat. Eff., 18 (1973) 231.
85 R. Laurent and G. Slodzian, Radiat. Eff., 19 (1973) 181.
86 E. Zwangobani and R.J. MacDonald, Radiat. Eff., 20 (1973) 81.
87 V.E. Yurasova, A.A. Sysoev, G.A. Samsonov, V.M. Bukhanov, L.N. Nevzorova and L.B. Shelyakin, Radiat. Eff., 20 (1973) 89.
88 H. Kerhow and M. Trapp, Int. J. Mass Spectrom. Ion Phys., 13 (1974) 113.
89 J.M. Fluit, P.K. Rol and J. Kistemaker, J. Appl. Phys., 34 (1963) 690.
90 C.E. Carlston, G.D. Magnuson, A. Comeaux and P. Mahadevan, Phys. Rev., 138 (1965) 3.
91 J.J.Ph. Elich, H.E. Roosendaal and D. Onderdelinden, Radiat. Eff., 10 (1971) 175.
92 R. Kelley and N.Q. Lam, Radiat. Eff., 19 (1973) 39.
93 P. Joyes, J. Phys. Chem. Solids, 32 (1971) 1269.
94 M. Leleyter and P. Joyes, Radiat. Eff., 18 (1973) 105.
95 M. Leleyter and P. Joyes, J. Phys. B, 7 (1974) 516.
96 R.F.K. Herzog, H.J. Liebl, W.P. Poschenrieder and A.E. Barrington, NASA Contract No. NASW-839, Tech. Rep., GCA-65-7-N, 1965.
97 R.F.K. Herzog, W.P. Poschenrieder and F.G. Satkiewicz, Radiat. Eff., 18 (1973) 199.
98 Z. Jurela, Radiat. Eff., 19 (1973) 175.
99 G. Blaise and G. Slodzian, Rev. Phys. Appl., 8 (1973) 105.
100 G. Carter, J.N. Baruah and W.A. Grant, Radiat. Eff., 16 (1972) 107.
101 R.K. Lewis, J.M. Morabito and J.C.C. Tsai, Appl. Phys. Lett., 23 (1973) 260.
102 F.G. Rudenauer, Vacuum, 22 (1972) 609.
103 C.A. Evans, Jr., Proc. Tutorial Session 8th Nat. Conf. Electron Probe Anal. Soc. Amer., New Orleans, La., August 1973, p. 29.
104 H. Liebl, Anal. Chem., 46 (1974) 22A.
105 GCA Corporation, GCA Technology Division, Bedford, Mass. U.S.A.
106 H. Nishimura and J. Okano, Jap. J. Appl. Phys., 8 (1969) 1335.
107 A. Benninghoven, Z. Phys., 230 (1970) 403.
108 A. Benninghoven, Surface Sci., 28 (1971) 541.
109 A. Benninghoven and E. Loebach, Rev. Sci. Instrum., 42 (1971) 49.
110 A. Benninghoven and E. Loebach, J. Radioanal. Chem., 12 (1972) 95.
111 R. Schubert and J.C. Tracy, Rev. Sci. Instrum., 44 (1973) 487.
112 Z. Šroubek, Rev. Sci. Instrum., 44 (1973) 1403.
113 K. Wittmaack, J. Maul and F. Schulz, Int. J. Mass Spectrom. Ion Phys., 11 (1973) 23.
114 J.V.P. Long, Brit. J. Appl. Phys., 16 (1965) 1277.
115 I.W. Drummond and J.V.P. Long, Proc. 1st Int. Conf. Ion Sources, Saclay INSTN, 1969, p. 459.

276

116 Applied Research Laboratories, P.O. Box 129, 9545 Wentworth Street, Sunland, Calif. 91040.
117 H. Liebl, Int. J. Mass Spectrom. Ion Phys., 6 (1971) 401.
118 Hitachi Ltd., Kokubunji, Tokyo, Japan.
119 H. Tamura, T. Kondo, H. Doi, I. Omura and S. Taya, Proc. Int. Conf. Mass Spectrosc., Kyoto, Japan, 1969 University of Tokyo Press, 1970, p. 205.
120 I. Omura, H. Tamura, T. Kondo and H. Doi, Proc. Int. Conf. Mass Spectrosc., Kyoto, Japan, 1969, University of Tokyo Press, 1970, p. 210.
121 H. Tamura, T. Kondo and H. Doi, Advan. Mass Spectrom., 5 (1971) 441.
122 AEI Scientific Apparatus Limited, Manchester, England.
123 J.M. Rouberol, J. Guernet, P. Deschamps, J.P. Dagnot and J.M. Guyon de la Berge, 16th Ann. Conf. Mass Spectrom. Allied Topics, Pittsburgh, Pa., May 12—17, 1968, p. 216.
124 CAMECA, Courbevoie (92) France.
125 R. Hernandez, P. Lanusse and G. Slodzian, C.R. Acad. Sci. Ser. B, 271 (1970) 1033.
126 R. Hernandez, P. Lanusse, G. Slodzian and G. Vidal, Method. Phys. Anal., 6 (1970) 411.
127 B.N. Colby and C.A. Evans, Jr., Appl. Spectrosc., 27 (1973) 274.
128 D.K. Bakale, B.N. Colby, C.A. Evans, Jr. and J.B. Woodhouse, Proc. 8th Nat. Conf. Electron Probe Anal. Soc. Amer., New Orleans, La., August 1973, p. 7A.
129 F.G. Ruedenauer, Int. J. Mass Spectrom. Ion Phys., 6 (1971) 309.
130 J.M. Morabito and R.K. Lewis, Anal. Chem., 45 (1973) 869.
131 J.A. McHugh, Pittsburgh Conf. Anal. Chem. Appl. Spectrosc., Cleveland, Ohio, March 5—9, 1973, paper no. 98.
132 G. Slodzian, Rev. Phys. Appl., 3 (1968) 360.
133 H. Liebl, Messtechnik, 80 (1972) 358.
134 J.A. McHugh and J.F. Stevens, Anal. Chem., 44 (1972) 2187.
135 C.A. Andersen, H.J. Roden and C.F. Robinson, J. Appl. Phys., 40 (1969) 3419.
136 J. Richards, R.L. Dalglish and J.C. Kelley, Radiat. Eff., 18 (1973) 207.
137 B. Blanchard, personal communication, 1973.
138 J.W. Coburn, Rev. Sci. Instrum., 41 (1970) 1219.
139 J.W. Coburn and E. Kay, Appl. Phys. Lett., 18 (1971) 435.
140 J.W. Coburn and E. Kay, Appl. Phys. Lett., 19 (1971) 350.
141 J.W. Coburn, E. Taglauer and E. Kay, J. Appl. Phys., 45 (1974) 1779.
142 H. Oechsner and W. Gerhard, Phys. Lett. A, 40 (1972) 211.
143 J.A. McHugh, 10th Nat. Meeting Soc. Appl. Spectrosc., St. Louis, Mo., Oct. 18—22, 1971, unpublished.
144 R. Kelley and H.M. Naguib, in D.W. Palmer, M.W. Thompson and P.D. Townsend (Eds.), Atomic Collision Phenomena in Solids, North-Holland, Amsterdam, 1970, p. 172.
145 M.L. Tarng and G.K. Wehner, J. Appl. Phys., 42 (1971) 2449.
146 S.D. Dahlgren and E.D. McClanahan, J. Appl. Phys., 43 (1972) 1514.
147 H.C. Feng and J.M. Chen, J. Phys. C, 7 (1974) L75.
148 G. Slodzian and J.F. Hennequin, C.R. Acad. Sci., Ser. B, 263 (1966) 1246.
149 M. Bernheim and G. Slodzian, Int. J. Mass Spectrom. Ion Phys., 12 (1973) 93.
150 B. Blanchard, N. Hilleret and J. Monnier, Mater. Res. Bull., 6 (1971) 1283.
151 B. Blanchard, N. Hilleret and J.B. Quoirin, J. Radioanal. Chem., 12 (1972) 85.
152 W.K. Hofker, H.W. Werner, D.P. Oosthoek and H.A.M. deGrefte, Radiat. Eff., 17 (1973) 83.
153 F. Schulz, K. Wittmaack and J. Maul, Radiat. Eff., 18 (1973) 211.
154 J.W. Guthrie and R.S. Blewer, Rev. Sci. Instrum., 43 (1972) 654.
155 R.S. Blewer and J.W. Guthrie, Surface Sci., 32 (1972) 743.

156 M. Croset, J. Radioanal. Chem., 12 (1972) 69.
157 J.A. McHugh, Radiat. Eff., 21 (1974) 209.
158 F. Brown and W.D. Mackintosh, J. Electrochem. Soc., 120 (1973) 1096.
159 H.L. Hughes, R.D. Baxter and B. Phillips, IEEE Trans. Nucl. Sci., NS-19 (1972) 256.
160 I.H. Wilson, Radiat. Eff., 18 (1973) 95.
161 J. Monnier, H. Hilleret, E. Ligeon and J.B. Quoirin, J. Radioanal. Chem., 12 (1972) 353.
162 A. Benninghoven, Chem. Phys. Lett., 6 (1970) 626.
163 V.I. Shvachko and Ya.M. Fogel', Kinet. Katal., 7 (1966) 722.
164 V.I. Shvachko and Ya.M. Fogel', Kinet. Katal., 7 (1966) 834.
165 V.I. Shvachko, B.T. Nadykto, Ya.M. Fogel', B.M. Vasyutinskiı and G.N. Kartmazov, Sov. Phys. Solid State, 7 (1966) 1572.
166 V.F. Rybalko, B.Ya. Kolot and Ya.M. Fogel', Sov. Phys. Solid State, 14 (1970) 1290.
167 A. Benninghoven, E. Loebach and N. Treitz, J. Vac. Sci. Technol., 9 (1972) 482.
168 S. Paletto, M. Perdrix, R. Goutte and G. Guilland, Surface Sci., 35 (1973) 473.
169 A. Benninghoven and A. Müller, Thin Solid Films, 12 (1972) 439.
170 A. Benninghoven and S. Storp, Appl. Phys. Lett., 22 (1973) 170.
171 B. McCarroll, J. Chem. Phys., 46 (1967) 863.
172 W.K. Huber and E. Löbach, Vacuum, 22 (1972) 605.
173 P.V. Fontana, J.P. Decosterd and L. Wegmann, J. Electrochem. Soc., 121 (1974) 146.
174 L. Habraken, V. Leroy and J.P. Servais, 5th Int. Conf. Electron Ion Beam Sci. Technol., The Electrochemical Society Inc., Princeton, N.J., 1972, p. 196.
175 J.W. Guthrie, J. Less-Common Metals, 30 (1973) 317.
176 J.M. Morabito, Anal. Chem., 46 (1974) 189.
177 H.W. Werner, H.A.M. deGrefte and J. Van Den Berg, Radiat. Eff., 18 (1973) 269. (1973) 269.
178 C.A. Evans, Jr. and J.P. Pemsler, Anal. Chem., 42 (1970) 1060.
179 J.M. Morabito and M.J. Rand, Proc. 8th Nat. Conf. Electron Probe Anal. Soc. Amer., New Orleans, La., August 1973, p. 8A.
180 R.E. Pawel, J.P. Pemsler and C.A. Evans, Jr., J. Electrochem. Soc., 119 (1972) 24.
181 C.A. Evans, Jr., personal communication, 1972.
182 P. Contamin and G. Slodzian, Appl. Phys. Lett., 13 (1968) 416.
183 B. Cox and J.P. Pemsler, J. Nucl. Mater., 28 (1968) 73.
184 D. Quataert and F. Coen-Porisini, J. Nucl. Mater., 36 (1970) 20.
185 J.R. Hinthorne and C.A. Andersen, Proc. 8th Nat. Conf. Electron Probe Anal. Soc. Amer., New Orleans, La., August 1973, p. 9A.
186 V. Cherepin, Advan. Mass Spectrom., 5 (1971) 448.
187 J. Maul, F. Schulz and K. Wittmaack, Phys. Lett. A, 41 (1972) 177.
188 W.K. Hofker, H.W. Werner, D.P. Oosthoek and H.A.M. deGrefte, Radiat. Eff., 17 (1973) 83.
189 R.P. Gittins, D.V. Morgan and G. Dearnaley, Brit. J. Phys. D, 5 (1972) 1654.
190 A.H. Kachare, W.G. Spitzer, A. Kahan, F.K. Euler and T.A. Whatley, J. Appl. Phys., 44 (1973) 4393.
191 R.D. Dobrott, F.N. Schwettmann and J.L. Prince, Proc. 8th Nat. Conf. Electron Probe Anal. Soc. Amer., New Orleans, La., August 1973, p. 10A.
192 A. Addamiano, G.W. Anderson, J. Comas, H.L. Hughes and W. Lucke, J. Electrochem. Soc., 119 (1972) 1355.
193 R. Castaing, Thesis, University of Paris, 1951; Publ. ONERA, No. 55.
194 H.W. Werner, Vacuum, 22 (1972) 613.
195 J.M. Rouberol, Ph. Basserville and J.P. Lenoir, J. Radioanal. Chem., 12 (1972) 59.
196 J.N. Coles and J.V.P. Long, Phil. Mag., 29 (1974) 457.

197 C.A. Andersen, J.R. Hinthorne and K. Fredrikson, Proc. Apollo 11 Lunar Sci. Conf. Vol. 1, Pergamon, New York, 1970, p. 159.

198 R.V. Gentry, S.S. Cristy, J.F. McLaughlin and J.A. McHugh, Nature, 244 (1973) 282.

199 C.A. Andersen and J.R. Hinthorne, Geochim. Cosmochim. Acta, 37 (1973) 745.

200 W.P. Poschenrieder, R.F. Herzog and A.E. Barrington, Geochim. Cosmochim. Acta, 29 (1965) 1193.

201 E. Gradsztajn, M. Salome, A. Yaniv and R. Bernas, Earth Planet. Sci. Lett., 3 (1968) 387.

202 J.W. Colby, Proc. 8th Nat. Conf. Electron Probe Anal. Soc. Amer., New Orleans, La., August 1973, p. 6A.

203 A.M. Huber and M. Moulin, J. Radioanal. Chem., 12 (1972) 75.

204 F. Yiou, M. Baril, J. Dufaure de Citres, P. Fontes, E. Gradsztajn and R. Bernas, Phys. Rev., 166 (1968) 968.

205 C.A. Evans, Jr., Anal. Chem., 42 (1970) 1130.

Chapter 7

THE USE OF AUGER ELECTRON SPECTROSCOPY AND SECONDARY ION MASS SPECTROMETRY IN THE MICROELECTRONIC TECHNOLOGY

J.M. MORABITO and R.K. LEWIS

I. Introduction

Fundamental concepts, the instrumentation used, modes of operation, and the general capabilities and limitations of both Auger electron spectroscopy (AES) and secondary ion mass spectrometry (SIMS) for surface and in-depth analysis have been discussed in previous chapters. Both analytical techniques have been demonstrated to be capable of localized, small selected volume, analysis from the material's surface into its bulk. The "ion image" and the "Auger image" provide chemical distribution maps of the impurities present on the surface of the sample and into its bulk. For in-depth analysis, energetic ion bombardment is used in both AES and SIMS to remove successive atomic layers while monitoring the emitted Auger electrons or secondary ions. Depth resolution is therefore fundamentally limited by the sputtering process when proper operating conditions are maintained, i.e. the absence of crater effects [1], redeposition [2], knock-on [3], etc.

This chapter will compare, by way of specific examples, the capabilities and limitations of the two techniques for surface analysis, depth profile analysis, and quantitative analysis on material systems of technological importance for the fabrication of hybrid integrated circuits. The analytical capabilities and limitations of both AES and SIMS are such that they are complementary techniques. The combined use of both on the same material system can often provide increased insight in the interpretation of both SIMS and AES data. There is, unfortunately, very little published data on the use of both AES and SIMS techniques on the same type of samples. This is because, in part, both are relatively new techniques and very few laboratories have both available. The authors are therefore limited in the number of examples which demonstrate the complementary nature of the two techniques.

II. Sample selection

A. TANTALUM THIN FILMS

There is considerable interest in both tantalum and silicon materials for use in the microelectronics industry. Tantalum nitride (Ta_xN_y) and tantalum oxynitride (TaO_xN_y) are important precision resistor materials [4] and can be conveniently trimmed to a desired resistor value by progressive anodization [5] or by laser trimming [6]. Tantalum thin films are also used to fabricate tantalum oxide capacitors [7].

The method of cathodic sputtering in an inert atmosphere, e.g. argon or in argon—reactive gas (N_2, O_2, CH_4) mixtures is used to deposit tantalum thin films. The latter is referred to as "reactive sputtering" or "doping." Since sputtering is used to deposit both the doped and undoped tantalum thin films, there is a possibility that contaminants in the sputtering chamber can be incorporated into the film and influence or control the electrical and structural properties of the deposited films. The effect of elements such as nitrogen, carbon, and oxygen on the electrical and structural properties of tantalum thin films (doped and undoped) has been the subject of extensive study [8—10]. Both AES and SIMS are capable of providing information on the amount and distribution of these light elements (N, C, O) in tantalum [11] which can establish the correlation between film composition and film properties necessary for the optimum in resistor and capacitor yield and reliability [12].

B. DOPED (B,P,As) SILICON

The importance of silicon in the fabrication of active electronic devices based on semiconductor properties has been well established.

Various combinations of thin metal films are used for making electrical contact to the silicon and for fabricating metallic conductor patterns. The possibility of interdiffusion and/or the formation of intermetallics during the deposition or subsequent high temperature processing of these multilayer thin films makes localized analysis of the surface, bulk, film—film, film—silicon interface absolutely necessary for a comprehensive chemical analysis. In addition, the amount and distribution of dopants such as boron, phosphorus, arsenic, etc., introduced into silicon play a major role in controlling the characteristics of active devices [13]. These dopants can be conveniently introduced into silicon from a chemical source [14] or by ion implantation [15], but the concentration of these dopants are typically below detection by conventional techniques such as the electron microprobe. Electrical methods [16,17] of obtaining the distribution of the dopants reveal only that fraction of dopant which is electrically active. Electrical profiling methods are also extremely tedious.

Although calculated profiles [18] for silicon implanted with boron or arsenic can be used for an estimate of the implantation and annealing conditions needed to satisfy a given set of transistor specifications, the calculations do not take into account changes in the profiles that might arise when dopant distributions overlap. For example, substantial perturbations of boron base-profiles are known to occur with the introduction of arsenic or phosphorus emitters from chemical sources [19]. Perturbations on base profiles are also suspected to occur when doping is done by ion implantation. Without knowledge of interaction effects that occur with overlapping dopant distributions, the accuracy of the calculated profiles is uncertain. Since high frequency transistors are highly sensitive to the base width and to the shape of the base impurity distribution, a precise characterization of the profiles for the overlapping distributions in these structures would contribute to a more detailed understanding of their electrical behavior.

Dopant profiling in silicon is possible by AES and SIMS, but the more sensitive SIMS technique has been found to be more useful for the profiling of dopants in the concentration range of highest interest for device fabrication (10^{19}—10^{17} atoms/cm^3) [20].

III. Selection of inert or reactive primary ion bombardment in AES and SIMS profiling

Inert argon or xenon are used in AES profiling. Xenon has the advantage of providing higher sputtering rates and must be used when the distribution of argon in a sample is of interest or when profiling boron in silicon since the boron transition at 179 eV interferes with one of the argon triplet peaks at 180 eV. Xenon is not a good choice when profiling Cr, O or I due to spectral interference. Primary ion energies in the range of 1—2 keV are typically used with maximum current densities of 250 μA/cm^2. The sputtering rates used in AES profiling are usually in the range of 50—100 Å/min depending on film thickness.

The primary ions most frequently used in SIMS are oxygen, argon, and nitrogen. Oxygen ions are used most often since they produce the highest secondary ion yield for the majority of elements, which results in the highest detection sensitivity. Both $^{16}O_2^+$ and $^{16}O^-$ ions can be extracted from the ion source. The advantage of using $^{16}O_2^+$ primary ions is that higher sputtering rates are possible, while the use of $^{16}O^-$ reduces charge buildup when analyzing insulators.

The use of reactive ion bombardment has the disadvantage of producing more interfering molecular ion species from the matrix. This is readily apparent in Table 1 which shows the molecular ions produced from oxygen bombardment of GaP. Argon, on the other hand, does not produce the M_xO_y species but does produce more intense polyatomic M_x species as shown in Table 1. The principal use for argon bombardment is in analysis for oxygen.

References pp. 326—328

TABLE 1

Relative ion intensities for GaP mass spectra*

Mass	Ion	I_p		Mass	Ion	I_p	
		O_2^+	Ar^+			O_2^+	Ar^+
10.3	P^{3+}		190	116	GaPO	9500	
15.5	P^{2+}		2600	118	GaPO	6400	
16	O	390		124	P_4	45	27
23	Na	63		131	GaP_2	7100	5000
27	Al	2300	43	133	GaP_2	4500	3400
31	P*	10,000	21,000	138	Ga_2	6700	7400
32	O_2	93		140	Ga_2	9000	9200
39	K	140		142	Ga_2	2800	3600
40	Ar		4100	147	GaP_2O	57	
47	PO	6400		149	GaP_2O	36	
52	Cr	70		154	Ga_2O	5900	
62	P_2	2900	2500	155	Ga_2OH	640	270
69	Ga	5.0×10^6	2.4×10^7	156	Ga_2O	7700	
71	Ga	3.3×10^6	1.6×10^7	157	Ga_2OH	750	340
78	P_2O	66		158	Ga_2O	2700	
80	Ar_2		30	159	Ga_2OH	240	120
85	GaO	840		162	GaP_3	410	210
87	GaO	570		164	GaP_3	250	140
93	P_3	1300	1200	169	Ga_2P	9100	8200
96.5	GaP_4^{2+}	75		171	Ga_2P	12,000	11,000
98.5	GaP_4^{2+}	52		173	Ga_2P	4100	3600
100	GaP	8600	110	185	Ga_2PO	2700	
102	GaP	5600	100	187	Ga_2PO	2600	
109	GaAr		160	189	Ga_2PO	1200	
111	GaAr		100				

*All intensities normalized to ^{31}P with oxygen bombardment. Intensities are also compensated for sputtering rate difference $\dot{Z}(O_2)$ 140 Å/sec, $\dot{Z}(Ar)$ 110 Å/sec.

Using oxygen primary ion bombardment does not preclude the analysis of oxygen because it is possible to take advantage of differences in emission energy from ions originating from the primary beam and ions originating from the sample. However, argon bombardment must be used to achieve high detection sensitivity for oxygen. In addition, inclusions present in a sample as oxides can be ion imaged with considerably more contrast with argon bombardment making them easily identifiable as oxides. It is also possible to produce higher detection sensitivity for certain reactive elements. For example, carbon has its highest ion yield as C_2^- and the highest yield for this species occurs with argon bombardment. The primary ions used in the SIMS and AES profiles discussed in this chapter are indicated in the figure captions.

IV. Sputtering rate measurements and depth resolution in AES and SIMS profiling

Since sputtering is used for both AES and SIMS depth profiling, accurate sputtering rate measurements are necessary to relate sputtering time to the depth sputtered. This rate is usually determined by measuring the crater depth and total time elapsed during sputtering. The depth measurement can be easily made on well polished large single crystal specimens such as Ge, Si, and the III—V semiconductors such as GaP, GaAs and most evaporated or sputtered thin films on smooth substrates. Light interferometers [21] or a mechanical stylus device such as a Talystep [22] are typically used for this measurement. The accuracy of the depth measurement with a Leitz microscope equipped with a Michaelson interferometer is about one-fifth of a fringe ($\lambda/2$ = 2700 Å) corresponding to ±300 Å. The accuracy of multiple beam interferometry [23] is <50 Å. The accuracy of the Talystep is in the range of 100 Å. When polycrystalline specimens are being analyzed, the crater bottom can often be uneven when sputtering to depths over several thousand angstroms making the light interference method less suitable. The Talystep method was used on these samples, but the error with the Talystep is higher than the light interference methods.

When it is not possible to measure the depth of a crater, it is necessary to have an accurate measurement of the beam density and knowledge of the sputter yield of the matrix sputtered to obtain the sputtering rate. Data on sputtering yields are available in the literature. However, a large fraction of the bombarded area will be at a lower and varying beam density than the selected area analyzed; consequently, it is difficult to derive a precise beam density measurement from primary current measurements. When this problem is encountered, an excellent technique for determining beam density is to sputter into a reference material. Transparent amorphous materials with a high light refractive index such as Ta_2O_5 or SnO_2 which develop interference colors in white light are excellent reference materials. The integral light optics available on the Auger [24] or SIMS instruments [25] can be used to observe the specimen while sputtering. The time when a specific color fringe develops and changes to another color fringe can then be recorded. The uniformity of the beam density over a given area can also be established with these color fringes. This technique compares favorably with the technique described by Guthrie and Blewer [26] and has the added advantage of providing depth information. The depth scales indicated on the figures of this chapter have been obtained by either the light interference or the Talystep methods.

Accurate depth profile measurements in both AES and SIMS depend on obtaining a flat bottomed crater and restricting the secondary ions or Auger electrons detected to this flat region, which avoids crater wall effects. Under proper operating conditions, the depth resolution possible with AES or SIMS on amorphous samples such as Ta_2O_5 using sputtering has been shown to be 5—10% of the thickness removed [27]. Any variation in sputtering rate, preferential sputtering effects, crystal orientation effects, etc. can lead to non-uniform crater bottoms and hence degradation in depth resolution. The depth resolution possible on polycrystalline samples, where there is a possibility of different sputtering rates at the grain boundaries, has not been established but is believed to be <30% of the thickness removed [28].

V. Chemical analysis of sputtered tantalum thin films by AES and SIMS

The Auger spectrum obtained prior to in situ argon ion sputtering of a Ta film sputtered at 3.8 kV and 420 mA on a glass substrate with a Ta_2O_5 underlay is shown in Fig. 1. Sulfur, tantalum, carbon, nitrogen, oxygen, iron, nickel, and copper were detected on the surface of this film. The carbon peak shape is similar to that of graphite [29] and indicative of carbonaceous surface contamination. Sulfur is a ubiquitous contaminant found on most metallic surfaces. The oxygen is due to the natural surface oxide present on tantalum thin films, and nitrogen either to adsorbed nitrogen or to nitrogen incorporated into the film during sputtering. The metallic impurities, iron and nickel, are due to the presence of stainless steel cathode support fixtures in the sputtering chamber. The presence of copper could result from impurities in the glass (Corning 7059) substrate. Fragments of the substrate are almost always present in the sputtering chamber due to accidental substrate fracturing.

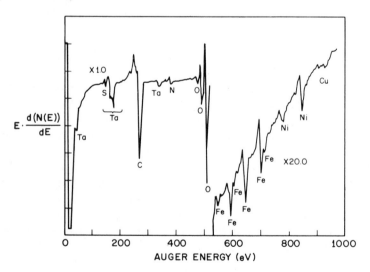

Fig. 1. Auger spectrum prior to in situ ion sputtering with argon of a β-Ta film (4000 Å) sputtered at 3.8 kV and 420 mA on a glass substrate with a Ta_2O_5 underlay (500 Å).

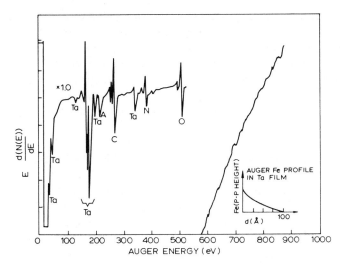

Fig. 2. Auger spectrum of β-Ta film (see Fig. 1) after removing ~150 Å by in situ ion sputtering. Insert, in-depth profile of iron within the first ~100 Å.

The Auger spectrum of the same film after removing ~150 Å of the film by in situ ion sputtering is shown in Fig. 2. The arrangement used to simultaneously ion sputter and Auger analyze the film is shown in Fig. 8 of Chapter 5. A low sputtering rate (~2 Å/min) was used to obtain the distribution of the surface impurities within the first 100—150 Å of the film. The sputtering rate in this case was calculated by removing material for a fixed period of time and measuring the crater depth with a Talystep [22]. The S, Fe, and Ni peaks were reduced after sputtering to a level below the detectability limits of the Auger technique (~0.1—0.5 at%). The carbon peak (Fig. 2) has changed shape to that characteristic of a carbide [29] rather than surface carbon (see Fig. 1). The removal of the surface carbon also resulted in an increase in both the amplitude and resolution of the low energy tantalum Auger peaks (Fig. 2). The nitrogen Auger peak also increased after ion sputtering, but the oxygen transition decreased. The reduced nitrogen peak height on the surface of tantalum prior to ion sputtering is due to the attentuation of the nitrogen Auger electrons by surface carbon and is not a true indication of the amount of nitrogen at the film surface. Although the presence of surface carbon attenuates the nitrogen peak, it does not interfere with or complicate the identification of nitrogen. The argon peak is due, in part, to the use of argon for in situ ion sputtering. A small fraction of these argon ions is implanted into the film as a result of the in situ ion sputtering, but argon is also introduced into the film during film deposition.

The SIMS spectrum obtained with primary argon ($^{40}A^+$) ions on an identical tantalum film is shown in Fig. 3. The impurities present in the mass

References pp. 326—328

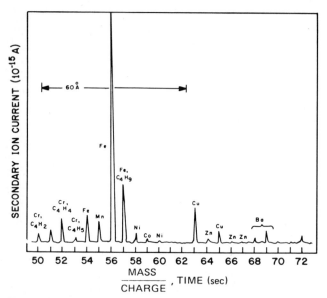

Fig. 3. Secondary ion mass spectrum of β-Ta film (see Fig. 1) in the mass range 50—72 recorded while removing 100 Å from the surface by argon sputtering.

range between 50 and 72 mass units on the surface (within the first 100 Å of the tantalum film) are Cr, Fe, Mn, Ni, Cu, Zn, Ba. Iron, nickel, and copper were also detected with AES on the surface of this film. Chromium (529 eV) and manganese (589 eV) were not within the detectability limits (~0.1 at.%) of AES. The barium present in the SIMS spectrum is a major constituent of the 7059 glass substrate. Trace amounts of copper and zinc are also present in 7059 glass which could explain the presence of these impurities on the surface of the film. The isotope ratios of Cr and Fe within the first 40—60 Å of the film are incorrect due to hydrocarbon interference effects [30]. The correct isotopic abundance of chromium (4.34, 83.8, 9.50) was most probably not obtained because of mass interferences from hydrocarbons (e.g. $C_4H_2^+$, $C_4H_4^+$, $C_4H_5^+$). Recent data obtained using high mass resolution ($M/\Delta m = 1000$) has confirmed the presence of hydrocarbons on an aluminum surface. The hydrocarbon ($C_2H_3^-$) interference with $^{27}Al^-$ rapidly decreased after sputtering ~50 Å into the aluminum foil as shown in Fig. 4. Hernandez et al. [31] have been able to identify nineteen hydrocarbon lines in the spectrum obtained on an aluminum surface using a MS7 mass spectrograph modified with a sputter ion source. They found that C_2^+, C_2H^+ and $C_2H_2^+$ interfered with the 24, 25, 26 isotopes of Mg. The silicon isotopes (28, 29, 30) had interference from CO^+, $C_2H_4^+$, $C_2H_5^+$ and $C_2H_6^+$. Hydrocarbons are present in varying amounts on the surface of most samples depending on the nature of the sample and of sample preparation prior to analysis. In the case of sputtered tantalum films, hydrocarbon interference effects disap-

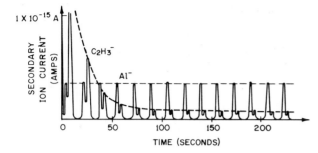

Fig. 4. Hydrocarbon interference ($^{27}C_2H_3^-$) with $^{27}Al^-$ for an aluminum foil using argon ions at 14.5 keV.

peared and did not return after the removal of ~40 Å from the tantalum film. The correct isotope ratio was obtained for copper, as shown in Fig. 3, after removing approximately ten monolayers (40 Å) from the tantalum film. However, hydrocarbon interference effects with $^{63}Cu^+$ and $^{65}Cu^+$ from $C_5H_3^+$ and $C_5H_5^+$ were a problem at the surface (< 40 Å).

Hydrocarbon interference effects definitely complicate quantitative and qualitative SIMS surface analysis. Therefore, the use of high mass resolution will often be essential if one wishes to perform surface analysis with SIMS over the first few hundred angstroms. High mass resolution is only achieved in SIMS with a decrease in detection sensitivity; however, the detection sensitivity available with SIMS is often more than is necessary for the detection of surface impurities. The mass resolutions necessary to resolve the common hydrocarbons from silicon, chromium, iron, and copper are listed in Table 2.

The AES in-depth profile of Fe within the first 150 Å is inserted into Fig. 2. A similar profile shape was obtained for carbon, oxygen, and nickel.

TABLE 2

Mass resolution necessary to resolve the common hydrocarbons from silicon, chromium, iron, and copper

Mass	Doublet	$M/\Delta m$
28	$C_2H_4 - {}^{28}Si$	524
52	$C_4H_4 - {}^{52}Cr$	572
54	$C_4H_6 - {}^{54}Fe$	499
56	$C_4H_8 - {}^{56}Fe$	438
63	$C_5H_3 - {}^{63}Cu$	670

References pp. 326—328

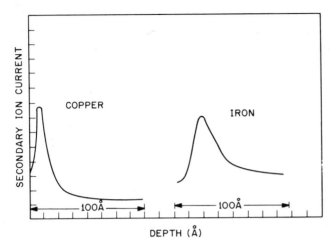

Fig. 5. SIMS depth profiles for iron and copper in β-Ta within $\sim$1000 Å of the surface using $^{16}O_2^+$ at 4.5 keV.

The SIMS depth profiles for iron and copper, shown in Fig. 5, were obtained using $^{16}O_2^+$ ions. The AES profile indicates a decrease in iron from the surface, but the SIMS profile shows an increase. The shape of the SIMS profiles within the first tens of angstroms is believed to be in error due to a complicated superposition of hydrocarbon interference effects [30], enhanced secondary ion yield of Fe^+ and Cu^+ attributed to the presence of surface oxide (chemical emission) [32] and the implantation range effect of primary oxygen ions [33]. Hydrocarbon interference and enhanced emission effects at the surface prevented the establishment of meaningful secondary ion intensity vs. surface concentration relationships within the first few hundred angstroms of the surface. Surface oxide and carbon also prevented the establishment of Auger electron intensity vs. concentration relationships for the low energy nitrogen (380 eV) Auger electrons.

Further material removal ($\sim$1000 Å) by in situ ion sputtering did not change the number of Auger transitions shown in Fig. 2, but the carbon and oxygen Auger peak heights were reduced. The sulfur, iron, nickel, and copper transitions did not reappear. The concentration of these metallic impurities in the bulk of the film is therefore below $\sim$0.1 at.% and/or non-homogeneously distributed throughout the film.

The SIMS spectrum obtained after sputtering through $\sim$1000 Å of the film is shown in Fig. 6. The major impurities found in the bulk of the film are Mg, Al, and Si, along with Ti, Cr, Co, Nb, Ba, B, K, Ca, Na, N, C, O, and As. The Al, Si, B, Ba, and As are major constituents of the glass substrate. Nb, Ca, Na, Mg are trace impurities in the tantalum cathode. The nonmetallic impurities, N, C, and O were also detected in the bulk of the film by Auger measurements (see Fig. 2). The oxygen in the film is most

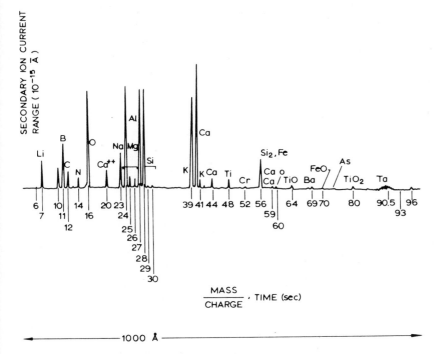

Fig. 6. Secondary ion mass spectrum of β-Ta (see Fig. 1) while removing $\sim$1000 Å of the film by $^{16}O_2^+$ bombardment. The mass range shown is 6—96 mass units. All species were singly charged except where indicated.

probably due to the presence of water vapor in the sputtering chamber, but oxygen from the ambient can also react and diffuse into the film after deposition. An oil diffusion pumped chamber was used to deposit the film which could explain the presence of carbon in the bulk of the film. The presence of low levels of nitrogen in the film suggests a small air leak. The rather large peak with a mass to charge ratio of 56 in the bulk of the film could be due to Fe^+ or Si_2^+. In order to resolve this difficulty, a secondary ion mass spectrum of 7059 glass was taken; the calculated Si^+ to Si_2^+ ratio was determined to be $\sim$160 but the measured ratio in Fig. 6 is 3.1. This peak is then a combination of $^{56}Fe^+$ and Si_2^+, but is predominantly due to iron. This also explains the peak at mass 70 which is $^{54}Fe^{16}O$.

Ion images (see Chapter 5) of the Al, Si, and Mg impurities in the tantalum film showed that these contaminants are present in particulate form. These impurities are due to the presence of glass chips on the 7059 substrate prior to formation of the Ta_2O_5 underlay or tantalum thin film deposition. Any mechanical abrasion of the slide can lead to the presence of these glass chips on the slides. Some of the particles can be blown off but smaller ones remain behind. Rubbing does not remove the smaller particles which in some cases project through the tantalum oxide underlay and sputtered film.

Glass chips on the substrate can result in a loosely held or non-adherent film and a film surface which follows the contours of these defects which have an average diameter of 300 Å. The removal of a non-adherent film by sputtering should expose the surface of the glass projections which explains the presence of Ba within the first 100 Å of the film (Fig. 3) by SIMS analysis.

VI. Quantitative analysis of sputtered tantalum films intentionally doped with nitrogen, carbon, and oxygen by AES and SIMS

Both the Auger and SIMS analysis detected the presence of nitrogen, carbon, and oxygen homogeneously distributed throughout the sputtered Ta films. These results suggested that the intentional and controlled doping of tantalum with nitrogen, carbon, and oxygen could be a convenient means of preparing homogeneous standards for quantitative AES and SIMS measurements.

The intentional doping or "reactive sputtering" was accomplished by adding a varying, but well-controlled, partial pressure of a single reactant or reactant mixture (N_2, O_2, CH_4) to an inert gas such as argon during sputtering from a pure metal cathode. The sputtering parameters used to deposit the doped films were 5 kV and 200 mA (single phase, unfiltered d.c. power supply) with a 9.5 cm cathode-to-anode spacing [12]. Methane—argon mixtures were used to prepare carbon-doped films [8]. The nitrogen [9] and oxygen [10] doped films were prepared by introducing small amounts of nitrogen or oxygen into the sputtering chamber. All the films were deposited on Corning 7059 glass substrates coated with a Ta_2O_5 etch stop layer. The films were optically smooth and flat, which minimized surface roughness effects on the Auger and secondary ion signal intensity and made accurate sputtering rate measurements possible. The depth profiles obtained by both AES and SIMS confirmed that the N, C, and O dopants were homogeneously distributed throughout the films, and it was possible to measure accurately (±10%) the amount of these dopants in the films by the electron microprobe. This analysis was then used to calibrate the AES and SIMS measurements for quantitative analysis. Details of sample preparation, instrument settings (accelerating voltage) etc., necessary for the quantitative analysis of light elements such as nitrogen, carbon, and oxygen in tantalum by the electron microprobe have been discussed [11,34].

Prior to the development of simultaneous ion sputtering—Auger analysis, quantitative AES by calibration methods had been restricted to submonolayer, uniformly distributed surface deposits prepared and analyzed under ideal vacuum conditions as first described by Weber and Johnson [35] and more recently by Meyer and Vrakking [36]. The in situ ion sputtering removes contamination such as surface carbon and surface oxides which can attenuate the intensity of low energy ($\leqslant$500 eV) Auger electrons and prevents the recontamination of the surface by contaminants present in the vacuum chamber such as water vapor and hydrocarbons. There is, however, a surface compositional change due to in situ ion sputtering. The magnitude of this compositional change (neglecting diffusion effects) for a binary system AB will be a function of the bulk concentration C ratio and the sputtering yield S ratio of the two components. For $S_A < S_B$, after steady state has been established, the ratio of the surface concentration of the two components $C_S^A / C_S^B = (S_B / S_A)(C^A / C^B)$ where C^A / C^B is the bulk concentration ratio. Since the Auger peak height I is a measure of surface concentration [35], $C_S^A / C_S^B = KI^A / I^B$ where K is a constant related to the Auger and sputtering yields of the two components. Therefore, for the Ta—N system, $C_S^N / C_S^{Ta} = KI^N / I^{Ta}$ or the magnitude of the peak height ratio I^N / I^{Ta} measured while sputtering is a relative measure of the actual bulk composition of the film which for a homogeneous sample can be accurately measured by techniques such as the electron microprobe [37], high energy ion backscattering [38] or nuclear reactions [39].

The *actual* steady state surface concentration ratio C_S^N / C^{Ta} could be calculated if the Auger and sputtering yields of both tantalum and nitrogen were accurately known. Calibration of Auger data for quantitative analysis is therefore possible for homogeneous samples of known composition by maintaining the same in situ ion sputtering conditions on samples with increasing amounts of a given dopant. The data [34] generated for nitrogen, carbon, and oxygen are shown in Fig. 7 where the log of concentration as determined by electron microprobe measurements is plotted versus the log of the

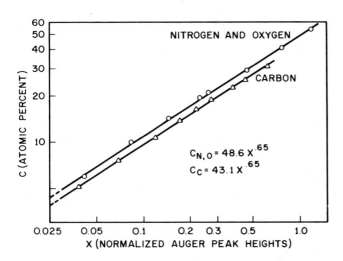

Fig. 7. Log C (at.%) determined by the electron microprobe vs. log of normalized N, O, C Auger peak height (X) in reactively sputtered tantalum thin films.

References pp. 326—328

normalized Auger peak heights, N/Ta, O/Ta, C/Ta. The non-linearity in the relationship is most probably due to changes in the escape depth d_e [40] and backscattering correction r [41] with increasing dopant concentration. X-ray analysis of the film also showed a change in structure with increasing dopant concentration. Transitions from the β-Ta structure to a mixture of β-Ta and b.c.c.-Ta, and finally to a single phase b.c.c. (bulk) region followed by oxide, nitride or carbide formation have been observed [12]. There is also a small change in the density (g/cm^3) of the films with increasing nitrogen, carbon, and oxygen concentration.

The nitrogen and oxygen data could be expressed by $C_{N,O} = 48.6X^{0.65}$ and the carbon data by $C_C = 43.1X^{0.65}$ where the concentration C is in at.% and X is the normalized Auger peak height (e.g. N/Ta). The above equations do not represent the data below 5 at.% [11].

Normalizing the data to the matrix (Ta) eliminated the effects of inadvertent changes in beam current, electron multiplier gain, etc. The similar magnitudes of the data for nitrogen, carbon, and oxygen is consistent with the measurements of Gerlach and Ducharme [42], and the similarity in the nitrogen and oxygen data suggests that the Auger yield and sputtering yield of nitrogen and oxygen must be the same from a tantalum matrix. This implies that the Auger peak height ratio of oxygen and nitrogen is a measure of the O/N ratio in the film. A comparison of the O/N ratio in TaO_xN_y films determined by electron microprobe analysis and from the oxygen-to-nitrogen Auger peak height ratio is made in Table 3. The observed agreement is quite good.

Although Auger peak heights measured while sputtering provide only a relative measure of the actual bulk concentration, SIMS provides a more direct measure of the bulk concentration since the fraction of ionized particles ejected from the sample while sputtering are detected. Andersen and Hinthorne [43] have described a quantitative model based on a thermal equilibrium model of the ion emission process and on an internal standard

TABLE 3

Comparison of the O/N ratio determined by electron microprobe analysis and the oxygen to nitrogen ratio determined by Auger analysis

Sample	Electron microprobe	Auger
1	0.50	0.53
2	1.16	1.31
3	1.08	1.12
4	0.425	0.43
5	0.620	0.53
6	0.560	0.51

(e.g. composition of matrix), but the general applicability of this model has not been demonstrated. The limitations of any generalized model due to the presence of chemical and orientational effects on secondary ion emission have been discussed by Castaing [44]. Calibration of SIMS data using homogeneous samples of known and accurate composition is at present the most reliable method for quantitative SIMS analysis provided such standards are available or easily prepared [30,45—47].

Morabito and Lewis [30] have recently described a method of quantitative analysis based on the use of standards and the measurement of parameters such as the secondary ion current of the dopant (i_{a_i}) and the secondary ion yield of the dopant relative to the matrix (K_{rel}). The equation which relates the concentration of a dopant in a given matrix to these parameters is

$$C = \frac{i_{a_i} 10^6 (100/a_i)S}{\eta K_{rel} K_m \, \sigma \dot{z} A (1.6 \times 10^{-19} C^\circ/\text{ion})} \tag{1}$$

where C is the concentration (ppm atomic), i_{a_i} the secondary ion current of dopant (amps), a_i the atomic abundance of dopant, S the sputtering yield of matrix (sputtered atoms/incident ion), K_m the secondary ion yield of matrix, σ the surface atom density of matrix (atoms/cm^2 monolayer), $\dot{z}$ the sputtering rate (Å/sec), A the area of analysis (μm^2), K_{rel} the secondary ion yield of dopant relative to the matrix, C° the number of coulombs and η the ion collection efficiency $\times$ transmission (%). The most important parameter in eqn. (1) is K_{rel}, which can be measured by the use of standards.

One can also show that

$$\frac{C_m}{10^6} \frac{S}{K_m \eta \sigma \dot{z} A (1.6 \times 10^{-9} C^\circ/\text{ion})} = \frac{1}{i_m (100/a_m)} \tag{2}$$

where C_m is the composition of the matrix, a_m the isotopic abundance of the matrix, and i_m the secondary ion current detected from the matrix.

Equation (1) can then be rewritten in the reduced form

$$C = \frac{i_{a_i}(100/a_i)C_m}{i_m K_{rel}(100/a_m)} \tag{3}$$

Normalizing the measured dopant secondary ion current i_{a_i} to the measured secondary ion current of the matrix i_m eliminates the untractable effects of changes in sputtering rate $\dot{z}$, sputtering yield S and surface atom density σ during a measurement or series of measurements. Normalization is of paramount importance for in-depth analysis since profile shape is particularly affected by changes in the secondary ion yield of the dopant and matrix. The observed changes in ion yields are often due to the presence of impurities such as oxygen [32,45] which enhances the yield of both the matrix and dopant.

The calibration of SIMS for N, C, and O quantitative analysis in tantalum is complicated by the change in crystallographic structure of the tantalum films due to the incorporation of these elements into the tantalum which occurs at concentration above 5 at.%. These reactive gases are accommodated into the sputtered tantalum lattice to form interstitial (random or ordered) solid solutions. The coordination number of these elements will be different in the β tetragonal structure [4] from the b.c.c. structure, which could affect the secondary ion yield of these elements. This matrix or coordination effect is shown for carbon in Fig. 8. The abrupt change in slope observed for the normalized secondary ion current of $^{12}C^+$ is thought to be due to a change in the secondary ion yield of carbon due to the crystallographic transitions (β-Ta $\rightarrow$ β-Ta + b.c.c.-Ta $\rightarrow$ b.c.c.-Ta) observed by X-ray analysis. This abrupt change also suggests a change in the mechanism of the secondary ion emission of $^{12}C^+$ from tantalum with an increase in the amount of carbon in the film. The possibility of chemical and matrix effects on ion yields make the calibration of SIMS data more restrictive than that of the AES calibration. The carbon calibration curve shown in Fig. 8 can only be used for tantalum—carbon films which are known to contain only small amounts of oxygen, since the presence of oxygen could change the secondary ion yield of the carbon and the tantalum. The presence of oxygen in the film is easily detected by the use of $^{40}Ar^+$ or $^{14}N_2^+$ primary ion bombardment prior to analysis for carbon. The calibration curves obtained for nitrogen and oxygen are shown in Figs. 9 and 10. Secondary ions of NO$^-$ were

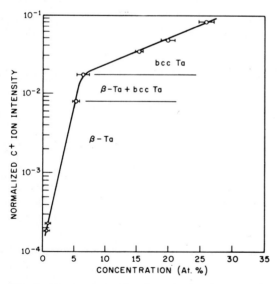

Fig. 8. Normalized C$^+$ ion intensity vs. concentration of carbon (at.%) in tantalum thin films (~1000 Å) reactively sputtered in (Ar + CH$_4$) mixtures. $^{16}O_2^+$ primary ions at 4.5 keV.

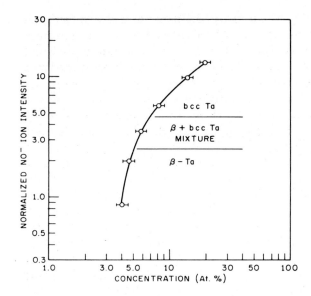

Fig. 9. Normalized NO⁻ ion intensity vs. concentration of nitrogen (at.%) in tantalum thin films (~1000 Å) reactively sputtered in mixtures. The primary ions were $^{16}O_2^+$ at 14.5 keV.

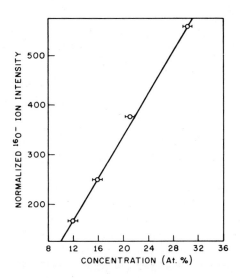

Fig. 10. Normalized $^{16}O^-$ ion intensity vs. concentration of oxygen (at.%) in tantalum thin films (~1000 Å) reactively sputtered in (Ar + O₂) mixtures. The primary ions were $^{40}Ar^+$ at 14.5 keV.

References pp. 326—328

monitored for nitrogen and $^{16}O^-$ for oxygen since the yield of NO^- is greater than N^+ and the yield of $^{16}O^-$ greater than $^{16}O^+$. Both signals were normalized to $^{181}Ta^-$. It was not possible to obtain oxygen data below ~13 at.% due to a lack of homogeneous samples in this concentration range.

In all three cases (Figs. 8—10) the normalized data were approximately a linear function of concentration above 13 at.%. This linear dependence demonstrates that the secondary ion measurement can be quantitative with one calibration point. The reproducibility of these results was better than 5%. Mass interference effects for $^{12}C^+$, NO^- and $^{16}O^-$ in a tantalum matrix were not a problem.

VII. Analysis of P-doped Ta$_2$O$_5$ films by AES and SIMS

Phosphorus-doped Ta_2O_5 samples have been used to compare the depth profiling capabilities of AES and SIMS [48]. The location of the P-rich zone can be controlled and varied by the anodization process [49], and the anodic oxide—metal interface is less diffuse than that of the thermal oxide—metal interface making estimates of depth resolution possible. The Auger spectrum of phosphorus-doped Ta_2O_5 prior to in situ ion sputtering is shown in Fig. 11. The phosphorus 120 eV Auger peak indicates that phosphorus is present at the surface of the anodic film. The in-depth profile of phosphorus, tantalum, and oxygen is shown in Fig. 12. The phosphorus, tantalum, and oxygen Auger transitions were monitored simultaneously by the use of a multiplexer [24]. The phosphorus rich zone was detected after sputtering ~500 Å into the anodic oxide film (~1000 Å in thickness) with a FWHM of 75 Å. The Ta_2O_5—Ta interface is not completely sharp, but is more indicative of a graded or oxygen-deficient region. A sharp interface between stoichiometric Ta_2O_5 and the metal is most probably thermodynamically unstable with the driving force (free energy) towards an oxygen deficiency, hence semiconduction in the oxide. Smyth et al. [50] have discussed the significance of the Ta_2O_5—Ta interface with regard to the conductivity profile of heat treated and unheat-treated anodic Ta_2O_5 films. The increase in the tantalum Auger peak at the oxide—metal interface is larger than that expected from the increase in density of tantalum atoms in the metal. This is due to changes in parameters such as escape depths d_e and back scattering factors r.

The SIMS in-depth profile of oxygen and phosphorus obtained on an identical phosphorus doped anodic film is shown in Fig. 13. The SIMS profile also indicates that phosphorus is present at the surface. The phosphorus-rich zone is located at 480 Å into the anodic oxide film with an FWHM of 98 Å. It is also quite clear from the oxygen profile that the secondary ion yield of oxygen has been affected by the presence of phosphorus and is highest at the maximum of the phosphorus ($^{31}P^+$) secondary ion intensity

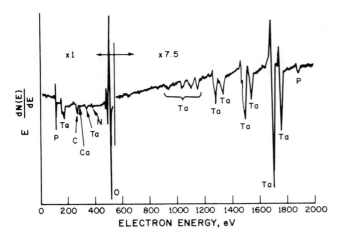

Fig. 11. Auger spectrum of phosphorus-doped Ta_2O_5 (1000 Å) prior to in situ ion sputtering.

(see Fig. 13). This gradient in the $^{16}O^+$ ion intensity is most probably associated with the presence of phosphorus in the anodic film. It is not due to a gradient in the oxygen concentration since the Auger analysis of an identical sample (see Fig. 12) and of a phosphorus-doped sample of twice the oxide thickness indicated a constant oxygen Auger peak height (hence concentration) in the bulk of the film. The SIMS analysis of this thicker (~2000 Å) film also showed a gradient in the oxygen ion intensity. Although

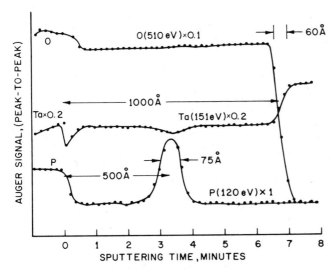

Fig. 12. In-depth profile of phosphorus, tantalum, and oxygen in phosphorus-doped Ta_2O_5 film (see Fig. 11). The primary ions were argon.

References pp. 326—328

298

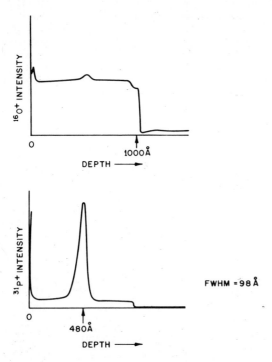

Fig. 13. SIMS profile of phosphorus and oxygen in phosphorus-doped Ta_2O_5 film (see Fig. 12) using primary ions of $^{16}O_2^+$ at 4.5 keV.

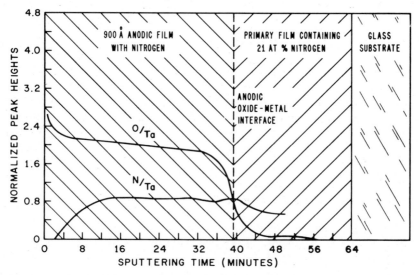

Fig. 14. Nitrogen distribution, determined by Auger spectroscopy using argon, in anodized Ta film formed by anodic oxidation of Ta—N alloy films.

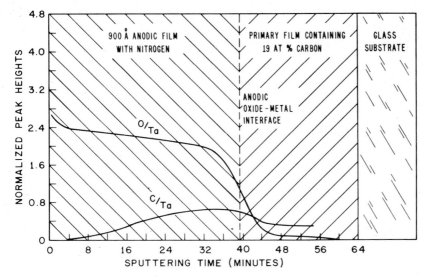

Fig. 15. Carbon distribution, determined by Auger spectroscopy using argon ions, in anodized Ta films formed by anodic oxidation of Ta—C alloy films.

the Auger oxygen peak height was not affected by the presence of phosphorus, the Auger tantalum peak height was affected as shown in Fig. 12. The change in the tantalum Auger peak due to the presence of phosphorus is suggestive of a direct substitution for tantalum by phosphorus on random sites or possibly to the formation of isolated regions (islands) of P_2O_5 in a Ta_2O_5 matrix. A local change of density due to the incorporation of phosphorus is also a possibility. The constant oxygen Auger peak height (Fig. 12) is consistent with a +5 valency for phosphorus which maintains the M_2O_5 stoichiometry.

It is also possible to profile light elements such as nitrogen and carbon incorporated into oxides anodized from nitrogen or carbon doped tantalum films. The results obtained using AES are shown in Figs. 14 and 15. The nitrogen and carbon Auger peaks have been normalized to the tantalum (matrix) Auger peaks. The absence of nitrogen at the surface of the anodized film and the nitrogen increase at the oxide—metal interface have also been observed by Simmons et al. [51]. The nitrogen-free region observed in the profile (Fig. 14) has been taken [51] as evidence that nitrogen remains stationary during oxide growth while the tantalum and oxygen ions are mobile.

Simmons [52] has also compared the nitrogen profiles obtained in Ta_2O_5 by AES and SIMS. Both the SIMS and AES profiles showed a nitrogen-free region, but the SIMS profile did not show an increase in nitrogen at the oxide—metal interface. The differences in the AES and SIMS nitrogen profiles are believed to be due to matrix and chemical effects on the secondary ion yield of nitrogen which complicates the interpretation of the SIMS data. The use of both techniques did, however, provide useful information on the growth mechanism of anodic films formed on nitrogen-doped tantalum films.

VIII. Analysis of platinum films containing phosphorus by AES and SIMS

Various combinations of thin films are used to make interconnections (metal—silicon) between the doped regions of silicon used in integrated circuit technology, and as terminations, i.e. contact pads. The metal—silicon (M—S) contacts often exhibit low resistance, linear (non-rectifying) I—V characteristics, ohmic contact. Platinum and aluminum are commonly used to make M—S ohmic contacts. Platinum forms platinum silicides which are chemically and metallurgically stable, but platinum is difficult to evaporate with a filament and must be deposited by sputtering or electron beam evaporation. These energetic processes can cause radiation damage in the oxide of field effect transistors such as metal oxide semiconductor (MOS) devices [53]. The result is a continuous drift in the gate voltage or changes in the source-to-drain current caused by a build-up of positive fixed charge in the oxide at the oxide—silicon interface [54].

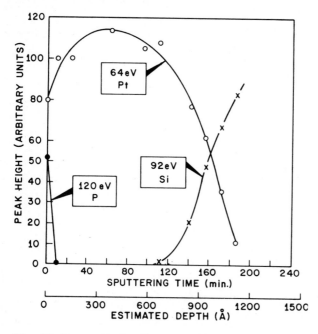

Fig. 16. Ion sputtering Auger profiles using argon ions of CVD Pt on Si as deposited at 225° C.

To avoid such an effect, a pyrolytic platinum chemical vapor deposition (CVD) from the volatile $Pt(PF_3)_4$ has been developed by Rand [55]. The possibility of incorporating phosphorus or fluorine into the deposited Pt film was of concern, since the amount and distribution of these impurities prior to and after silicide formation could have a pronounced effect on the mechanical, electrical, and chemical properties of Pt and its silicides [28].

The P, Pt, and Si profiles [28] obtained on the as-deposited (CVD) Pt by AES and SIMS are shown in Figs. 16 and 17. Fluorine was below detection by both techniques. The AES profile of phosphorus indicated that all the phosphorus prior to silicide formation is within the first ~60 Å of the Pt film. The amount of phosphorus at the Pt—Si interface (~400 Å) was below AES detection (≤1 at.%) under the conditions used for the analysis. However, the more sensitive SIMS technique could detect phosphorus not only at the surface (not shown in Fig. 17), but in the bulk of the film and at the Pt—Si interface (see Fig. 17). The phosphorus signal at the surface is not shown since surface oxide greatly enhances the secondary ion yield of phosphorus at the surface. The rather large phosphorus signal at the Pt—Si interface is also due in part to a chemical enhancement effect since a comparison of this signal normalized to $^{30}Si^+$ with the normalized ($^{31}P^+/^{30}Si^+$) signal from phosphorus-doped silicon standards [20] indicates that the phosphorus

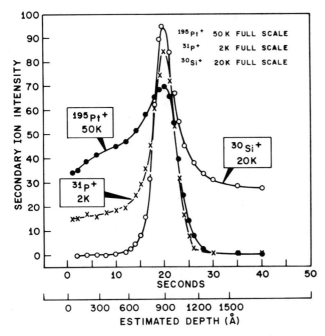

Fig. 17. SIMS profiles of CVD Pt on Si as deposited at 225°C using primary oxygen ions ($^{16}O_2^+$) at 4.5 keV.

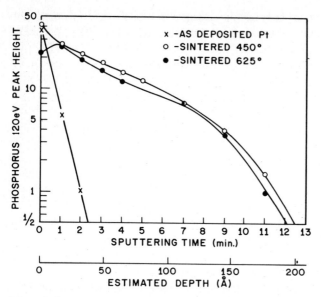

Fig. 18. Ion sputtering Auger phosphorus profiles in PtSi films from CVD Pt using argon ions.

concentration is ~10 at.%. This high concentration would have been detectable by AES.

The AES profiles [28] of phosphorus in the as-deposited Pt and sintered (450°C, 625°C) platinum silicide is shown in Fig. 18. Silicide formation has resulted in the migration to the surface of the previously undetectable phosphorus in the bulk and interface of the film. A similar result is observed in the SIMS profile [28] shown in Fig. 19. Note the absence of phosphorus at

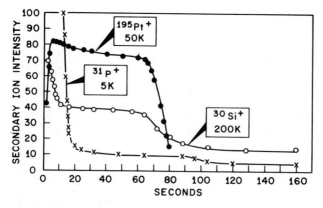

Fig. 19. SIMS profiles of 2000 Å of PtSi after a 450°C sinter using primary argon ions ($^{16}O_2^+$) at 4.5 keV.

the Pt—Si interface. The low energy silicon Auger peaks shifted from 92 eV to 78 eV. This shift is characteristic of Si in the form of SiO_2 [56], and was not observed below ~200 Å of the surface.

A PtSi single crystal was used to determine the composition of platinum silicide made from CVD Pt sintered in hydrogen at 450°C and 625°C. The Pt(64 eV)/Si(92 eV) Auger ratio was found to be constant throughout the film at both temperatures. This was also the case for the $^{195}Pt^+/^{30}Si^+$ SIMS ratio. The measured Auger and SIMS Pt/Si ratios at both temperatures were the same as those obtained on the PtSi standard. Similar results have been obtained using high energy ion back scattering [57].

IX. Analysis of alumina ceramic substrates by AES and SIMS

Alumina ceramic substrates are commonly used in the hybrid integrated circuit (HIC) technology to provide support for active (i.e. transistors) and passive thin film (resistors and capacitors) components. Thin film adherence to ceramic substrates is therefore of concern and has been studied extensively [58]. Recently, Sundahl [59] and Conley [60] have used AES to study the relationship between the relative Ca and Si surface impurity concentration on ceramics and adherence characteristics. In Fig. 20, the correlation obtained by plotting the average 90° peel strength (a measure of adherence) of tantalum nitride resistors [4] on ceramic versus the log of the sum of the

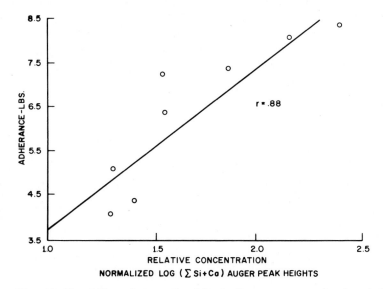

Fig. 20. The 90° peel strength of leads thermocompression bonded to Ta_2N resistors on ceramic vs. log of the summation of surface calcium and silicon normalized to the aluminum Auger peak height.

References pp. 326—328

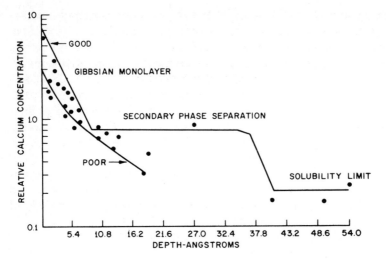

Fig. 21. In-depth profiles of Ca on ceramic characterized to have very good and very poor adherence.

surface calcium and silicon normalized to the aluminum Auger signal is shown [60]. The $90°$ peel tests were performed on leads which were thermocompression bonded to AuPdTi—Ta_2N. The observed adherence failures were at the Ta_2N—Al_2O_3 interface. The in-depth profiles [60] of Ca obtained on ceramic substrates characterized to have very good and very poor

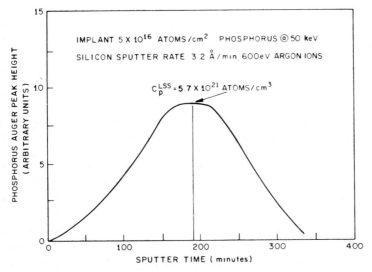

Fig. 22. Phosphorus depth profile in silicon ion implanted to 5×10^{16} ions/cm^2 at 50 keV using primary ions of $^{40}Ar^+$ at 600 eV primary energy.

adherence are shown in Fig. 21. The exponential drop can possibly be associated with a surface energy driven Gibbsian type monolayer of calcium oxide. The first plateau possibly represents the presence of a second phase, and the second plateau the solubility limit concentration of Ca in alumina. A similar profile [60] was obtained for silicon. The major source of Ca and Si impurities is from the grain boundaries of the ceramic itself and their presence appears to be a necessary requirement for good adherence [59]. Sundahl [59] found that etched ceramic had low impurity (Ca, Si) levels and poor adhesion, while annealing of etched samples led to recovery of surface Ca and Si impurity levels and adhesion.

Conley [61] has also studied the surface chemistry of ceramic substrates by SIMS. SIMS in-depth profiling showed that Ca and Si were concentrated at the surface, but this could be due to the enhanced secondary ion yield at the surface of the ceramic. Both the AES and SIMS analysis, however, did detect the presence of impurities such as Si, Ca, Mg, and K.

X. Chemical analysis of P-, As-, and B-doped silicon by SIMS and AES

A. PHOSPHORUS

SIMS and AES measurements on phosphorus-doped silicon by the technique of ion implantation have shown that ion implantation is a useful means of calibrating AES and SIMS for quantitative analysis [20].

The phosphorus in-depth profile using AES in combination with in situ ion sputtering is shown in Fig. 22. The Auger data are in good agreement with the LSS [62] prediction for the location and distribution of the implanted phosphorus ions. These measurements indicated that the detectability limit for an in-depth profile analysis of phosphorus in silicon is 5×10^{19} atoms/cm^3 by AES. The in-depth AES detectability limit is determined by the analysis time (time constant on lock-in-amplifier) and sputtering rate used for optimum depth resolution.

The *bulk* AES detectability for phosphorus was determined by measurements obtained on bulk-grown phosphorus-doped samples. The phosphorus profile is constant in these samples and a longer analysis time can be used since depth resolution is not a consideration. The results [63] obtained for phosphorus are shown in Fig. 23. The phosphorus data can be represented by $C_{Phos} = 1.9 \times 10^{21}$ (P/Si)$^{1.16}$ where P/Si is the phosphorus 120 eV Auger peak normalized to the silicon high energy 1619 eV transition.

The SIMS profile obtained by monitoring $^{31}P^+$ was in poor agreement with both the AES profile and LSS theory due to a mass interference from $^{31}(SiH^+)$. Profiles in better agreement with both the AES profile and LSS theory were obtained by monitoring $^{31}P^-$ or $^{47}(PO^-)$ as shown for $^{47}(PO^-)$ in Fig. 24. The vertical bars represent the minimum and maximum

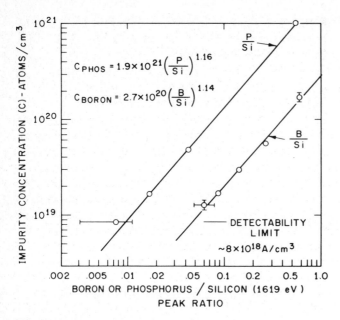

Fig. 23. Auger calibration curves for B and P in silicon.

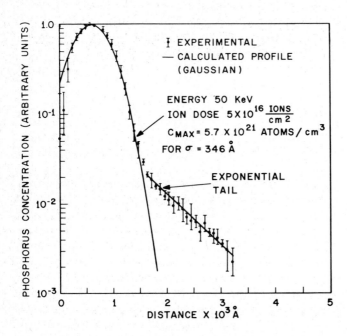

Fig. 24. SIMS profiles of phosphorus in silicon ion implanted to 5×10^{16} ions/cm² at 50 keV using primary ions of $^{16}O_2^+$ at 14.5 keV.

readings of five channels from an AS 200 multichannel analyzer. The exponential tail in the profile occurs at a concentration of $\sim 5 \times 10^{18}$ atoms/cm^3. This asymmetry in the phosphorus profile has also been observed by radioactive tracer [64] and electrical methods [16,17].

SIMS measurements on samples implanted to a total phosphorus dose of $3.2 \times 10^{+14}$ atoms/cm^2 indicated that the *in-depth* detectability limit for typical operating conditions and background is in the 10^{18} atoms/cm^3 range. At the expense of depth resolution, the *bulk* detectability limit could be improved to $\sim 10^{17}$ atoms/cm^3 which is about two orders of magnitude better than AES.

B. ARSENIC

The integral of the SIMS in-depth profiles (Fig. 25) from silicon samples ion implanted with arsenic [65] doses of 1×10^{14} atoms/cm^2, 1×10^{15} atoms/cm^2, and 1×10^{16} atoms/cm^2 at 50 keV, were found to be nearly proportional to the ion dose. The peak concentration at 360 Å is in good agreement with the theoretical projected range (324 Å). The profiles (shown in Fig. 25) were obtained with primary oxygen ions ($^{16}O_2^+$) at 14.5 keV and the secondary AsO$^-$ ions were collected from a 70 μm diameter area. The AsO$^-$ species provided the highest detection sensitivity for arsenic which was found to be in the 2×10^{18}–3×10^{18} atoms/cm^3 range for the operat-

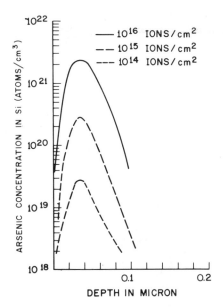

Fig. 25. SIMS profiles of arsenic ion implanted to 10^{16} ions/cm^2, 10^{15} ions/cm^2, and 10^{14} ions/cm^2 at 50 keV using primary ions of $^{16}O_2^+$ at 14.5 keV.

References pp. 326—328

ing conditions used. AES in-depth detection sensitivity for As has been estimated [20] to be in the 10^{+20} range.

The arsenic profile obtained from the electrical (differential conductivity) and the SIMS data for a sample annealed for 15 min at 1050°C in nitrogen are shown in Fig. 26. The shapes of the electrical and SIMS profiles are in good agreement. The apparent tails in the SIMS arsenic profile has been attributed [64] to mass interference from a hydrogen-containing polyatomic ion species such as $^{91}(Si_2O_2H)$. The suspected mass interference is expected to be sample-dependent (e.g. hydrogen content in the samples, etc.) and a function of the partial pressure of contaminants such as water vapor in the sample chamber of the SIMS instrument. Profile tails were not present in the profiles obtained on the as-implanted samples (see Fig. 25).

The bulk detectability limit for arsenic by SIMS has also been determined [66] by use of bulk-doped silicon. Three samples were analyzed which had bulk concentrations of 1.4×10^{20} ions/cm^3, 2.5×10^{18} ions/cm^3, and 4.5×10^{16} ions/cm^3 as determined from resistivity measurements and Irvin's curves [67]. The sample with 4.5×10^{16} As ions/cm^3 was found to be

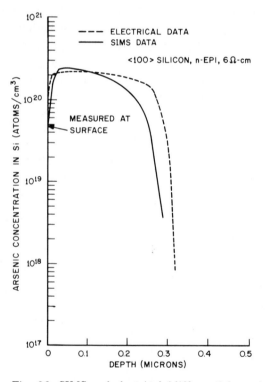

Fig. 26. SIMS and electrical (differential conductivity) profiles of arsenic ion implanted to 5×10^{15} ions/cm^2 at 50 keV, and annealed at 1050°C for 15 min in nitrogen.

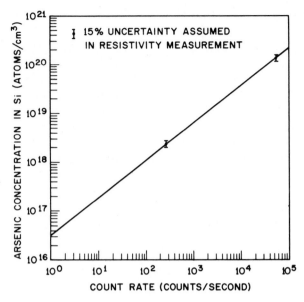

Fig. 27. SIMS bulk detectability for arsenic using primary ions of $^{16}O_2^+$ at 14.5 keV.

below detection with SIMS. In Fig. 27, a straight line has been drawn through the two data points assuming a power dependence for the AsO^- signal versus concentration. By defining the detectability limit as the point at which the arsenic signal becomes $\approx 2/1$ of the background signal, the detectability limit for a given background can be estimated from Fig. 27. Typically, the background count is in the range of 30—300 c/s, probably due to a mass interference from the $^{91}(Si_2O_2H^-)$ complex. For example, for the background count of 184 c/s, the arsenic concentration detectability is 1.0×10^{18} ions/cm^3. The arsenic concentration detectability at a background count of 30 c/s is $\sim 4.2 \times 10^{17}$ ions/cm^3.

C. BORON

Gittins et al. [68] Croset [69] and Hofker et al. [70] have profiled boron in silicon by SIMS. Profile tails have also been observed on boron-implanted samples. The reason for the observed profile tails in boron appears to be associated with a "knock on" phenomenon [1]. The SIMS bulk detectability limit for boron in silicon is in the 10^{14} atoms/cm^3 range, whereas AES detectability limits (see Fig. 23) are in the 10^{19} atoms/cm^3 range.

A boron profile from a silicon sample implanted at 275 keV to a total dose of 9.0×10^{12} ions/cm^2 followed by a second implant at 50 keV to a total dose of 8.9×10^{13} ions/cm^2 is shown in Fig. 28. The boron profiles were obtained by monitoring the BO^- secondary ions using primary oxygen

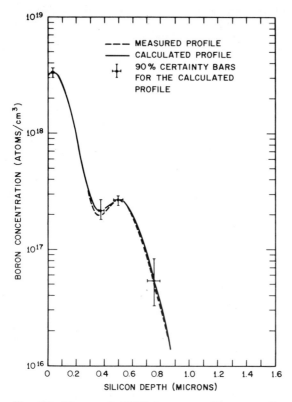

Fig. 28. Measured SIMS boron profile normalized to the calculated boron profile in silicon using primary ions of $^{16}O_2^+$ at 14.5 keV.

ion bombardment [66]. The implanted boron profile exhibits two peaks whose calculated separation is given by R_p (275 keV) $-R_p$ (50 keV) $\equiv \delta R =$ 472 Å where R_p is the sputtered depth at the peak. If the silicon is assumed to sputter at a constant rate between the boron peaks, the sputter rate $\dot{z}$ is given as $\dot{z} = \delta R/n\tau = 2360/n$ Å/sec where n is the number of channels in the multichannel analyzer between the boron peaks and τ is the dwell time per channel. For the boron profile shown in Fig. 28, $n = 46$, corresponding to a $\dot{z}$ of 51 Å/sec. A comparison between the calculated boron profile and the measured profile, also shown in Fig. 28, by arranging the ordinates of the two profiles so that the calculated and measured peaks of the higher energy (250 keV) boron implants matched, thus calibrating counts/channel in terms of ions/cm^3.

It is also possible to profile boron and arsenic when both are present as is the case when fabricating a high frequency, high gain transistor [18]. The comparison, achieved by normalizing the profiles to the integrated implanted dose as discussed above between the calculated boron and arsenic profiles

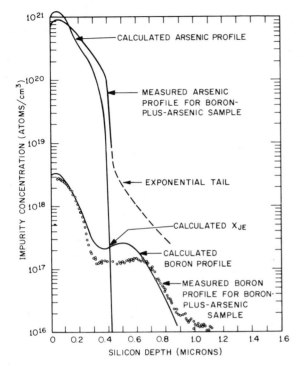

Fig. 29. SIMS profiles of arsenic and boron in silicon ion implanted with both boron and arsenic in silicon using primary ions of $^{16}O_2^+$ at 14.5 keV.

and those of the measured boron and arsenic profiles, is shown in Fig. 29. Fitting the profiles in this manner causes the peak concentration of the arsenic profile to occur at a concentration of $\sim 10^{21}$ ions/cm^3, with the distribution falling off into the background at a concentration of $\sim 2 \times 10^{17}$ ions/cm^3. The minimum detectable arsenic concentration for the arsenic profile agrees well with the predicted sensitivity from Fig. 27, where 15 c/s corresponds to a sensitivity of 1.5×10^{17} ions/cm^3.

The exponential tail (see Fig. 29) which is observed at a depth of ~ 0.43 μm, however, indicates the measured arsenic concentration is greater than the boron concentration for all values of x. This, in fact, cannot be so since a junction (X_{je}) is known to occur at a silicon depth of ~ 0.41 μm. In addition, there appears to be a substantial redistribution of boron in the vicinity of the emitter—base junction (X_{je}), which is illustrated in Fig. 29. The redistribution is substantial and amounts to as much as 40% at the second boron peak. Measurements [71] by the nuclear reaction method on silicon doped with boron and arsenic from a chemical source have also indicated a depletion of boron extending over a range of 2000—4000 Å in the region of the boron—arsenic junction.

Blanchard and coworkers [72] and more recently Dobrott et al. [73] have also observed a similar redistribution for boron- and arsenic-doped silicon by SIMS profiling. This redistribution of boron could be due to both the arsenic implantation and subsequent anneal, and is not believed to be an artifact of the SIMS technique.

XI. In-depth, bulk, and surface sensitivity comparison of AES and SIMS

It is very difficult to compare the sensitivities of AES and SIMS since the SIMS signal depends on the amount of material volatilized by sputtering but Auger does not. The minimum volume of material which has to be removed to measure a given concentration will depend on the ionization yield K of the element, the ion collection efficiency multiplied by the transmission of the instrument η, and on the precision p of the measurement. The relationship between these parameters is given by the expression [30]

$$\delta = \frac{1}{CK\eta} \frac{M}{\rho N} \frac{10^4}{p} \frac{10^2}{a} \tag{4}$$

where $\rho N/M$ is the atom density (atoms/cm^3), with ρ the density, N Avogadro's number, and M the atomic weight, $100/a$ is the isotopic abundance correction, and η is of the order of 10%. Using this expression, the depths at concentrations of 10^{-2}, 10^{-4}, and 10^{-6} have been calculated (Table 4) for $^{11}B^+$, $^{31}P^+$, $^{75}As^+$ in silicon corresponding to analyzed areas of 20 μm^2 (5 μm beam diameter), 10^4 μm^2 and 9×10^4 μm^2, or to rastered area sizes of 100 μm $\times$ 100 μm and 300 μm $\times$ 300 μm. The precision selected was 3%.

TABLE 4

The minimum sputtered depth versus concentration for boron, phosphorus, and arsenic in silicon

Element	C	Depth corresponding to analyzed areas of		
		20 μm^2	10^4 μm^2	9×10^4 μm^2
$^{11}B^+$	10^{-2}	250 Å	5 Å	0.55 Å
$^{11}B^+$	10^{-4}	2.5 μm	500 Å	55.5 Å
$^{11}B^+$	10^{-6}	250 μm	5×10^4 Å	5.5×10^3 Å
$^{31}P^+$	10^{-2}	370 Å	7.4 Å	0.82 Å
$^{31}P^+$	10^{-4}	3.7 μm	740 Å	82 Å
$^{31}P^+$	10^{-6}	370 μm	7.4×10^4 Å	8.2×10^3 Å
$^{75}As^+$	10^{-2}	2500 Å	50 Å	5.5 Å
$^{75}As^+$	10^{-4}	25 μm	5000 Å	555 Å
$^{75}As^+$	10^{-6}	2500 μm	5×10^5 Å	5.5×10^4 Å

The secondary ion yields using primary oxygen ions are K (boron) = 5.2 × 10^{-6}, K (phosphorus) = 3.5 × 10^{-6} and K (arsenic) = 5.2 × 10^{-7}. These yields were measured from the number of counts obtained from a known volume of ion-implanted silicon and assuming η to be 10%. These calculations do not correct for mass interference effects, hence background. From the results in Table 4, it would be very difficult to measure concentrations of boron, phosphorus, and arsenic in the 10^{-4} range consuming material corresponding to a depth less than 50 Å. The depth must be increased further for the analysis of lower ion yield elements, but it is decreased for the analysis of elements with higher yields such as aluminum.

The required sputtered depth (Table 4) places a lower limit on the depth resolution possible at a given concentration with the SIMS technique. AES and SIMS sensitivity must therefore be compared at the same depth resolution and amount of sample consumed. This can be done with ion-implanted samples as discussed by Morabito and Tsai [20]. When depth resolution is not a consideration, i.e. in the case of bulk analysis, the detection of elements at the ppb level for some elements is possible with SIMS. A comparison of the bulk sensitivity limits for P, B, and As in silicon and for N, O, C in Ta using AES and SIMS is shown in Table 5.

Although SIMS is, in principle, also more surface sensitive than AES, mass interference effects (particularly from hydrocarbons) and primary ion implantation effects [33] can often severely complicate the interpretation of SIMS surface analysis. The surface analysis of the sputtered tantalum films [30] and the chemical vapor deposited (CVD) platinum films [28] by AES and SIMS were in good qualitative agreement, but SIMS surface analysis alone would have been ambiguous. Primary ion implantation effects on SIMS analysis within the first few hundreds of angstroms of the surface are discussed in the next section.

TABLE 5

Comparison of bulk sensitivity limits for P, B, and As in silicon and for N, O, C in tantalum using AES and SIMS

Element	Technique	Detectability limit (atoms/cm^3)
N, O, C	AES	10^{19}
N	SIMS	10^{19}
C, O	SIMS	10^{17}
P, B	AES	10^{19}
As	AES	10^{20}
B	SIMS	10^{14}
P	SIMS	10^{17}
As	SIMS	10^{18}

XII. Anomalous ion yield effects produced at the surface in SIMS depth profiles

To avoid serious error during depth profile measurements resulting from secondary ion yield enhancement, it has become common practice to obtain a normalization signal [11]. This signal is obtained by monitoring some suitable matrix ion species. The technique unfortunately is not always reliable since the ion species selected may be enhanced differently from the ion species being monitored. This has been shown to be true particularly when sputtering from one phase into another [28] or when steep chemical gradients are being followed. In addition, two anomalous surface effects are apparently occurring in SIMS measurements over the first few hundred angstroms of depth [33]. These effects cause a variation of secondary ion yield preventing establishment of meaningful secondary ion intensity versus concentration relationships over this range.

The first effect is an enhancement due to the presence of a surface oxide. This oxide results from ambient oxygen which has adsorbed on the surface and has reacted over the first few angstroms of depth. The ion yield enhancement resulting from the surface oxide usually disappears after the first few monolayers have been sputtered away. The second effect is an enhancement of the secondary ion yield which is directly related to the implantation range of the primary bombarding ion. The second effect is observed with reactive primary ions (e.g. $^{16}O_2^+$, $^{16}O^-$, $^{14}N^+$) which are used to enhance the secondary ion yield by the "chemical emission" mechanism [32].

Selected area depth profiles of oxygen [33] and nitrogen using argon ion sputtering Auger and SIMS from craters formed by $^{16}O_2^+$, $^{16}O^-$, and $^{14}N^+$ primary ions showed that the implanted primary reactive ions have a gaussian type distribution consistent with LSS theory [62]. However, the range predicted by LSS theory is lower than that actually measured by Auger and SIMS profiling. This is possibly due to the high sputtering rate used while implanting the ions, which modify the assumptions used in LSS calculations, or to channeling effects.

Depth profiles obtained on pure silicon displaying the superposition of these two surface effects are shown in Fig. 30(a) and (b), where $^{16}O^-$ primary ion bombardment has been used and the $^{30}Si^+$ species monitored. The surface oxide effect is delineated better by the expanded scales in Fig. 30(b), where the $^{30}Si^+$ yield is shown enhanced over the first few angstroms of depth sputtered. The initial valley for $^{30}Si^+$ in Fig. 30(b) bears no resemblance to the true nature of the composition of the surface since the sample was pure silicon. The phenomenon can be described as a dynamic implantation range effect. At the initiation of the sputtering process, primary oxygen ions are implanted with a gaussian distribution below (~ 100 Å) the surface. As a result, the concentration of primary oxygen is not high enough to enhance sufficiently the secondary ion yield near the surface so that the ion

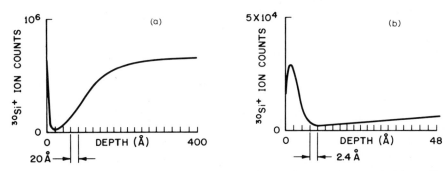

Fig. 30. Silicon depth profiles of bulk-grown single-crystal silicon with (a) profile range of 0—400 Å, and (b) profile range of 0—48 Å using primary ions of $^{16}O^-$ at 14.5 keV.

production mechanism is principally a kinetic process. The chemical and kinetic secondary ion emission mechanisms have been described previously [32]. Significant enhancement of the secondary ion yield cannot occur until the sputtering front reaches that point in the implanted oxygen distribution where the oxygen concentration is high enough to cause chemical emission to predominate over kinetic emission. When the sputtering depth finally reaches the peak concentration of the initial primary oxygen implant distribution, the measured secondary ion signal then reaches a steady state, oxygen enhanced value. At this point, the secondary ion signal follows concentration variations accurately. It is necessary, of course, to achieve this steady state condition when using any reactive gas as a bombarding species for quantitative analysis in SIMS. A profile identical to the profile in Fig. 30(a) was also obtained when pure silicon was bombarded with $^{16}O_2^+$ and $^{16}O^-$ monitored as a function of depth as shown in Fig. 31. This demonstrates that there is a one-to-one correspondence between the enhancement of the $^{30}Si^+$ signal and the concentration gradient produced by the implanted primary oxygen ion beam.

The effect of these anomalous ion yields on a typical depth profile is

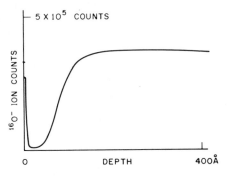

Fig. 31. Profile of $^{16}O^-$ from a silicon sample ion implanted with $^{16}O_2^+$ ions at 14.5 keV.

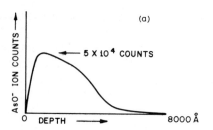
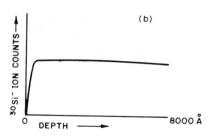

Fig. 32. Arsenic and silicon depth profiles in silicon ion implanted with arsenic at 50 keV and diffused at 1050°C for 15 min in nitrogen (a) AsO$^-$ ion counts vs. depth and (b) ^{30}Si$^-$ ion counts vs. depth.

shown in Fig. 32. The sample selected was silicon which had been implanted with ^{75}As$^+$ at 50 keV to a dose level of 5×10^{15} atoms/cm^3. The sample was analyzed by bombarding with O$_2^+$ and monitoring AsO$^-$. The artifact introduced in the AsO$^-$ profile shown in Fig. 32(a) resulting from the implantation of primary oxygen ions is apparent. The AsO$^-$ signal should have started at about 5×10^4 counts at the surface and continued at about this level over the next 400 Å. The decreased depth over which the primary ion implantation effect is observed when compared with Fig. 31(a) is due to the fact that the O$_2^+$ ion splits on impact and divides the 14.5 keV bombarding energy between the two oxygen atoms. Similar profiles (not shown) were observed with boron and phosphorus implant-diffused samples when bombarding with ^{16}O$_2^+$ and monitoring BO$^-$ and PO$^-$ secondary ions.

The surface oxide has been observed to modify the depth profiles for all the reactive metallic systems studied, and the primary ion implantation effect has also been observed in these systems whenever reactive primary ion bombardment has been used. However, in most metals the effect displays more variation and is not as reproducible as it is in silicon. This variation is observed in the ^{63}Cu$^-$ profiles obtained in pure polycrystalline copper as shown in Fig. 33(a) and (b) from two grains of random orientation. Differences in channeling undoubtedly account for much of the differences seen in

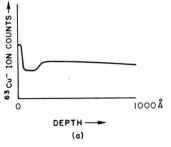
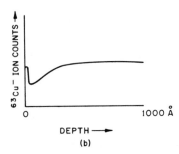

Fig. 33. Primary ion implantation effect on the emission of secondary copper ions for polycrystalline copper from grains of different (a and b) preferred orientation.

Fig. 33(a) and (b). The effects of channeling on secondary ion yield have been described by Slodzian [74] and Castaing [44].

It is obviously highly desirable to eliminate or reduce the primary ion implantation effect. Lewis et al. [33] have recently reported that for silicon, at least, the technique of introducing a stream of oxygen at the surface while bombarding with oxygen does effectively eliminate the initial valley resulting from the primary ion implantation. This has been demonstrated by repeating the same profile shown in Fig. 31(a) except with the residual oxygen pressure set at 10^{-4} torr. In this profile (shown in Fig. 34) the ion yield of silicon is seen to remain constant from the surface, i.e. the ion yield one would expect from SiO is observed within a few monolayers of sputtering.

The effects on secondary ion emission resulting from the presence of a high residual pressure of oxygen in the ambient atmosphere during ion bombardment were first described by Slodzian and Hennequin [75]. They demonstrated that increasing the ambient oxygen level during argon bombardment considerably enhances the secondary ion emission. The ion yield increases with increasing ambient oxygen pressure and levels off at a pressure of about 10^{-4} torr. At this pressure, the ion yield is very nearly the same as the yield obtained from the respective pure metal oxides. Blanchard et al. [76] have also shown that a high ambient oxygen level can improve

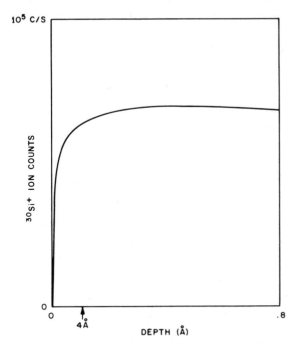

Fig. 34. SIMS depth profile of bulk grown single-crystal silicon in the presence of an oxygen ambient using primary ions of $^{16}O^{-}$.

References pp. 326—328

depth profiles, and were able to obtain a constant Si signal when sputtering from SiO_2 into Si by bombarding with oxygen and maintaining a residual oxygen pressure of about 10^{-4} torr. Apparently, from the very onset of sputtering, a continual breaking of metal/oxide "bonds" causes "chemical emission" to predominate. At an oxygen partial pressure of about 10^{-4} torr, the rate of forming surface oxide layers appears to be faster than the rate of destructive sputtering of the primary beam. This rapid renewal of surface oxide maintains, in effect, a continuous surface coverage for sputtering rates up to at least a few tens of angstroms per second. The advantages of performing SIMS analysis in the presence of an oxygen ambient are currently being studied in several laboratories.

XIII. The use of high energy and low energy secondary ion discrimination in SIMS

Energy interference effects for the detection of phosphorus, boron, and arsenic in silicon and for the light elements (N,C,O) in tantalum were not a problem in the AES technique. The possibility of mass interference effects in SIMS, however, was an important consideration for the detection of both phosphorus and arsenic in silicon. In fact, one of the most serious problems encountered in SIMS is the problem of mass interference from molecular ion species produced from the matrix in the ion bombardment process. These interferences often reduce the detection sensitivity considerably by raising, at the mass position of interest, the background signal to abnormally high values. In addition to polyatomic mass interference problems, another type of interference problem is also frequently encountered. This is the low energy tail that occurs with certain mass species particularly when the species has a high sputter yield. This low energy tail can sometimes show considerable intensity extending over several mass units below the low energy side of the peak, seriously limiting the detection sensitivity of impurities occurring in this mass region.

The yield versus energy relationship of various polyatomic species has been studied by Herzog et al. [77]. Their measurements clearly show that discriminating against ions released at energies below a few electron volts can dramatically reduce the intensity of polyatomic ion species in the mass spectra with very little loss of the parent ion species. To achieve low energy discrimination (LED) with the Castaing—Slodzian design [78], a grid has been introduced in the ion optical path at the exit crossover beyond the second magnet deflection where the stigmator is located as shown in Fig. 35. This grid complements the electrostatic mirror which provides a high energy cut-off. The electrostatic mirror is polarized at a voltage $V_a + V_m$ where V_a is the accelerating potential of the secondary ions, and V_m is a voltage variable from 0 to 50 V. Ions then having energies in excess of $V_a + V_m$

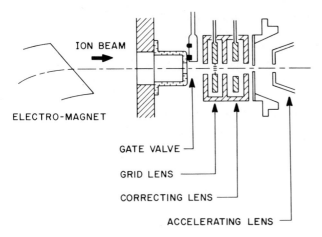

Fig. 35. Schematic of a low energy discriminator.

collide with the mirror and are lost from the beam. This high energy discrimination has been used effectively to reduce chromatic aberration improving mass resolution. Similarly, the grid is placed at a voltage $V_a + V_g$ where V_g is a voltage variable from 0 to 20 V. Ions having energies greater than this value are allowed to pass through and be detected while ions of lower energy are deflected and lost from the beam. It becomes possible then with the grid and mirror controls to achieve a very precisely defined voltage window (V_g to V_m) through which the secondary ions, of a particular charge to mass ratio selected by the mass spectrometer, can be detected.

The data shown in Fig. 36 show the effectiveness of the LED in attenuating the polyatomic ion species when bombarding silicon. This figure not only shows the attenuation of the positive polyatomic ion species, but also that attenuation of the negative polyatomic ion species is possible. The ion yield of the various species has been plotted as a function of positions of the LED threshold. This threshold was selected to give an attenuation of 2, 4, 8 and 16 respectively of the parent ^{28}Si species. In Fig. 36, the dimer and single oxide species are attenuated only slightly more than the silicon species, while the doubly charged species are attenuated the same amount. However, the more complex species are attenuated quite drastically.

The data obtained on a typical analytical problem are shown in Table 6. Here, the attenuation is listed for the various ion species interfering with the detection of arsenic in silicon by using LED. For the peak of interest at mass 91 corresponding to AsO^-, the mass interference results from the Si_2O_2 species. The abundance yields are shown in Table 6 together with the attenuations achieved with and without the LED. When the AsO^- (91) peak is attenuated 2.5 times the main Si_2O_2 (88) peak is attenuated 20 times.

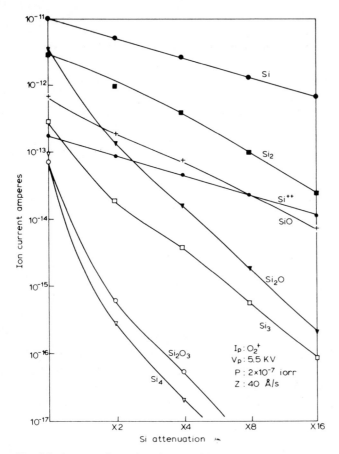

Fig. 36. Attenuation of positive poly-atomic ion species by low energy discrimination.

Therefore, the contribution of this species at mass 91 will also be attenuated 20 times. This results in almost an order of magnitude lowering of the background signal. In fact, the photomultiplier dark current now becomes the limiting factor. An arsenic bulk detectability in silicon of about 4×10^{16} atoms/cm^3 can be obtained by the use of the LED.

XIV. Summary and conclusions

A number of examples representative of typical problems encountered in thin film and silicon device technology have been studied by AES and SIMS.

Both AES and SIMS have unique analytical capabilities which make selected area three dimensional (x,y,z) microchemical analysis possible. Two

TABLE 6

LED attenuation of polyatomic matrix ion species interfering with the detection of As in Si

Mass no.	Relative abundances				Peak counts		Attn.
	Si_3^-	$Si_2O_2^-$	SiO_4^-	AsO^-	With LED off	With LED on	
84	0.784				43,000	6900	6.2
85	0.120				7000	1000	
86	0.085				4700	700	
87	0.008				500	80	
88	0.003	0.850			2300	115	20
89		0.087			300	18	
90		0.059			200	15	
91		0.003		1.00	10,000	4000	2.5
92		0.001	0.927		8000	900	8.8
93			0.047		500	70	
94			0.031		400	40	

dimensional (x,y) chemical analysis is possible by means of the "ion image" or by the recently developed "Auger image", which provide chemical distribution maps of the impurities present on specific areas of a surface and into the bulk. In-depth concentration profiles, the z dimension, can also be obtained with good depth resolution, i.e., 5—10% of the depth sputtered, in the absence of preferential sputtering effects from grain boundaries, knock-on, redeposition and crater effects. The comparison of data obtained by both methods on the same sample show clearly that the individual analytical capabilities are extremely complementary. The combined use of both has, in fact, had a synergistic effect on the final characterization achieved compared with that possible by AES or SIMS analysis alone.

A. SPUTTERING ION BEAM

Both argon and xenon ion bombardment are used in AES. Xenon is used when higher sputtering rates are desirable. In SIMS, oxygen and argon are used most extensively. Oxygen has been highly publicized as the most useful ion bombardment species for SIMS because it generally produces the highest secondary ion yield and both positive and negative ions may be extracted from the ion gun. Argon provides many unique advantages that often dictate its use. These include, for example, no interference from polyatomic M_xO_y species, higher detection sensitivities for some electronegative elements (e.g. C and S) and the highest detection sensitivity for oxygen.

B. SPUTTERING RATE

Depth measurements for calibrating sputtering rates are difficult because the distances involved typically range from only a few hundred to a few thousand angstroms. Crater depths are usually determined after completing an analysis by measuring a deep crater specifically produced for the depth calibration. The depth is usually determined by light optical interferometry or mechanical stylus techniques with accuracies to a few hundred angstroms, which are compatible with the control of other parameters affecting the accuracy of calibrating the depth scale, namely the flatness of the crater bottom and the uniformity of the sputtering rate.

When the material being analyzed is not amenable to subsequent crater depth measurement, internal methods of estimating the amount of material removed are used such as beam densities and sputter atom yields. Beam uniformity and density can be measured most conveniently by including a standard such as Ta_2O_5 and noting the interference color change. No depth calibration technique, however, has been found completely satisfactory. The greatest errors occur with depth calibration for a few hundred angstroms below the surface and through metal/metal oxide interfaces where the sputtering rate changes rapidly because of the change from metal to chemical bonding.

It would be highly desirable to have a continuous monitoring of the sputtering rate by measuring the re-deposition of the sputtered material. Incorporation of this type of instrumentation would appear to be quite feasible, but is not yet available.

C. MASS AND SPECTRAL INTERFERENCES

Serious mass interferences are frequently encountered in SIMS analysis. These interferences typically result from matrix polyatomic species, hydrocarbons, hydrides, and matrix compound formation. In addition, post ionization beyond the zero potential plane in the ion acceleration space can produce a high background for peaks on the low energy side of an intense matrix peak. Low energy discrimination with a sharp cut-off grid filter is very effective in eliminating this low energy tailing and effectively attenuates heavy polyatomic ion species emitted at low energies. A mass resolution of 1000 effectively eliminates hydrocarbon interferences. However, the mass resolution needed to eliminate most hydrides and compound interferences is usually greater than 3000. This resolution currently is not available in SIMS without severely limiting the spectrometer transmission efficiency. More sophisticated mass spectrometric techniques should overcome this difficulty.

Spectral interferences can also be a problem in AES, but fortunately electrostatic analyzers with high energy resolution are available. Also, most elements have several Auger transitions and this can often eliminate spectral interference difficulties, e.g. the detection of silicon in lead using the low energy 92 eV Si Auger electrons would be impossible since lead has transitions at 90 and 94 eV, but silicon also has high energy (1515—1620 eV) transitions which do not interfere with any of the lead transitions. However, the detection of nitrogen in titanium is difficult with AES since the major nitrogen transitions (357, 367, 381, 389, 405 eV) interfere with titanium transitions at 354, 364, 383, and 387 eV. Nitrogen has no transitions above 405 eV and is therefore only detectable by observing the increase in the 387 eV Ti transition compared with that of pure titanium.

D. SURFACE ANALYSIS

Surface analysis by AES and SIMS can be divided into ranges of 0—10 Å and 10—500 Å.

In the first range, quantitative analysis by both AES and SIMS can be in error because of effects of adsorbed oxygen and surface contaminants such as hydrocarbons. SIMS is also particularly sensitive to the presence of reactive elements, e.g. H, C, O and N, since they enhance secondary ion yields. Obtaining good quantitative analysis in this range appears to be unlikely with either technique at the present time, but qualitative surface analysis is definitely possible. Qualitative surface analysis is less ambiguous by AES than SIMS and pre-characterization by AES prior to SIMS surface analysis is often necessary due mostly to hydrocarbon interference effects and surface oxide (enhanced secondary ion yield) effects.

The second range (10—500 Å) is amenable to the AES technique for concentrations $\geqslant 0.1$ at.%. The AES technique has as its principle advantage an excellent one-to-one relationship between concentration and signal level with only limited interference effects. SIMS, on the other hand, while considerably more sensitive, suffers from ion yield artifacts. In particular, there is an anomalous ion yield "self-induction" effect in this range with reactive gas ion bombardment. Fortunately, this effect can be effectively reduced by using a high ambient pressure of oxygen. This is best accomplished by directing a controlled stream of oxygen against the area under ion bombardment. Interdiffusion studies in thin film conductor systems (Ti—Pd—Au, NiCr—Au, etc.) or conductor—silicon systems such as Pt—Si, Au—Si, Al—Si are conveniently studied by AES, but the SIMS profiles obtained from multilayer thin films can often be difficult to evaluate quantitatively. However, since SIMS is more sensitive than AES, it should be used to detect the presence of elements below the surface region prior to the less sensitive AES analysis.

E. IN-DEPTH ANALYSIS (>500 Å)

The technique of sputtering into the bulk of a sample and analyzing the chemical variation in depth is defined as depth profiling. By sputtering to distances of several thousand angstroms, it is possible to correlate surface and bulk chemistry directly in one experiment. This feature makes depth profiling very useful in many areas of materials characterization.

Depth profiling is inherent in SIMS analysis because of the sputtering action of the primary ion beam. SIMS depth profiles can be obtained rapidly with high detection sensitivity. The AES depth profiling technique has recently been developed by including a sputter ion gun as an accessory to standard AES instrumentation. However, the sputtering technique is secondary to the AES method of analysis and the more sophisticated ion guns utilized in SIMS have not been developed for AES, and consequently depth profiling with AES lacks the speed and more localized sputter control available in SIMS.

One of the main advantages of AES for depth profiling is the small depth dimension ($\leqslant 20$ Å) from which the excited Auger electrons are emitted. This results in a superior depth resolution for AES as compared with SIMS profiling. Furthermore, AES analysis is a non-destructive method; SIMS is not. That is, it is possible to stop the sputtering action and obtain a complete spectral analysis correlating as many elements as desired at a specific depth level. SIMS depth profiles are limited to the measurement of one mass at a specific depth level for magnetic mass analyzers. The depth reached by the sputtering front before detecting a second mass depends on the number of mass units that must be scanned by the magnetic field before reaching the second peak. For example, there can be a time lapse of as much as ten seconds switching from one mass peak to another, which can easily correspond to a depth of several hundred angstroms. The actual difference in depth will depend on the sputtering rate used. It is not possible to arbitrarily reduce sputtering rate to any low level because it is necessary to develop counting statistics required for a particular precision on the measurement. In addition, below a sputtering rate of about one angstrom per second, the recontamination rate of the surface from the ambient gases in the source region can raise the background and cause loss in detection sensitivity and serious mass interference problems. This lower sputtering rate limit applies to a vacuum of about 10^{-7} torr where most commercial SIMS instruments operate. The inclusion of a cold surface in the source region minimizes the contribution from condensible vapors. However, it does not eliminate the recontamination problem and a cold surface impedes the interchange of samples which can seriously affect the analysis time.

It is theoretically possible to overcome the switching problem in SIMS by using a mass spectrometer with multiple ion detection. Including multiple electronic ion detection on a mass spectrograph of the Mattauch–Herzog design does present some formidable instrumentation problems because of the intense magnetic field at the exit slit and the variable angle of incidence of each mass at the focal plane. Faster scanning between peaks is another possibility and can be accomplished with electrostatic deflection. However, there is a mass scan range limit of about 10% of the mass with this technique. The best scanning system appears to be the use of magnetic scanning with primary beam blanking between peaks.

F. DEPTH RESOLUTION

It is very difficult to achieve the theoretical depth resolution available with AES or SIMS because of the difficulty in achieving a flat bottomed crater. In addition, serious errors in peak intensity measurements with SIMS or AES can occur if the peak position shifts during sputtering. This shift can occur if voltage differences develop which is possible at high sputtering rates on insulators. The problem is most pronounced with SIMS at the low mass region where the mass dispersion is highest, and it is sometimes necessary to sacrifice mass resolution to minimize this problem. The best solution to the problem is to scan a SIMS mass or AES energy range that always includes the peak during a particular burn and integrate the total counts. This integrated count is then used instead of the peak intensity. This technique is handicapped because it requires more time than peak-to-peak switching and, as a result, the correlation in depth becomes more of a problem. Rastering the primary ion beam overcomes the problem of having a uniform beam density over the area being analyzed, but differences in sputter yield can often occur in depth, e.g. changes in grain orientation in polycrystalline samples or large differences in chemical composition. These differences tend to be emphasized with increasing depth, consequently the depth resolution decreases with sample thickness. In the absence of sputtering artifacts the depth resolution possible is 5—10% of the depth sputtered with AES or SIMS. However, artifacts due to non-uniform crater bottoms, crater wall effects, redeposition, and knock-on can often limit the accuracy of the depth profiles obtained by AES or SIMS. When possible, comparisons of AES and SIMS profiles with the profiles obtained by high energy ion backscattering should be made. In some cases, it is possible that sputtering artifacts could limit AES and SIMS profiles to the 10—1000 Å range.

G. QUANTITATIVE ANALYSIS

Quantitative analysis by both techniques is in the early stages of development, but is definitely possible. At present, AES appears to be the less ambiguous technique for quantitative analysis. However, its detection sensitivity is 1—0.1 at.%, whereas SIMS can detect concentrations two to five orders of magnitude lower depending on the secondary ion yield. Consequently, only SIMS can determine concentration distributions at most dopant levels of interest in semiconductor device technology provided the area available for analysis is sufficient.

Reliable quantitative analysis in both SIMS and AES requires the use of calibration standards. Equations relating the various parameters needed to compare standards against an unknown have been established for both AES and SIMS. However, securing suitable standards in SIMS can often be very difficult in complex materials because neighboring impurities and crystal orientation can alter ion yields unpredictably. It has been possible to produce uniform reactively doped tantalum for use as thin film standards. The concentrations of impurities in these films were calibrated using the electron microprobe. Accurate standards can be prepared for semiconductors by using ion implantation techniques.

It is expected that once escape depths, back-scattering correction factors, and Auger currents can be conveniently and accurately measured, then the quantitative capabilities of AES will approach that possible with the electron microprobe. It will be more difficult to make the SIMS technique as quantitative as the electron microprobe due to the presence of chemical and orientational effects on secondary ion yield and the possibility of mass interference effects. However, high mass resolution SIMS is possible and the use of low energy secondary ion discrimination does minimize molecular and polyatomic interferences. In addition, a controlled oxygen jet on the sample during primary reactive or inert ion bombardment will reduce secondary ion yield variations due to chemical and orientational effects in many, but unfortunately not all, cases.

Acknowledgments

The authors acknowledge with pleasure the cooperation of a number of colleagues, in particular, D. Gerstenberg, J.C.C. Tsai, R.J. Scavuzzo, M.J. Rand, and J.H. Thomas. P. Palmberg (Physical Electronics, Edina, Minnesota) made Figs. 11 and 12 available and D.K. Conley (Western Electric, Allentown) provided Figs. 20 and 21.

References

1 C.A. Evans, Jr. and J.P. Pemsler, Anal. Chem., 42 (1970) 1130.
2 B. Blanchard, N. Hilleret and J.B. Quoirin, Pittsburgh Anal. Chem. Appl. Spectrosc. Conf., 171 (1973) 131.
3 F. Schulz, K. Wittmaack and J. Maul, Radiat. Eff., 18 (1973) 211.
4 R.W. Berry, P.M. Hall and M.T. Harris, Thin Film Technology, Van Nostrand—Reinhold, Princeton, 1968, Chap. 7.
5 W.H. Jackson and R.J. Moore, Proc. Electron Components Conf., Washington, D.C., (1965) 45.
6 L. Braun and D.E. Lood, Proc. IEEE, 54 (1966) 1521.
7 R.W. Berry and D.J. Sloan, Proc. IRE, 47 (1959) 1070.

8 D. Gerstenberg and J. Klerer, Proc. Electron Components Conf., Washington, D.C., (1967) 77.
9 D. Gerstenberg, J. Electrochem. Soc., 113 (1966) 542.
10 L.G. Feinstein and D. Gerstenberg, Thin Solid Films, 10 (1972) 79.
11 J.M. Morabito, Anal. Chem., 46 (1974) 189.
12 R.D. Huttemann, J.M. Morabito and D. Gerstenberg, IEEE Trans. Parts Mater. Packag., 11 (1975) 67.
13 A.S. Grove, Physics and Technology of Semiconductor Devices, Wiley, New York, 1967, Chap. 3.
14 C.J. Frosch and L. Derick, J. Electrochem. Soc., 104 (1957) 547.
15 J.W. Mayer, L. Eriksson and J.A. Davies, Ion Implantation in Semiconductors, Academic Press, New York, 1970.
16 R.A. Moline, J. Appl. Phys., 42 (1971) 3553.
17 T. Tokuyama, T. Ikeda and T. Tsuchimoto, 4th Microelectronics Congr., Munich, Oldenbourg Verlag, Munich, 1970, p. 36.
18 B. Fair and G.R. Weber, J. Appl. Phys., 44 (1973) 273.
19 K.H. Nicholas, Solid State Electron., 9 (1966) 35.
20 J.M. Morabito and J.C.C. Tsai, Surface Sci., 33 (1972) 422.
21 O.S. Heavens, G. Hass and R.E. Thun (Eds.), Physics of Thin Films, Vol. 2, Academic Press, New York, 1964, p. 193.
22 N. Schwartz and R. Brown, Trans. 8th AVS Symp. and 2nd Int. Congr. Vacuum Sci. Technol., Pergamon, New York, 1961, p. 836.
23 S. Tolansky, Surface Microtopography, Interscience, New York, 1960.
24 Physical Electronics Industries, Inc., Edina, Minnesota.
25 Applied Research Labs., Sunland, Calif., Cameca Instruments, Elmsford, New York.
26 J.W. Guthrie and R.S. Blewer, Rev. Sci. Instrum., 4 (1972) 654.
27 P.W. Palmberg, J. Vac. Sci. Technol., 10 (1973) 274.
28 J.M. Morabito and M.J. Rand, Thin Solid Films, 22 (1974) 293.
29 T.W. Haas and J.T. Grant, Appl. Phys. Lett., 16 (1970) 172.
30 J.M. Morabito and R.K. Lewis, Anal. Chem., 45 (1973) 869.
31 R. Hernandez, P. Lanusse, G. Slodzian and G. Vidal, Method. Phys. Anal., 6 (1970) 411.
32 R. Castaing and J. Hennequin, Advan. Mass Spectrometry, 88 (1972) 419.
33 R.K. Lewis, J.M. Morabito and J.C.C. Tsai, Appl. Phys. Lett., 23 (1973) 260.
34 J.M. Morabito, Thin Solid Films, 19 (1973) 21.
35 R.E. Weber and A.L. Johnson, J. Appl. Phys., 40 (1969) 314.
36 F. Meyer and J.J. Vrakking, Surface Sci., 33 (1972) 271.
37 J. Philibert, in R.F. Bunshah (Ed.), Modern Analytical Techniques for Metals and Alloys, Vol. III, Part 2, Interscience, New York, 1970, pp. 419--531.
38 M.A. Nicolet, J.W. Mayer and I.V. Mitchell, Science, 177 (1972) 841.
39 G. Amsel and D. Samuel, Anal. Chem., 39 (1967) 1689.
40 P.W. Palmberg and T.N. Rhodin, J. Appl. Phys., 39 (1968) 2425.
41 T.E. Gallon, J. Phys. D, (1972) 822.
42 R.L. Gerlach and A.R. Ducharme, Surface Sci., 32 (1972) 329.
43 C.A. Andersen and J.R. Hinthorne, Anal. Chem., 45 (1973) 1421.
44 R. Castaing, Proc. 8th Nat. Conf. Electron Probe Anal., New Orleans, 1A (1973).
45 H.W. Werner, Develop. Appl. Spectrosc., 7A (1969) 239.
46 H.W. Werner and H.A.M. deGrefte, Surface Sci., 35 (1973) 458.
47 A. Benninghoven, Surface Sci., 35 (1973) 427.
48 1972 Fall Workshop on Surface Analysis and Secondary Ion Mass Analysis, Tarrytown, N.Y., Nov. 2—3, 1972.
49 R.E. Pawel, J.P. Pemsler and C.A. Evans, Jr., J. Electrochem. Soc., 119 (1972) 24.
50 D.M. Smyth, G.A. Shirn and T.B. Tripp, J. Electrochem. Soc., 111 (1964) 1331.

328

51 R.T. Simmons, P. Morzenti, D.M. Smyth and D. Gerstenberg, Thin Solid Films, 23 (1974) 75.
52 R.T. Simmons, M.S. Thesis, Lehigh University, 1972.
53 P. Richman, Characteristics and Operation of MOS Field-Effect Devices, McGraw-Hill, New York, 1967.
54 J.P. Mitchell and D.K. Wilson, Bell Syst. Tech. J., 46 (1967) 1.
55 M.J. Rand, J. Electrochem. Soc., 120 (1973) 686.
56 J.T. Grant and T.W. Haas, Appl. Phys. Lett., 15 (1969) 140.
57 A. Hiraki, E. Lugujjo and J.W. Mayer, J. Appl. Phys., 43 (1972) 3643.
58 D.S. Campbell, in L. Maissel and R. Glang (Eds.), Handbook of Thin Film Technology, McGraw-Hill, New York, 1970, Chap. 12.
59 R.C. Sundahl, J. Vac. Sci. Technol., 9 (1972) 181.
60 D.K. Conley, Proc. 8th Nat. Electron Probe Anal., New Orleans, 21A (1973).
61 D.K. Conley, Proc. 6th Nat. Electron Probe Anal., Pittsburgh, 9A (1971).
62 J. Lindhard, M. Scharff and H. Schiott, Mat. Fys. Medd. Dan Vid. Selsk, 33 (1973) 1.
63 J.H. Thomas and J.M. Morabito, Surface Sci., 41 (1974) 629.
64 G. Dearnaley, M.A. Wilkins and P.D. Goode, 2nd Int. Conf. Ion Implantation in Semiconductors, Springer-Verlag, Berlin, 1971, p. 439.
65 J.C.C. Tsai, J.M. Morabito and R.K. Lewis, 3rd Int. Conf. Ion Implantation in Semiconductors, Pergamon, New York, 1973, 89—98.
66 R.J. Scavuzzo, R.S. Payne and J.M. Morabito, unpublished Bell Telephone Laboratories data, 1971.
67 J.C. Irvin, Bell Syst. Tech. J., 41 (1962) 387.
68 R.P. Gittins, D.V. Morgan and G. Dearnaley, J. Phys. D, 5 (1972) 1654.
69 M. Croset, Rev. Tech. Thompson-CSF 3, 1 (1971) 19.
70 W.K. Hofker, H.W. Werner, D.P. Oosthoek and H.A.M. de Grefte, in B.L. Crowder (Ed.), Ion Implantation in Semiconductors and Other Materials, Plenum Press, New York, 1973, p. 133.
71 J.F. Ziegler, G.W. Cole and J.E.E. Baglin, Appl. Phys. Lett., 21 (1972) 177.
72 M. Baunis, B. Blanchard, M. deBrebisson and J. Monnier, Electrochem. Soc. Meet., Miami, Florida, 1972, Abstr. 266.
73 R.D. Dobrott, F.N. Schwettmann and J.L. Prince, Proc. 8th Nat. Electron Probe Analysis, New Orleans, 10A (1973).
74 G. Slodzian, Abst. Pittsburgh Anal. Chem. Appl. Spectrosc., 95 (1973) 113.
75 G. Slodzian and J.F. Hennequin, C.R. Acad. Sci. B, 263 (1966) 1246.
76 G. Blanchard, N. Hilleret and J. Monnier, Mater. Res. Bull., 6 (1971) 1283.
77 R.F. Herzog, W.P. Poschenrieder and F.G. Satkiwicz, NASA Rep. CR-683, 1967.
78 R. Castaing and G. Slodzian, J. Microsc. Paris, 1 (1962) 395.

Chapter 8

THE ATOM-PROBE FIELD ION MICROSCOPE

E.W. MÜLLER

I. Introduction

The atom-probe field ion microscope is a microanalytical tool of ultimate sensitivity to be used in basic investigations of surface phenomena as well as in metallurgical applications. The operator may view his specimen in full atomic resolution, pick up at his discretion a single surface atom or a few atoms or molecules from a selected surface area and identify these particles by their mass. Following the introduction of the field emission microscope [1] the basis of field ion mass spectrometry was laid in 1941 with the discovery of the new physical effect of field desorption [2]. For the first time, a well-defined electric field exceeding 10^8 V/cm or 1 V/Å was applied to a surface with the result that an adsorbate could be removed in the form of ions essentially without thermal activation and at a perfectly controlled rate. Ten years later, the experimental verification of the so far only theoretically known effect of field ionization was achieved with the introduction of the field ion microscope [3]. At this time, the gradual dissolution by a high field of the surface lattice of the emitter tip itself, later named field evaporation, was also first noticed, although at $T > 800$ K. In a lecture at the University of Chicago, Müller [4] suggested the use of a field ion emitter as a source for a mass spectrometer, and in 1954 the first quite exciting experimental results of field ion mass spectrometry were published by Inghram and Gomer [5]. Mass spectrometry of externally supplied gases with a field ionization source has since then developed to a productive branch of physico-chemical analysis, particularly by the efforts of Beckey [6] and Block [7]. The investigation of products of field evaporation proved to be experimentally more difficult where, because of the limited supply of particles, the emitter tip itself is consumed in the process. Similarly, adsorbed species could be detected only by a train of repetitive desorption pulses between which the adsorbate was replenished from the gas phase, with the complications by superimposed effects due to free-space field ionization.

By imaging a field ion emitter through a wide aperture einzellens and a magnetic deflection field onto a phosphor screen, Müller [8] found tungsten to field evaporate at room temperature as a multiply charged ion. Thomson

[9] observed that copper field evaporated preferentially as a hydride in the presence of hydrogen. Later, Vanselow and Schmidt [10] were able to increase the sensitivity of their mass spectrometer to detect singly charged ions of Pt field evaporating at $T > 850$ K. The first analysis of low-temperature field evaporation became possible with a $60°$ magnetic deflection second order focusing mass spectrometer by Barofsky and Müller [11]. In this instrument, the slowly field evaporating tip was focused onto an exit slit which was followed by a resistance strip electron multiplier as a detector. The spectrum was scanned by varying the magnetic field, with the practical problem of having enough of an evaporation rate at the instant when one "spectral line" just passed the exit slit, while evaporating ions of other species were wasted. With this apparatus, field evaporation products of Be, Fe, Ni, Cu and Zn tips were successfully investigated, and particularly the change of multiplicity of ionic charge with temperature was observed. In the presence of evaporation promoting hydrogen, the abundant formation of metal hydride ions, except for zinc, was also well established. The sensitivity and the signal-to-noise ratio of this mass spectrometer required the evaporation or desorption of the order of 3 monolayers/sec, so that with the given small transmission at least 100 ions would be received while one ion species passed the exit slit.

Since almost twenty years field ion microscopy has been a well established technique uniquely permitting the viewing of metal surfaces in atomic resolution. A finely etched tip of the metal to be investigated is the specimen. Its surface, perfected by field evaporation to form a hemisphere of typically 200—2000 Å radius, is radially projected onto a phosphor screen, using preferably field ionized helium, neon or hydrogen. The many crystal facets exposed at the tip surface are revealed in million times magnification and a resolution of 2.5 Å to show the individual atoms as building blocks of the crystal lattice. The direct visualization of lattice imperfections such as vacancies, interstitials, radiation damage, dislocations slip bands and grain boundaries, as well as access to the depth of the specimen by controlled layer-by-layer field evaporation was well established by the present author's work [8] before 1960. The obvious applications to surface physics and metallurgical problems led to major advances in the following decade. Yet, in spite of detailed investigation of the imaging process, a limitation remained in the inability to identify by its appearance the chemical nature of an atom species, be it a constituent of an alloy or an impurity atom at a particular surface location. This atomic surface analysis has now been made possible by the author's idea of the combination of a field ion microscope with a mass spectrometer of single ion sensitivity, employing the selection of a test site of atomic dimension by a probe hole in the image screen.

II. Principles of atom-probes

A basic advance leading to the identification of the species associated with one individual atomic image spot as selected by a probe hole in the screen was achieved with the overcoming of the noise problem using the idea of correlating the field evaporation event with the detection event. Such a device [12,13], for which the name atom-probe field ion microscope has been accepted, may employ various mass spectrometric principles. In the time-of-flight version, the field evaporation pulse is used to trigger the time-read-out of an oscilloscope sweep or of an electronic clock, thereby gating the detector for a period of $10-20$ μs within which the occurrence of a noise signal is unlikely [14]. In a magnetic sector mass spectrometer [15], the correlation is spatial when individual ion signals are displayed on a screen at the known location of a "spectral line," or it may be combined with a time correlation by gating the detector display screen electronically or optically through a photographic shutter synchronized with a "slow" evaporation pulse at the specimen tip. When the concept of a field desorption microscope became a reality with the availability of the channel plate, the entire tip cap could be imaged by surface ions from a single desorption pulse [16]. A "10 cm atom-probe" has been developed by Panitz [17] employing photoelectric pick-up of the screen image or pulse-gating of the channel plate [18] for obtaining a one-shot picture of the distribution of an atomic species over the entire tip area.

III. Models of field ionization and field evaporation

Comprehensive theoretical treatments of the basic effects of field ionization and field evaporation are available in the literature [19,20]. In the context of the present subject of atom-probe field ion microscopy, we need to consider only briefly the elementary concepts of the theory. In the present review, a number of fundamental discoveries made with the atom-probe will be described which eventually will require considerable modifications and refinements of the theory.

Field ionization occurs when an atom or a molecule in a very high electric field loses an electron by quantum mechanical tunneling through the barrier formed by the potential trough of the ion core and the external field. This process was conceived by Oppenheimer [21] in 1928, and verified for hydrogen by Müller [3] when the required field of the order of 3 V/Å became experimentally accessible. Subsequently, Inghram and Gomer [5] suggested the one-dimensional WKB treatment of field ionization in free space and near a metal surface. The barrier penetration probability is

$$D(E, V_{(x)}) = \exp\left\{-\left(\frac{8m}{\hbar^2}\right)^{1/2} \int_{x_1}^{x_2} [V_{(x)} - E]^{1/2} dx\right\} \tag{1}$$

where $V_{(x)}$ is the potential and E the total energy of the electron, while the barrier lies between x_1 and x_2. Near a metal surface, the barrier is reduced by the image potential, and a significant boundary condition is that the electron's energy must be above the Fermi level of the metal (Fig. 1). Thus field ionization can occur only beyond a minimum critical distance x_c as given by

$$eFx_c = I - \phi - \frac{e^2}{4x_c} + \tfrac{1}{2}(\alpha_a - \alpha_i)F^2 \approx I - \phi \tag{2}$$

where I is the ionization energy of the atom, ϕ the work function of the metal surface, and α_a and α_i the polarizabilities of the atom and the resulting ion, respectively. For more advanced treatments using three-dimensional quantum mechanical tunneling and for further refinements such as field penetration into the metal surface and the complexities of the gas supply to a surface site as affected by atomic polarization which are essential for field ion microscopy but less so for atom-probe work, we must refer to the literature [20].

There is no basic difference between field desorption and field evaporation, the former referring to the field-induced removal of an adsorbed atom, the latter of a lattice atom. Compared with field ionization, the process is much more difficult to handle theoretically because the essential electronic transition occurs much closer to the surface than in field ionization. Any one-dimensional calculation cannot be more than an approximation, while a three-dimensional treatment is forbiddingly complicated.

A useful first approximation is the image force model introduced by Müller [22]. The metal ion of charge ne escapes over a barrier formed by the superposition of the potential $-neFx$ as provided by the applied field to the image potential energy $-(ne)^2/4x$ which holds the ion at the surface. Thus, the barrier is reduced at the Schottky saddle to $(n^3e^3F)^{1/2}$ below the zero field value (Fig. 2). The activation energy $Q_n(F)$ for field evaporating an n-fold charged ion by the field F is given by

$$Q_n(F) = Q_0 - (n^3e^3F)^{1/2} = \Lambda + \sum_n I_n - n\phi - (n^3e^3F)^{1/2} \tag{3}$$

where Q_0 is the energy required to remove a neutral surface atom to infinity as an n-fold charged ion in the absence of a field, Λ being the sublimation energy, the $\sum_n I_n$ the total ionization energy, and ϕ the work function. The time required to provide the activation energy is $\tau = \tau_0 \exp(Q_n/kT)$, where τ_0 is the vibrational time of the considered surface atom. Solving for the field required for evaporation as an n-fold charged ion, we obtain

$$F_n = m^{-3}e^{-3}(\Lambda + \sum_n I_n - n\phi - kT \ln\tau/\tau_0)^2 \tag{4}$$

After it had been realized that barium desorbs from tungsten most likely as a

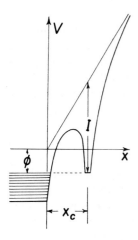

Fig. 1. Potential energy of an electron near a metal surface, with a field ionizing gas atom at the critical minimum distance x_c.

doubly charged ion [22], Brandon [23] affirmed that for most metals the evaporation field should be the lowest for double rather than single charges. The order of magnitude of experimental evaporation fields agrees well with the results of the simple theory, as uneasy as one may feel about the validity of the image force model so very close to the three-dimensional surface. For the case of field desorption, Gomer [24] and Gomer and Swanson [25] had suggested a charge exchange model which requires knowledge of localized bonds and well-defined energy levels of a metal atom and an ion close to the surface about which a considerable uncertainty again prevails. Even after advanced theoretical and experimental investigations including field evaporation rates, it is still not possible to decide definitely which of the two models should be preferred [20], but the image force seems to describe the data somewhat better [26,27].

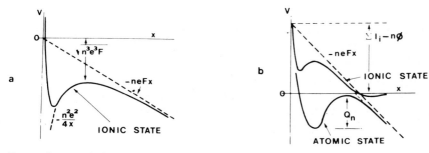

Fig. 2. Potential diagrams for field evaporation. (a) The image force model, (b) the charge exchange model.

References pp. 375—378

334

IV. The TOF atom-probe

A. DESIGN CONSIDERATIONS

The straight time-of-flight (TOF) atom-probe field ion microscope conceived by the author [12] in 1967 and repeatedly improved is schematically shown as the 1973 version in Fig. 3 [14]. The direction of the FIM tip may be manipulated to shift the ion image over the screen so that the operator can place a selected atomic image spot over a central probe hole. To the image voltage V_{dc}, a pulse voltage V_p of a few nanoseconds duration is superimposed. Multiply ionized field desorbing or evaporating surface atoms will acquire the kinetic energy $neV_e = ne(V_{dc} + V_p)$ very near the tip, and those admitted through the probe hole will travel at constant velocity through the drift path of total length l until they arrive at a detector of single ion sensitivity. The time of flight t, typically in the $1-20$ μs range, is indicated by a spike on an oscilloscope trace, or measured by an electronic clock. The correlation of the evaporation and detection events is provided by triggering the oscilloscope sweep or the clock by the evaporation pulse. The identifying mass-to-charge ratio, m/n, is finally obtained by solving the energy equation $mv^2/2 = neV_e$, with the velocity known by the measured time of flight, to be

$$m/n = 2e(V_{dc} + V_p)t^2/l^2 \tag{5}$$

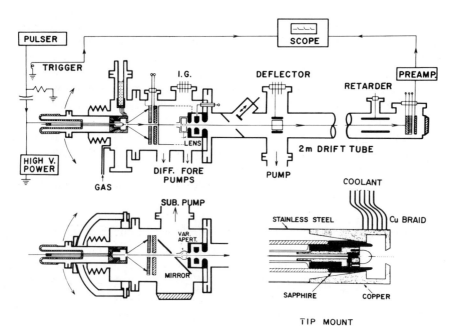

Fig. 3. TOF atom-probe [14].

As n is a small integer from 1 to 4 and is usually known by experience, the atomic or molecular ion is unambiguously identified by its mass in atomic mass units (amu).

After a simple glass—metal design of a prototype TOF atom-probe had been operated successfully [13], several other modified instruments with all-metal vacuum systems have been constructed in the author's laboratory [28—33] and by other investigators [34]. A well-engineered design was described in detail by Brenner and McKinney [35]. The tip direction is manipulated via an internal gimbal system, and the tip is connected to the cryogenic cold finger by a flexible copper braid, which simultaneously serves as the electrical lead for the d.c. and pulse voltage. The screen with the probe hole is actually a micro-channel plate with a proximity focused phosphor screen, which allows a comfortable viewing of the faint field ion image by converting it into an about 1000 times intensified electron image. At the end of a 70 cm drift path is a funneled curved channeltron as the detector. This instrument performed very well in metallurgical applications where only modest mass resolution sufficed.

Turner et al. [36—38] have constructed an atom-probe with a similarly designed tip manipulation. The channel plate ion electron image converter uses magnetic focusing, which gives a somewhat sharper and brighter image of the specimen surface. The small electric field between the channel plate and the screen is thought to cause less defocusing of the ion beam than in a proximity channel plate-screen assembly. However, the disturbing field in the probe hole of the channel plate itself remains the same, and it appears that pulsing is done better with the image intensifier turned off, although in the ordinary use of the atom-probe it is convenient and desirable to have the image gas present during pulsing. The published ion image contains a dead spot ten times larger in diameter than the effective probe hole. The various potentials over relatively long sections of the total 160 cm drift path and the lack of differential pumping of the drift tube may significantly contribute to the large number of unidentifiable random time signals, appearing as "grass", of up to 10% of the largest peak height in the automatically recorded mass spectra. This situation has been somewhat improved by the incorporation of a 15° electrostatic prism system into the drift tube, which deflects from the detector the disturbing lower energy ions created in the gas by free-space ionization and collision processes.

The TOF atom-probes in the author's laboratory have been repeatedly improved as the need for a better mass resolution became more obvious. The design used until the end of 1973 is shown in Fig. 3 [14,20]. The tip is moved by an external bellows-sealed gimbal system. Although more bulky than an internal manipulator, it allows for the provision of a fixed geometry of constant impedance for the pulse lead [33] rather than a loop of posi-tion-dependent inductance. The 2.5 mm probe hole in the proximity focused channel plate-screen assembly can be effectively reduced to 1 mm diameter by a variable aperture behind the viewing mirror. The narrow communication between the microscope chamber and the separately pumped drift tube permits the maintenance of a two orders of magnitude pressure differential. The two instrument chambers made of stainless steel with metal gasketed flanges are pumped by separate liquid nitrogen trapped oil diffusion pumps. In addition, there is a large liquid nitrogen cooled titanium sublimation pump attached directly to the microscope chamber in order to retain a low background pressure of residual hydrogen and water vapor when the main diffusion pump is valved off while the noble imaging gases, usually helium or neon at 10^{-5} to 10^{-6} torr, are leaked in. The system is occasionally baked at $250°$C. The background vacuum, as measured by a nude Bayard—Alpert gauge, reaches the high 10^{-9} torr range without baking after a tip specimen has been replaced, and at least an order of magnitude better after baking. The long total drift path of 250 cm is favorable for a high resolution, but it requires the use of an einzellens for focusing the ions onto the distant detector. Auxiliary electrodes in the drift tube are x—y deflection plates, sometimes used to direct the beam exactly onto the center of the detector, and a cylindrical retarder [33], which has been employed to measure energy deficits of field evaporated ions.

B. DETECTORS

A critically important part of a TOF atom-probe is the ion detector, which is an electron multiplier of suitable design. After several years of experimentation with various types such as venetian blind multipliers, curved channeltrons and flat strip-magnetically focused multipliers, the Penn State instrument was finally equipped with a stack of two microchannel plates (Chevron—CEMA of Galileo Electro-Optics Corp.) followed by a conductively coated phosphor screen as the output electrode. With a gain of the order of 10^7, single ions can easily be detected by coupling the oscilloscope via an FET preamplifier. Moreover, the ion arrival can be seen or photographed as a bright light flash on the screen and, by using the strong image gas ion beam emerging from an individual atomic site on the specimen, the alignment and focusing condition of the instrument can be judged. The great disadvantage of the channel plate of only 60% efficiency must be considered acceptable until a better detector becomes available.

The obvious requirements for an atom-probe detector are a gain of the order of 10^7, a high efficiency approaching 100%, a narrow pulse height distribution, a low "dark current" counting rate or noise and a rise time in the nanosecond range. The primary dynode should be flat to define an exact path length for the incoming ions. This is not the case for the venetian blind type [13,29,30] and is particularly bad for the curved channeltron (Bendix CEM 4028). The latter has a funnel-shaped opening with an active depth of the order of 7 mm, which when used in a short (70 cm) drift tube [35] represents an uncertainty of the path length. This results in a 2% statistical error of mass determination, wiping out the possibility of metal hydride ion detection. A flat primary impact dynode, of a fairly large area is offered by the Bendix M306 magnetically focused strip electron multiplier, which has been extensively used in atom-probes [31—33,36—38]. However, all of these detectors suffer from the occurrence of afterpulse signals which appear some 50—1000 ns after a primary ion impact event and cannot be distinguished from a real ion signal. Brenner and McKinney [39] have studied the afterpulse artifacts of the curved channeltron and the magnetic strip detector, finding an intolerable occurrence of up to 10 or 20% of the primary pulses when the dynode voltage is increased for improving the gain. The afterpulse spectrum of the channeltron peaks below a time gap of 100 ns and extends to 250 ns, while the magnetic multiplier has most of its afterpulses following the primary pulse by 200—800 ns, with a peak at 350 ns. The authors conclude that to reduce the afterpulse incidence to less than 3%, the multipliers must be operated at a low gain which necessitates additional external amplification, a practice that has been employed for some time with the Penn State instruments [31—33]. Brenner and McKinney also studied the afterpulse performance of a home-built double channel plate-screen assembly for which they found a useable performance with 2% afterpulses by optimizing the individual plate voltages. Subsequently, Müller et al. [40] demonstrated that a modest incidence of afterpulses in the M306 did not exclude the possibility of statistically assuring the existence of closely spaced real ion time signals such as those obtained from Rh^{2+} and $RhHe^{2+}$ ions, provided that the atom-probe offers a high resolution by the use of a long (235 cm) drift path.

Fortunately, the problem of afterpulses has become irrelevant with the finding that a well-built double channel-plate detector does not seem to have any afterpulses at all. The limit of no afterpulses following 400 neon ions of 15 kV within a time span of 20—1000 ns reported in the first communication [40] has since been extended to no afterpulses after more than 1000 heavy metal ions. This detector is now exclusively used in the advanced Penn State instruments [14,41—43] which will be described later.

So far, no satisfactory explanation has been suggested for the origin of detector afterpulses. Most observations seem to be compatible with the release of a negative ion by the impact at the first dynode of the primary positive ion. The afterpulse time gap represents the hopping time of the slow negative ion, which upon returning to the primary dynode neutralizes, sending its electron into the multiplication channel. The reduction of channeltron afterpulse incidence to 2%, as observed by Brenner and McKinney [39] to occur with a negative bias of the primary dynode, may result from preventing some of the negative ions to fall back onto the dynode. With our afterpulse-free double channel plate detector-screen assembly, we have observed visually the appearance of about 10% of spurious output events surrounding a focused beam of image gas ions as soon as the front surface was biased positively by 20—60 V. While under ordinary operating condition with a highly negative front plate (output screen near ground potential), the negative ions are emitted away from the front plate but they will return to it after a small hop when a retarding field of the order of 10—30 V/cm is acting at the front plate. This explanation also suggests that such negative secondary ion spots from a channelplate are the most likely cause of the extra "atoms" seen in the center of the actually empty (011) plane shown in the "10 cm atom-probe" developed by Panitz [18]. Here, the small retarding field that returns the negative secondary ions is provided by the highly negative electrode near the tip, which "sees" the channel plate through the wide opening in the ground potential wall of the microscope chamber.

C. PULSERS

To the imaging voltage V_{dc}, which may be taken from any well-regulated d.c. power supply (6—25 kV), a pulse voltage V_p must be superimposed to initiate field evaporation. The pulse amplitude must be reproducible, while its absolute value may be determined by calibrating the instrument with a known mass [44]. Signals from a field desorbed noble gas or from field evaporating a tip of a single isotope metal such as Rh, Nb or Ta may be used. A fast rise time, preferably less than a nanosecond, and a slight overshoot to assure evaporation at the beginning of the pulse only are desirable. To avoid late evaporation, the pulse duration should be just long enough to let the heaviest ion travel through the acceleration zone from the tip to the first grounded electrode. In the Penn State instruments, this distance is 1 mm, and thus a 10 ns pulse length is sufficient. The maximum pulse amplitude should be at least 3—4 kV, as it is desirable to view a tungsten tip at best image voltage, while its evaporation voltage is about 25% higher.

Such pulses are most conveniently obtained by suddenly connecting a length of a charged coaxial cable with an output cable of the same impedance. The pulse propagates along the output cable with an amplitude of one half of the voltage at the charged line, and the pulse duration equals the time an electromagnetic wave takes to propagate back and forth the charged line at about 2/3 of the velocity of light, or 1 ns for every 20 cm cable length. A short rise time requires an effective high voltage switch, a mercury whetted, electromagnetically triggered reed contact in a high pressure inert atmosphere. In their early work, Brenner and McKinney [35] also experimented with a triggered gaseous discharge tube as a switch or with a power-tube amplified pulse, which both do not seem to give a sufficiently short rise time. A severe problem is the impossibility to terminate the pulse line at the tip end with a low impedance. Thus, overvoltages of unknown amplitude may occur, which are constant only when a geometrically fixed connection to the tip is provided such as in the concentric lead of the various Penn State designs [29—33,41—43]. The consequences of a less than ideal shape of the pulse front will be discussed later.

D. TIME-OF-FLIGHT READ-OUT

Since the introduction of the atom-probe, the time signal read-out has been most reliably made by a fast oscilloscope, the sweep of which was triggered by the evaporation pulse. The 150 MHz Tektronix 454 or, better, the 350 MHz Tektronix 485 with its sharper beam, allows a time determination to ±20 or 40 ns when the entire range of the mass spectrum is to be covered with a 10 or 20 μs sweep. For high resolution work, particularly with the new energy focused atom-probe to be discussed below, a delayed onset of the sweep may be used to display a selected time slot around the mass range of special interest. At a sweep rate of 50 ns/div. a read-out of ±1 ns can be attained and may sometimes be needed for fully utilizing the potential mass resolution. The time traces are photographically recorded with up to 60 traces on one sheet of Polaroid film for economy and an easy survey of the abundance of various ion species in repeated shots (Fig. 4).

For less exact demands in time read-out precision, a storage oscilloscope [35] may be useful, although the beam trace is somewhat blurred. This mode of operation is more economical when very low evaporation rates are used so that a time signal is rarely found on a single sweep.

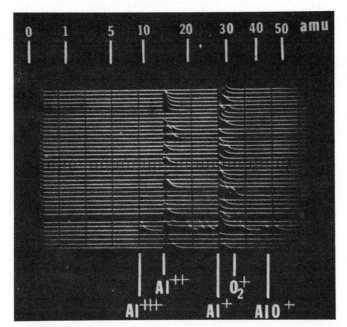

Fig. 4. Oscilloscope traces from an aluminum tip with argon as the image gas.

References pp. 375—378

340

Although direct oscilloscopic time read-out gives the most precise results, the operation is tedious and the accumulation of data for bulk analysis by extended evaporation sequences is slow. A computer-drawn plot of 1360 time signals from a rhodium tip is shown in Fig. 5. With a relatively modest investment in electronic circuitry, the time signals can be measured by an electronic clock working with a frequency commensurate with the required time resolution [45,46]. In addition, controls may be added that repeat the evaporation pulses at a desired rate until a preset number of ions arriving at the detector have been collected. Such a timer and event accumulator was first constructed by Johnson [47] for use with the Brenner—McKinney instrument. It provided a time resolution of ± 40 ns and could register up to five arrival events following an evaporation pulse. A similar but more elaborate timer and evaporation control has been developed by Turner et al. [37,38], in which the time-of-flight is measured by counting the number of pulses from a 82 MHz clock between the evaporation pulse and the detector signal, thereby achieving a resolution of 12 ns. The system counts only two subsequent arrivals following the evaporation pulse, thus requiring very low evaporation rates and weighing the distribution of various ion species in favor of light masses. Also, the accessible time window starts beyond a dead time of 1 μs after the evaporation pulse, so that revealing signals from H^+

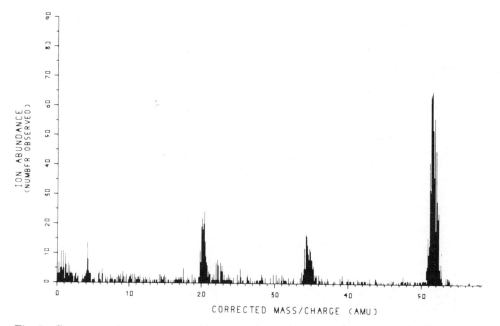

Fig. 5. Computer-plotted time signals of 1360 evaporation events from a rhodium tip with He—Ne as the imaging gas, taken with a 1968 atom-probe [30]. Shown are He^+, Ne^+, Rh^{3+}, Rh^{2+} and $RhHe^{2+}$.

and H_2^+ may be missed. The very practical features of this device are registration of the time data on a paper tape which may subsequently be fed into a computer, which calculates the mass-to-charge ratios and plots a mass spectrum.

Independently, Berger [48] has developed a circuit to be used for an atom-probe presently under construction at Cornell University. Employing a 100 MHz crystal oscillator as a clock, the time discrimination is ± 10 ns and thus sufficient for the resolution of a conventional TOF atom-probe. Up to eight arrival times following the evaporation pulse can be recorded as 4-decade decimal numbers. Another circuit with a still higher time resolution has been described by McLane [49]. A 5 ns resolution digital timer can record four sequential detector signals, and the device may be switched to act as an ion rate counter to be used for aligning and centering an ion image in the probe hole and for comparing relative brightnesses of atomic images at various surface sites.

E. MASS RESOLUTION

Assuming that the evaporation voltage $V_e = V_{dc} + V_p$ is accurately known, the mass resolution $\Delta m/m$ is essentially limited by the error in time determination Δt, as can be seen by differentiating eqn. (5).

$$\frac{\Delta m}{m} = \frac{(8ne V_e)^{1/2}}{lm^{1/2}} \Delta t \tag{6}$$

This result also illustrates the usefulness of a long drift path l. In various designs by Brenner and by Turner, lengths of 65—163 cm have been employed, while the Penn State instruments extended the drift paths to 235 and 250 cm.

While the d.c. voltage, the path length and the read-out time can be measured with sufficient accuracy, the effective pulse voltage is not precisely known because of a pulse shape deformation at the unterminated end of the pulse line at the tip. In addition, there may be a constant electronic time delay in triggering the oscilloscope and a transit time in the detector. Thus, it was found necessary to calibrate [44] the atom-probe with ions of known m/n. Suitable ions are those of the field desorbed imaged gases such as H^+, H_2^+, He^+, Ne^+ and Ar^+ as well as field evaporated ions from single isotopes metal tips such as Be, Al, Co, Nb, Rh, Ta and Au. With these metals, the possibility of hydride ions formed in the presence of a small background pressure of residual hydrogen must be considered. When the time signals are taken from counters or delayed oscilloscope sweeps, it is advantageous to read the time interval t_i between signals at t_1 and t_2 of two known masses $m = M/n$, from which one obtains the two times and the effective evaporation voltage from eqn. (5) as follows [32].

$$t_1 = t_i \left[1 + \frac{1}{(m_1/m_2)^{1/2} - 1} \right]; \quad t_2 = t_1 - t_i; \quad V_e = \frac{m_1 l^2}{2 e t_1^2} \tag{7}$$

The resolution desirable for practical applications is at least 1/200, which would marginally allow the separation of ions from metals adjacent in the periodic table, such as ^{196}Pt and ^{197}Au, of the heavy metal isotopes, and the discrimination between metal hydride ions and metal ions.

The existing atom-probes have been laid out for a resolution of this magnitude with Brenner and McKinney [35] reporting $\Delta m/m$ of about 1/100, Turner et al. [37] choosing design data to give a potential resolution of 1/180, and Müller [33] of 1/250. An inspection of published mass spectra (Fig. 5) shows that the resolutions actually achieved are in most cases much poorer, although a superior performance ($\Delta m/m < 1/500$) has also been demonstrated [33,40]. Experimental resolutions are generally derived from distribution curves obtained from a large but still very limited number of repeated shots. Thus, the practical definition of resolution used in conventional mass spectrometry of a 10% dip between two adjacent mass peaks is not applicable for the operation of the atom-probe with only a statistically insufficient small number of ions. The half-width of the distribution curve is more useful, but still unsatisfactory for judging the possibility of separating, for instance, a metal hydride ion of 1% abundance from the parent metal ions. In such a case, not one ion signal must fall beyond a ± 0.5 amu wide slot of the metal ion's mass. Obviously, the performance of an atom-probe is limited when the distribution curve has a broad base which does not clearly stand out from the general noise background or "grass" over the entire mass scale. In order to describe the resolution performance more realistically, at least the width at 10% peak height should be given in addition to the usually quoted width at half peak. If a sufficiently large number of ion signals have not been recorded, the total spread of all ions reasonably assigned to one species must be considered.

Practical results show that, usually, the expected resolution of the design is not at all achieved. A computer-plotted histogram by Turner et al. [38] shows for ^{27}Al$^+$ a half width of about 0.7 amu, representing $\Delta m/m$ of about $\pm 1/80$, while 90% of all Al$^+$ ions fall within a band of 3 amu, giving an effective resolution of $\pm 1/18$. Such an instrument thus cannot resolve the seven isotopes of molybdenum, requiring a $\Delta m/m$ of about 1/100, nor can it safely indicate the presence of an iron-hydride ion, for which a resolution of 1/56 would be required. Similarly, a recent paper by Goodman et al. [50] does not show clearly the separation of the two copper isotopes at 63 and 65 amu, and the signals are tailed by unresolved multiple hydride ions. Similarly, in a set of oscilloscope traces by Müller [33] the signals of ^{16}O$^+$ obtained as a decomposition product of adsorbed CO scatter by about one half amu. Obviously, the attainable resolution also depends upon the specimen

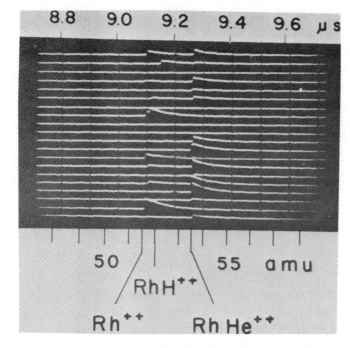

Fig. 6. High-resolution oscilloscope traces from a rhodium tip in helium.

conditions. From the same Penn State instrument, Müller et al. [40] also show traces of Rh^{2+} and $RhHe^{2+}$ taken at a sweep rate of 100 ns/div. which scatter by less than 5 ns, thereby representing a mass resolution of better than $\pm1/900$ (Fig. 6). Recent work revealing the major cause of unsatisfactory resolution and leading to a breakthrough in atom-probe development will be described in the following two sections.

F. ION ENERGY DEFICITS

The basic presupposition for achieving the potential resolution is that all field evaporated ions attain the same full energy neV_e as determined by the d.c. and pulse level voltages while passing through the acceleration zone between the tip and the aperture in the first grounded electrode. It turned out that this is not the case and that statistically distributed ion energy deficits of up to several hundred eV are the major cause for the resolution limiting width of the peaks in histograms of repeated evaporation shots. A few years ago, Lucas [51] suggested that an ion rapidly accelerated from the tip surface should lose kinetic energy by exciting, through its long range Coulomb field, collective surface modes. Excitation of several and up to 100 surface plasmon quanta of 10—15 eV was foreseen, thus placing a severe limit to the ultimate resolution of the atom probe of the order of a few percent. The advent of this theory prompted an experimental investigation [33] of possible energy deficits of field evaporated ions. A cylindrical electrode held at a potential some 200—1000 V below the tip potential retarded the ions over a section of the drift path. The arrival of ions bearing an energy deficit would be relatively more delayed than of ions with the full energy. This scheme made it possible in experiments with tantalum to discriminate unambiguously between signals delayed by the formation of TaH^{3+} from delayed signals of energetically deficient Ta^{3+} ions. In addition, small deviations in the internal consistency of precisely measured time signals from multiply charged ions of known masses such as Ne^{2+} and Ne^{+}, Ar^{2+} and Ar^{+} as well as Ta^{4+} and Ta^{3+} were investigated using eqn. (7). Evidently there were energy deficits of the order of up to several hundred volts, statistically spreading for one ion species, and distinctly different between differently charged species. However, the results showing larger deficits with increasing m/n are in disagreement with Lucas' prediction of an increase of ΔV_e with n^2. Based on our atom-probe results, Lucas and Sunic [52] have more recently concluded that field desorption should have a lesser effect on plasmon excitation than previously estimated, but they continue to see the periodic structure in the energy distribution of field ionization as discovered by Jason [53] to be due to plasmon excitation losses via long range Coulomb interaction. This interpretation, too, is in disagreement with experimental results by Müller and Krishnaswamy [54] who measured the energy of field ionization of various gases at a single atomic surface site, utilizing an atom-probe modified with the addition of a Möllenstedt energy analyzer. In particular, the absence of a periodic structure in H_3^+ ions, which can only be created at the surface and not by free space ionization, contradicts the plasmon excitation hypothesis. The real cause of the resolution limiting energy deficits in the atom-probe lies in the less than ideal pulse shape in the sub-nanosecond region of the pulse front [33]. Field evaporation occurs before the pulse voltage has reached its plateau. It had been realized earlier that an ion travels a large distance in the high acceleration zone near the tip within a time much shorter than the typical rise time of a good fraction of a nanosecond. As an example, the travel time of a Rh^{2+} ion required to reach a distance d from the tip surface is plotted in Fig. 7 when a constant voltage $V_e = 11$ kV is applied to the tip which is 0.1 cm away from the grounded electrode. The space potential V_d is calculated for a field between confocal paraboloids. It is seen that the ion falls through 10% of the total voltage just about one radius away from the tip surface within a time of 10^{-12} s, and it has already picked up one half of its final energy at a distance of about 100 tip radii or 5 μm within a travel time of 0.5 ns.

In the real situation, the accelerating voltage changes when the pulse matures during the critical first nanosecond [14]. In order to calculate the resulting final energy of the ion, Krishnaswamy and Müller [55,56] consider

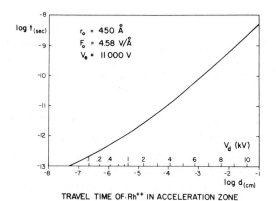

TRAVEL TIME OF·Rh⁺⁺ IN ACCELERATION ZONE

Fig. 7. Calculated travel time of a Rh^{2+} ion through the 1 mm deep acceleration zone.

the distance $r*$ reached by the ion after the time t_m at which the pulse has matured. From there on the final accelerating tip voltage V_f is constant. Knowing the space potential $V_{sp}(r*)$ due to V_f the total energy when the ion has traveled to the ground electrode at distance R is simply $1/2\, m\dot{r}*^2 + ne(V_f - V_{sp}(r*))$. The location $r*$, the velocity at this point $\dot{r}*$, and the potential of this point, $V_{sp}(r*)$ can be found with a computer solution of the non-linear, second-order differential equation of motion. Two different pulse shapes were considered, in which the final tip voltage level is approached exponentially from below, or exponentially from an overshoot. The time constant τ was chosen as 1.25×10^{-10} s corresponding to the measured rise time of 3×10^{-10} s from 10% to 90% of the pulse voltage. The axial potential distribution between the tip and the ground electrode was again represented by the paraboloid approximation

$$V_{sp}(r,t) = \frac{2V_{tip}(t)}{\ln(2R/r_0)} \{\ln(2R)^{1/2} - \ln(2r + r_0)^{1/2}\} \tag{8}$$

The equation of motion of the ion is

$$m\ddot{r} = neF(r, t) \tag{9}$$

where $F(r,t)$ is the field in the acceleration region

$$F(r,t) = -\frac{d}{dr}[V_{sp}(r,t)] \tag{10}$$

The tip voltage $V_{tip}(0) = V_e$ is an independent parameter at the instant of evaporation, where the time is zero, so that for a rising pulse

$$V_{tip}(t) = V_e + \Delta V_e[1 - \exp(-t/\tau)] \tag{11}$$

References pp. 375—378

and for a pulse decreasing after an overshoot

$$V_{\text{tip}}(t) = V_{\text{f}} + \Delta V_{\text{e}} = V_{\text{e}} \qquad 0 \leqslant t \leqslant t_1 \tag{12a}$$

and

$$V_{\text{tip}}(t) = V_{\text{f}} + \Delta V_{\text{e}}[\exp(-t/\tau)] \qquad t > t_1 \tag{12b}$$

The results of the deficit computation are given in tables in the original paper. For a typical case of Ta^{3+} ions evaporating at an instant when the tip voltage is below the nominal pulse by ΔV_{e}, the final energy deficit is $V_{\text{L}} = 0.56\ \Delta V_{\text{e}}$. Also, assuming a fixed $\Delta V_{\text{e}} = 0.055\ V_{\text{f}}$, the loss increases linearly with the tip radius, and $V_{\text{L}}/V_{\text{F}}$ is nearly constant at 0.11 for tip radii from 300 to 1000 Å. The energy loss decreases by about 40% when the mass increases from 4 to 100 amu.

Pulse evaporation may also occur at the crest of an overshoot ΔV_{e} from which the pulse diminished exponentially to the plateau as given by eqns. (12a) and (12b). Again, the relative loss $V_{\text{L}}/\Delta V_{\text{e}}$ is nearly constant at 0.37, but it is more strongly mass dependent; for 20 keV ions and a $\Delta V_{\text{e}} = 500$ V, it increases from 68 eV at mass 1 amu to 207 eV at mass 103 amu. The sign of the mass dependence of the energy deficit thus is opposite for the two shapes of the pulse considered. The shape actually occurring in the subnano-second front of the pulse cannot be studied directly by connecting the probe of a fast oscilloscope with the tip. There is a possibility of deriving the pulse shape via the field emission current obtained from a negatively pulsed tip and sampling the current in repeated, selected time intervals through a fast oscilloscope, but this experiment has not yet been carried out.

Because of the fundamental importance of energy deficits in the atom-probe, Müller and Krishnaswamy [43,57] measured the actual energy distri-bution of the field evaporated ions directly by incorporating an electrostatic prism energy discriminator (Fig. 8). First, a 90° cylindrical and later a 90° concentric-sphere deflector was used, the latter providing stigmatic imaging of the emitter tip, with the focusing on the detector double channel plate-screen assembly assisted by the einzellens. The potential distribution $\phi(r)$ in a radial field between concentric spheres of radii R_1 and R_2 at potentials V_1 and V_2 is given by

$$\frac{d\phi(r)}{dr} = \frac{1}{r^2}(V_2 - V_1)\frac{R_1 R_2}{R_2 - R_1} \tag{13}$$

The potential difference at the electrodes for guiding an ion of energy neV_{e} along the central trajectory of radius $r = R_0$ is obtained by the condition that the centrifugal and the electrostatic forces are balanced.

$$\frac{mv^2}{r} = -ne\frac{d\phi(r)}{dr} \tag{14}$$

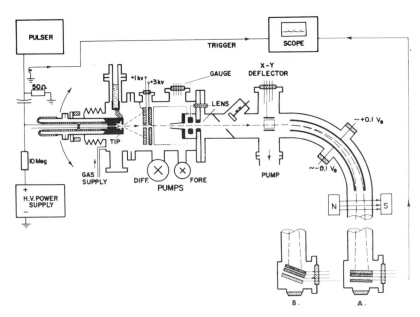

Fig. 8. Atom-probe equipped with a 90° spherical sector as an energy analyzer. For operation as an energy-deficit compensated atom-probe, the detector plane is tilted as shown in the insert, and the cross-magnetic field N → S removed.

which yields

$$V_2 - V_1 = V_e \left(\frac{R_2}{R_1} - \frac{R_1}{R_2} \right) \tag{15}$$

In addition, the condition that the central trajectory be at ground potential determines the slightly asymmetric deflection voltages to be

$$V_1 = V_e (1 - \frac{R_2}{R_1}) \text{ and } V_2 = V_e \left(1 - \frac{R_1}{R_2} \right) \tag{16}$$

Using calculations by Purcell [58], the deflection Δ^*_f from the central trajectory of an ion with an energy deficit $-e\Delta V_e$ can be calculated to be $\Delta x_f = 80 \ \Delta V_e / V_e$ cm for the given dimensions of $R_0 = 50$ cm and an additional straight path to the detector of 30 cm. The sensitivity of the device can also be calibrated by using the nearly monoenergetic ion beam obtained by field ionization of He and varying the applied d.c. voltage in steps of 100 V. The spot diameter of 0.5 mm on the screen allows an energy resolution of better than 0.05%, as seen by the separation at $V_e = 20$ kV of the first Jason peak 8 V below the main peak of the helium ion energy distribution. While the energy spectrum of the ions is displayed horizontally, a mass discrimination in the vertical direction is provided by a magnetic crossfield just behind the 90° deflector. With a proper pulse voltage, the field evaporation or desorption rate is sufficient to send 1—10 ions through the 2.5 mm probe hole. Each ion impact on the screen appears as a bright flash that can be easily recorded photographically. Spurious weak continuous emission spots on the detector screen are eliminated by using an exposure time of 1/60 s and triggering the evaporation pulse with the flash contact of the camera shutter.

A number of field evaporation and field desorption experiments were carried out in which a variety of ion masses and charge states could be compared directly under identical pulse conditions. The distribution of energy deficits of He^+, Ne^+, W^{4+}, W^{3+} was obtained from pulsing tungsten tips in helium—neon image gas mixtures. With V_e = 15—20 kV and V_p = 3 kV, He^+ has an energy spread of up to 300 eV, peaking at —130 eV below the fastest He^+ observed. Ne^+ peaks at —270 eV, W^{4+} at —330 eV and W^{3+} at —400 eV. In order to ascertain that the considerable difference between the deficits of the image gases and the tip metal ions is a matter of m/n only and not due to the apex versus lattice sites of the ions' origins, tips of two alloys, Cu—Be and Pt_3Co, were also investigated. These tips present for analysis Be^{2+} and Be^+ at 4.5 and 9 amu, Cu^{2+} at 31.5 and 32.5 amu, Cu^+ at 63 and 65 amu, Co^{2+} at 29.5 amu and Pt^{2+} from 97 to 99 amu, where the adjacent metal isotopes are not resolved by the small magnetic cross field dispersion. All these metal ions show the same general spread of energy deficits and increase with m/n. These results combined with the theoretical consideration above demonstrate that, at least in the Penn State atom-probe, the evaporation event occurs near the peak of an over-shoot and the essential acceleration takes place in the first nanosecond during which the voltage at the tip decreases. The relatively large spread of deficits of each ion species indicates a certain range within which evaporation takes place, affected by the availability of thermal activation at the instant of evaporation and by a varying effect of locally lowering the evaporation field which most likely depends upon the instantaneous position of a noble gas or impurity gas adsorbate.

It is now clear that, even at a low evaporation rate, no completely consistent mass calibration is possible in the straight TOF atom-probe. For instance, when a tungsten tip is pulsed in the presence of He and Ne as imaging gases, and the lowest (abundant) W^{3+} isotope is set at 182/3 = 60.67 amu, the location of $^4He^+$ comes out at 3.75 amu, and of $^{20}Ne^+$ at about 19.8 amu, simply because the lighter ions carry more energy by their predominant acceleration earlier in the diminishing pulse. The atom-probe resolution, as determined by the peak width in a histogram of repeated shots, is obviously very much dependent upon the sharp definition of local surface conditions at the subnanosecond instant of evaporation. There may be some fortunate situations prevailing as those that led to a resolution of 1/900 in the case of Rh^{2+} and $RhHe^{2+}$ shown in Fig. 6. The common situation prevailing at technically more interesting specimens with imperfect lattice structures as well as surfaces covered with evaporation field reducing adsorbates is such that the field binding energy, and thus the instant of evaporation in repeated shots, is poorly defined. This explains the quite unsatisfactory actual resolution limit of the order of 1/100 for the half width and 1/30 for the 10% of the peak height distribution width frequently experienced with all straight atom-probes described so far.

G. ENERGY DEFICIT COMPENSATION

The serious limitation of the resolution was easily relieved by a simple modification of the existing 90° deflector energy discriminating system (Fig. 8(b)). The plane of the detector on which the energy dispersed ion spectrum is displayed is tilted through an appropriate angle to give the slower ions of lesser energy a shorter total flight path [14,41,42]. From the basic energy equation, eqn. (5), we obtain by differentiation the path difference Δl for an ion of energy $e(V_e - \Delta V_e)$ to be

$$\Delta l = -l_c \Delta V_e / 2 V_e,$$ (17)

where l_c is the total path from the tip through the central trajectory of the deflector to the center of the detector. With the deflector's lateral energy dispersion $\Delta x_f = K(\Delta V_e / V_e')$ the tilting angle δ for compensating the velocity difference by a path difference is then $\delta = \arctan \Delta l / \Delta x_f = -l_c / 2K$. This angle is an instrumental constant and is independent of the ion mass and the applied evaporation voltage. A small correction is needed because an ion with lesser energy travelling inside the central path of the deflector in a region of lower potential will be accelerated and reduce its time loss with respect to an ion on the central path. The time of flight t_s in the 90° spherical sector field is found to be in first approximation [59—61]

$$t_s = R_0 \left(\frac{m}{2e V_c} \right)^{1/2} \left[\frac{\pi}{2} + 2(\tfrac{3}{8}\pi - 1)\Delta V_e / V_e \right]$$ (18)

while the total flight time including the straight sections l_{st} is

$$t = t_s + l_{st} \left[\frac{m}{2e(V_e + \Delta V_e)} \right]^{1/2} \simeq t_s + l_{st} \left(\frac{m}{2e V_e} \right)^{1/2} (1 - \Delta V_e / 2 V_e)$$ (19)

The energy deviation ΔV_e causes a time difference

$$\Delta t = \left(\frac{m}{2e V_e} \right)^{1/2} \Delta V_e / V_e [2R_0(\tfrac{3}{8}\pi - 1) - l_{st}/2]$$ (20)

which requires a compensating path length

$$\Delta l \approx \Delta t \left(\frac{2e V_e}{m} \right)^{1/2} = \Delta V_e / V_e [2R_0(\tfrac{3}{8}\pi - 1) - l_{st}/2]$$ (21)

The energy deficit compensation attained simply by tilting the detector plane dramatically improves the resolution performance of the atom-probe [14,41]. Although the optimum focusing of the einzellens can be made for only a median ion energy, experiments with tungsten field evaporation at 20 kV gave a time scatter of not more than ±5 ns, representing a mass resolution of ±1/1000. For the first time, the tungsten isotopes could be clearly resolved and their associated helium compound molecular ions definitely identified [62].

While the 90° spherical sector and the einzellens can focus a fairly wide aperture beam, the time-of-flight compensation by tilting the detector plane for achieving isochronous arrival of energy deficient ions works well for narrow apertures only. For attaining the resolution of 1/1000 in the present atom-probe, the effective probe hole diameter had to be limited to 1 mm. However, a probe hole of 2.5 mm is often desirable in order to have a larger transmission for identifying not only one selected atomic site but also the composition of its neighorhood by one evaporation shot. With a wide angular aperture, ions of one species travelling outside the deflector's central trajectory will be delayed, and those moving closer to the inside electrode will be accelerated throughout the entire sector. Thus, in spite of nearly perfect spatial focusing, the ions of one species will not arrive isochronously at the detector. The solution of this problem is to employ an energy discriminating deflector system in which there is a crossover of the off-central path beams half-way through the deflector, so that the delay encountered extra-centrally in the first half of the sector is exactly compensated by a gain over the intra-central path in the second half section, and vice versa for the opposite beam.

H. THE ENERGY FOCUSING ATOM-PROBE

Isochronous arrival at a detector plane normal to the central trajectory could be achieved with a 90° spherical sector by choosing a straight path section $l_{st} = 0.712 \, R_0$ as can be seen by setting $\Delta l = 0$ in eqn. (21). However, this would have required two einzellenses for focusing and a very large sector radius for matching the existing atom-probe FIM section. A first-order theory of several ingenious energy focusing configurations for TOF mass spectrometers with conventional ion sources has been presented by Poschenrieder and Oetjen [63], for which a 163.2° toroidal deflector system seemed to be adaptable to the existing Penn State atom-probe. In such an electrostatic sector field, the path length is always shorter for an ion of lesser initial energy. As this ion also picks up velocity near the inner electrode, it can regain the time loss which it had suffered relative to a full energy ion in the straight drift section. Thus, with a proper combination of a straight path length and an energy discriminating sector field, first order energy focusing may be achieved. In addition, the stigmatic imaging by the toroidal sector eliminates the need for an adjustable einzellens and reduces the setting of the instrument to choosing the proper deflection voltages V_1 and V_2, which are fixed fractions of the evaporation voltage. The schematic of this configuration is shown in Fig. 9, and its incorporation in the existing atom-probe is shown in Fig. 10. The deflector plates have principal radii of $R_1 = 31.1$ cm and $R_2 = 35.2$ cm, while the axial radii are 139 and 143 cm. The curved plates are 10 cm wide and their spacing of 4.10 cm is held to within ±0.01 cm. The tip and the detector plane are 2.35 $R_0 = 77.9$ cm away from the effective ends of the sector, as defined by Herzog fringefield shunts. The total flight path is 250 cm, after which the ions of one species are time- and energy-focused into a spot of 1 mm diameter.

The evaporation voltage range within which the time focusing is better than ±2 ns extends to ±3.5% of the median V_e for which the deflection voltages are set. For the present instrument, these voltages are $V_1 = -0.1295$ V_e and $V_2 = 0.1185$ V_e, as obtained either by calculation or by focusing with field ions from the image gas. The voltages are taken from two precision power supplies which are controlled by two ganged 10-turn potentiometers coupled via a voltage divider network to a digital voltmeter. On this, the operator needs only to dial the estimated evaporation voltage, within about ±2%, in order to set the instrument before applying the evaporation pulse. The energy focusing is so perfect that when, in a sequence of 40 shots, the tip voltage was gradually increased by up to 1400 V in order to maintain a good evaporation rate, the deflector voltages remaining constant, all ions of one species arrived within ±2 ns as recorded on the oscillogram.

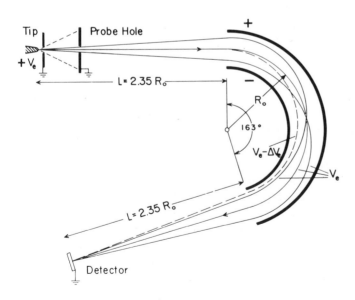

Fig. 9. Schematic of a 163° deflector, energy focusing TOF atom-probe.

References pp. 375—378

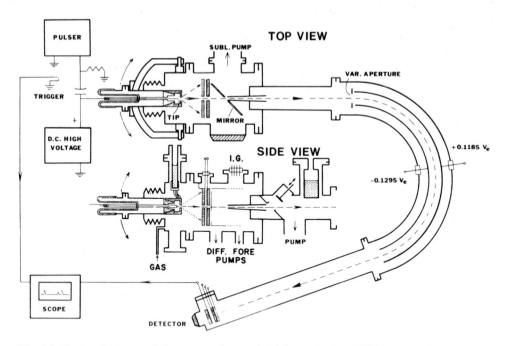

Fig. 10. Design features of the energy-focused, high-resolution TOF atom-probe.

In order to utilize fully the time resolution of the energy-focused atom-probe, read-out by one of the electronic time clocks described earlier would be inadequate. Oscilloscopic time recording with delayed triggering of a 50 ns/div. sweep using the Tektronix 454 (the 485 is better) permits a read-out to ±1 ns. The histogram of Fig. 11, taken from a rhodium tip at 18 kV, shows a width at half peak height of 2 ns, at 10% peak height of 10 ns, and that all ions are contained within a 11 ns wide time slot. Thus, the mass resolutions are

$$(\Delta m/m)_{50\%} = \pm 1/4834 \text{ and } (\Delta m/m)_{10\%} = \pm 1/967$$

while all Rh^{2+} ions appear at 51.50 ± 0.06 amu.

A mass resolution of the order of 1/1000 is also seen in an oscillogram and a histogram obtained from the field evaporation of molybdenum in the presence of 2×10^{-6} torr helium (Figs. 12 and 13). With the earlier straight atom-probes, the molybdenum isotopes could never be resolved.

A welcome additional advantage of the energy focused atom-probe is the rejection of artifact signals of ions with random energies of less than about 4% of V_e, stemming from free space ionization of the imaging and residual gases by the d.c. holding field, as well as from charge exchange on the flight path. Thus, the spectra are free from the background "grass" that made the

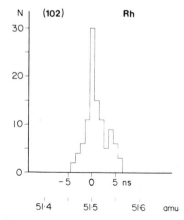

Fig. 11. Histogram of 102 time signals from a rhodium tip taken with the energy focusing atom-probe. Half width ±1 ns representing a mass resolution of 1/5000.

identification of rare ion species difficult or impossible. Together with the elimination of artifact signals from afterpulses by the double channel plate detector, the new energy-focused atom-probe design, with a resolution improved by at least an order of magnitude, seems to represent a real breakthrough in atom-probe analysis.

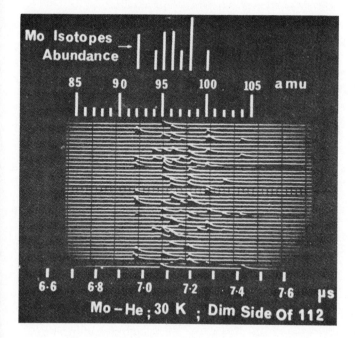

Fig. 12. Time traces at a sweep of 100 ns/div. from a molybdenum tip evaporating in helium, taken with the energy focused TOF atom-probe.

References pp. 375—378

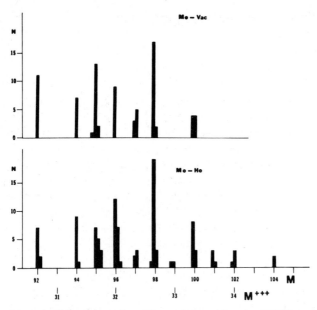

Fig. 13. Histogram of oscillographic read-out of TOF mass spectra of molybdenum evaporating in ultra-high vacuum, and in helium as an image gas, the latter showing the formation of $MoHe^{3+}$ ions.

A remaining problem of minor importance is the aiming at one particular atomic site. The field-evaporating ion does not follow exactly the trajectories of the imaging noble gas ions. The latter originate at or beyond the critical zone about 4.5 Å above the surface, giving a local over-magnification of a small net plane. The field evaporating ion's initial trajectory is much more influenced by the local atomic structure of the evaporation site. It was realized early [12,35] that, for instance, the probe hole must be placed well inside the ion image of the W (011) net plane ring to receive the field evaporating ions from the net plane edge. More recently, the multilayer field evaporation pattern of the tip as seen in the field desorption microscope (Section V) also reveals certain dark zone lines (ref. 19, Fig. 12) characteristic for each metal, which are not reached by the metal ions as a result of near-surface ion trajectory effects. If the atom probe hole is aimed at such directions, the yield is very small. However, these characteristic zones have an angular width of the order of one degree only, and the aiming is no problem when, as usual, the probe hole is larger.

V. A 10 cm TOF atom-probe

In some applications, it would be desirable to obtain a mass identification of all the surface atoms visible in an entire field ion micrograph, rather than only the few atoms within the narrow area selected by the probe hole. Imaging the tip surface by field desorption, which was first thought to be the mechanism of the FIM [3] and was later attempted by a train of successive desorption pulses between which the imaging gas was replenished by adsorption from the gas phase [15], became an experimental possibility with the advent of the single-ion detection capability of the channel plate. Walko and Müller [64] recorded the first desorption images of a tungsten tip. Because the surface features are shown by the individual field evaporated surface atoms themselves, rather than by some thousand slightly scattered image gas ions, the image is sharp even at room temperature. Müller et al. [65] suggested the use of the field desorption microscope as an atom-probe by delay-pulsing the channel plate screen for imaging only one selected ion species in order to obtain its distribution over the entire tip surface. The poor resolution, limited by the technical difficulties of sufficiently fast pulsing appropriate for the very short flight path, discouraged the Penn State group. However, Panitz [17] pursued this idea and made it a workable proposition by adding several technical features: the time-of-flight was extended by decelerating the ions over the main length of the 10 cm drift path to 3 or 4 kV. Originally, Panitz used two modes of ion detection: by picking up the signal from the conductive backing of the phosphor screen, the sequential arrival of all ions over the entire image area can be oscillographically recorded. The second detection mode is to pick up the light output of a single ion impact flash through a flexible glass fiber light pipe connected to a fast photomultiplier and an oscilloscope. The light pipe may be set at a selected image point which the operator wishes to identify. The trouble with recording the total screen output is the broadening of the signal by the different path lengths to the central and peripheral screen areas. With point recording through the light pipe, the large radial shift of the image in the instant of pulsing due to the change of FIM magnification with the voltage ratio of the three-electrode system makes the aiming questionable. Because of the short flight times, of the order of 1 μs for W^{3+}, the resolution obtained is only 1/15, and the hydrogen ions at 1 and 2 amu are lost in the initial trace transients. In a subsequent paper, Panitz [18] reports several technical improvements, such as the use of a curved double channel plate to eliminate the earlier radial path differences. He also employs the method of sequential pulsing of the channel plate, as originally suggested by Müller et al. [65], to gate the total screen output pulse for one narrow time slot of the arrival of one selected species. The gating pulse width of 10– 30 ns is a contributing factor to limiting the mass resolution, which is not explicitly mentioned except that no distinction can be made between masses 1, 2 and 3 of the hydrogen ion species. Also not discussed are energy deficits of the ions due to premature field evaporation in the long 5 ns rise time of the desorption pulse which should have a large effect on the resolution, as the energy deficits are no more insignificant in comparison with the low ion energy after retardation.

References pp. 375—378

Finally, the field desorption image seems to be marred by artifacts due to emission of secondary negative ions, which return to the front plate after a short hop by the repelling negatively biased electrode. This makes ions appear in the central region of the (011) plane, among other places, and led the author to conclusions about the "statistical nature" of the evaporation process, which are not in agreement with the well-founded experiences of field ion microscopists with perfectly controlled field evaporation. As in this system the long evaporation pulse diverges the trajectories of particles of different mass in the three-electrode field during their flight, confusion in the image is unavoidable. This situation does not occur in the standard TOF atom-probe, and drawing conclusions on the latter's aiming capability on the basis of the experience with the 10 cm atom-probe seems not to be justified. In summary, unless a considerable improvement in resolution can be achieved, the usefulness of a 10 cm drift path atom-probe appears to be quite limited. No new information on tip surface composition or processes has been obtained so far.

VI. A magnetic sector atom-probe

The novelty of the magnetic sector atom-probe [14,66] lies in the display of the mass spectrum on a screen, making it possible to record instantaneously the momentum spectrum of ions coming off the tip in d.c.- or slow pulsed evaporation, rather than scanning the spectrum sequentially through the conventional exit slit. A high mass resolution is achieved while the capability of discerning often revealing energy distributions is retained, in contrast to the high-resolution energy focused TOF atom-probe. As a number of significant results have been obtained with the new instrument, which is in several aspects complementary to the TOF atom-probe, some details of the design may be discussed.

As shown in the schematic diagram Fig. 14, the tip orientation can be manipulated by rotating the ring-shaped and bellows-sealed cold finger support over an external hemispherical steel shell centered to the tip. Thus, a selected small area of the field ion image of the tip surface may be positioned over the probe hole at the center of the 50 mm diameter channel plate-screen assembly below the tip. The screen image is viewed or photographed via a 45° inclined mirror. The probe-hole selected ion beam passes through an elliptical hole in the mirror and through the 60° magnetic sector lens. The spectrum is second-order focused by properly curved magnetic pole shoe edges to be displayed on the front plate of a 75 mm diameter double channel plate-screen assembly in the lower chamber of the instrument.

For obtaining the desired mass and energy resolution, the ion energy is usually held at 1000—2000 eV, and the necessary field at the emitter is obtained by an intermediate electrode at some 5—15 kV negative voltage.

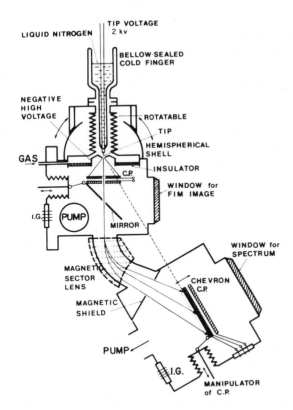

Fig. 14. Schematic of a magnetic sector atom-probe.

The variation of image magnification with voltage is not disturbing within the voltage range used, and is too insignificant in the central probe hole to cause an aiming error. The radius of the central trajectory within the magnetic sector field is 80 mm. The magnetic field in the 10 mm gap can be varied between 500 and 12,000 Gauss and thus can cover a mass range from 1 to 270 amu for 2 keV ions.

The front plane of the double channel plate on which the spectrum is displayed may be adjusted in situ to a best fit with the only very slightly curved focusing plane by rotation about two independent axes normal to the direction of dispersion.

The atom-probe is evacuated by two trapped 2 in. oil diffusion pumps and a titanium sublimation pump. The ultimate pressure is below 10^{-10} torr after a 200°C bake-out. Differential pumping was considered desirable as the double channel plates in the lower chamber should be operated below 10^{-5} torr, while up to 10^{-3} torr of an image or reactant gas may be applied dynamically around the tip. The gas supply is fed into the tip space while the pressure in the upper chamber is already low enough for the safe operation

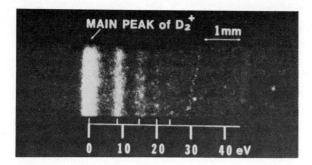

Fig. 15. Periodic peaks of the energy spectrum of deuterium.

of the single channel plate for viewing the tip surface.

The mass resolution can be demonstrated with the display of a neon spectrum, in which the isotopes 20, 21 and 22 are separated by about 10 mm and have a width of 0.1 mm, indicating a $\Delta m/m$ = 1/2000. Relative intensities may be measured by counting the number of individual ion spots that make up a spectral line. Up to 20 or 30 spots can be reliably counted in one photographed spectral line, and increased statistical significance is obtained by taking a sequence of photographs. A certain limitation is the range of the mass spectrum accessible over the width of the screen by one setting of the magnetic field, as the extreme trajectory radii in the sector are 65 and 92 mm. For example, ions in the mass range from H_3^+ to He^+, or W^{4+} to W^{3+} can be seen at one setting. The energy resolution equals the mass resolution, so that for 2000 eV ions, 1 eV energy difference can be discerned, as may be seen in Fig. 15 from a photograph of the Jason periodic ion energy distribution of deuterium. The various experimental results obtained so far with the magnetic sector atom-probe and its uniquely different behavior when compared with the TOF version will be discussed below.

VII. New phenomena observed with the atom-probe

A. MULTIPLY CHARGED IONS

While the atom-probe was originally conceived as a micro-analytical tool for identifying the atomic constituents of the specimen tip, it also revealed a number of unexpected phenomena in field evaporation and field ionization. These discoveries have extended our knowledge of the physical processes basic to field ion microscopy, and the seemingly established theories of these effects are now seen to require new approaches.

A surprise already experienced with the first operation of the prototype atom-probe [13] was the occurrence of three- and four-fold charged ions.

The more advanced first stainless steel instrument [28,67] showed tungsten and tantalum to field evaporate with three and four positive charges, and iridium with two or three charges. The bright single atoms representing the characteristic [100] zone decorations on tungsten and the [110] zone decorations on iridium were identified as triply charged tip metal ions [68]. The multiple charges for W and Ir were reported at the same time by Brenner and McKinney [34]. Subsequently, the higher refractory metals Re, Mo and Nb were also found to form up to four-fold charged ions, while the other metals used in field ion microscopy showed multiple charges decreasing with their respective evaporation fields [29—32]. Thus, Pt and Rh produce some three-fold and mostly doubly charged ions, while Pd, Ti, V, Fe, Ni, Co and Be are predominantly doubly charged [69,70]. Cu, Au and Al have double as well as single charges. Generally, at larger pulse overvoltages and concommittant higher evaporation rates, the ratio of abundance of higher to lower charge states increases, so that even for Au and Al some triply charged ions are seen at high evaporation rates. Accurate data for relative abundances of the various charges are not yet available, as this ratio depends upon the specific conditions of field evaporation, not only the rate but also the tip temperature, and the presence of imaging as well as adsorbed residual gases. The latter is probably responsible for the dependence of the charge ratio upon the d.c. voltage or rather the holding field F_0. Müller and Krishnaswamy [70] studied a Cu—Be alloy and found at 78 K for $F_0 > 2.8$ V/Å the ratio of $Cu^+/Cu^{2+} \approx 1$, while for lower fields the ratio increased to 1.5. Surprisingly, from exactly the same surface, the ratio of Be^+/Be^{2+} was about 0.5 at high holding fields, and 0.25 at low fields. Presumably residual low ionization potential gases, capable of approaching the tip surface at lower holding fields only, affect the relative charge ratio in a different way for the two constituents of the alloy. Generally, it is experienced that the presence of hydrogen, which is known to reduce the evaporation fields of all metals, also reduces the charge number, although a closer inspection with sufficient mass resolution may show that the mostly singly charged ions are, in fact, metal hydrides.

Neither the image force nor the charge exchange theories can explain the high positive charges since, at most, double charges were expected [23]. While some uncertainty remains in the application of the theories because of the lack of exact data for the higher ionization potentials and polarizabilities of surface atoms, detailed reviews of the situation showed that no conclusive justification for the high charges, including the possibility of free-space post ionization, can be given at the present time [20,26]. Certainly the interpretations of the carefully performed field evaporation experiments by Ehrlich and Kirk [71] as well as by Plummer and Rhodin [72], which were designed to determine atomic binding energies of single atoms planted by vapor deposition onto an FIM tip, are no longer valid because of the uncertainty of the charge number of the field evaporated atom, a neglect of the effect of the imaging gas and an underestimate of the polarizability of the adatom. These data may become available from further investigations with the atom-probe.

B. FIELD ADSORPTION OF THE IMAGING GASES

Equally unexpected as the high charges were ion signals representing the noble imaging gases helium, neon and argon at their appropriate masses [29,30]. The FIM image of a surface atom is made up by some 10^2-10^3 gas ions/sec at a gas pressure of 10^{-5} torr typical for an instrument equipped with a channel plate. Thus when a nanosecond evaporation pulse is applied, the chance of having a gas ion just in position at the critical zone above a metal surface atom is of the order of 10^{-6}. Actually, when a refractory metal is pulsed to evaporation in the presence of helium or neon, almost every shot may produce a signal at mass 4 or 20, respectively, besides the metal ion. Thus, the noble gas atom must have been waiting at the surface site in an adsorbed state to be available at the nanosecond instant of the pulse.

In the absence of a field, helium and neon are bound to a metal surface by van der Waals forces only, which provide a binding energy of the order of 0.013 or 0.025 eV, respectively. Thus, no appreciable degree of coverage is to be expected for a metal surface in contact with these gases at temperatures above 10 or 20 K. By contrast, atom-probe experiments show the presence of the noble gases under the following conditions.

(1) Field adsorption of He and Ne occurs up to 180 K.

(2) Field adsorption of He requires a holding field $F_0 > 2.0$ V/Å at 20 K and $F_0 > 3.3$ V/Å at 78 K.

(3) Field adsorbed Ne at 78 K is seen in the field range from $2.3 < F_0 < 5.0$ V/Å.

(4) Argon is also field adsorbed at 78°, but no boundaries have as yet been determined.

(5) In most cases, the pulsed field must be high enough to also field evaporate the metal.

Field adsorption of the noble gases as well as the observed temperature and field ranges can be explained as being due to polarization forces leading to a binding energy of the order of $1/2 \ \alpha F_0^2$. A realistic short range attraction, however, can be understood only by also considering a localized dipole at the protruding metal surface atom. Tsong and Müller [73] used the polizability α_M of an imaged kink site surface atom, as obtained from relevant field evaporation rate measurements, and derived the potential energy of an imaging gas atom of polarizability α_A directly above the surface atom to be

$$U_A = -\tfrac{1}{2}\alpha_A f_A F_0^2 \tag{22}$$

where

$$f_A = \frac{(1 + 2\alpha_M/d^3)^2}{(1 - 4\alpha_M\alpha_A/d^6)} \qquad (23)$$

is an enhancement factor due to the interaction of the two dipoles at a distance d. The field binding energy H thus obtained for adsorption in the apex position at a minimum distance $d = r_M + r_A$, the sum of the radii of the two atoms, is

$$H = \tfrac{1}{2}\alpha_A(f_A - 1)F_0^2 \qquad (24)$$

Using atomic radii of r_W = 1.36 Å, r_{He} = 1.22 Å, r_{Ne} = 1.60 Å, r_{Ar} = 1.92 Å, and a surface polarizability for a tungsten atom α_M = 4.6 Å^3, we obtain, at the approximate evaporation fields of tungsten in the three gases of 5.7, 5.4 and 4.1 V/Å, field binding energies of 0.30, 0.36 and 0.67 eV, respectively, far exceeding the van der Waals forces and in good agreement with atom-probe experiments. In the upper temperature range at which field adsorption can still be detected, the thermal evaporation rate of the adsorbate must be comparable to the pressure-, temperature-, and field-dependent gas supply rate. Tsong and Müller [74] have considered the statistical degree of coverage or the probability of having an apex atom adsorbed at the instant of pulsing, a quantity which can be experimentally determined with the atom-probe

$$p = [1 + (\nu_0 T/Zp_{gas}F)\exp(-H/kT)]^{-1} \qquad (25)$$

where ν_0 is the vibrational frequency of the adsorbate, and Z the gas supply function at the surface site. Adsorption probabilities calculated for various assumed binding energies H agree well with experimental data for two holding fields F_0 = 4.9 and 5.6 V/Å by corresponding binding energies H = 0.23 and 0.25 eV. Conversely, experimental atom-probe data of $p(T)$ may be used to determine surface atom polarizabilities α_M.

C. METAL—NOBLE GAS COMPOUND IONS

The field induced apex adsorption of the imaging gas at kink site surface atoms is of fundamental importance for the mechanism of field ionization, as the electron must tunnel through an already adsorbed gas atom. Field desorbed noble gas ions are not always seen to accompany field evaporation, and the exact conditions of the adsorption as well as the field desorption still need to be determined. Following the atom-probe discoveries of the high charge of field evaporated metal ions and of the adsorption of the noble imaging gases, a third surprise was the occurrence of metal—helium compound ions [29—33], such as WHe^{3+}, $RhHe^{2+}$ and $TaHe^{3+}$, and also some rare metal—neon compound ions. While the original observations of these species was difficult because of the limited resolution of the early atom-probes, the occurrence of afterpulses, and in the case of neon of other artifact signals, there can now no longer be any doubt since the availability of high resolution, artifact-free instruments. It is now definitely established that a kink site tungsten atom in the presence of helium at a partial pressure as low as 10^{-8} torr predominantly field evaporates as a WHe^{3+} ion from all crystallographic regions that are dimly imaged in the FIM [62]. In previous experiments with limited resolution, He^+ ions were obtained from the brightly imaged region of tungsten around (111), (112), (114) and (100) only, together with W^{3+} ions, while the evaporation products of the dimly imaged regions in the wide vicinity of (011) were thought to be W^{3+}, and almost no He^+ was found. It is now evident that actually He is field adsorbed everywhere. The experiment further suggests that a triply charged molecular ion is coming off during the evaporation process that may dissociate in the high-field, brightly imaged regions following post ionization according to $WHe^{3+} - \epsilon \rightarrow W^{3+} + He^+$. This post ionization process should occur as soon as the WHe^{3+} ion reaches the critical ionization zone above the surface, but attempts to localize the origin more accurately by energy distribution measurements will most likely be frustrated by the large random deficits due to premature evaporation. The probability of dissociation of WHe^{3+} by postionization is supported by the absence of a WHe^{4+} species and particularly by the observed temperature dependence of the ratio $WHe^{3+}/(WHe^{3+} + W^{3+})$, which peaks around 80 K. At a constant evaporation rate, the required evaporation field is decreased when the temperature is increased, so that in the lower temperature range an increasing fraction of WHe^{3+} may escape without postionization. The gradual disappearance of the metal—helide ions beyond 100 K reflects the decreasing degree of coverage of apex adsorbed helium as calculated above. The prevalence of the WHe^{3+} species at liquid nitrogen temperature calls for caution when field evaporated W^{3+} ions are employed for mass calibration. Most likely the species with the lowest mass/charge ratio is the helide at 62.0 rather than the metal ion at 60.7 amu. Using the new energy focused atom-probe, the dependence of the abundance of metal helide ions upon crystallographic orientation and temperature was also investigated for Mo, Ir, Pt and Rh, each exhibiting a specific behavior which still needs to be correlated to the chemical and band structure properties of the substrate.

In order to appear at the appropriate place on the mass scale, a metal helide molecular ion has only to be stable for the time it requires for passing through the acceleration zone of the TOF atom-probe, that is of the order of 5 ns. It was therefore important to confirm the existence of these compounds with the magnetic atom-probe [66]. This has been done with $RhHe^{2+}$, where it was found that at 80 K at the edge of the (100) plane all metal ions evaporate as helides when $p_{He} > 5 \times 10^{-5}$ torr. Their relative abundance decreases with the pressure, so that below 5×10^{-7} torr, Rh^{2+} alone is observed. These experiments were further confirmed with the use of the isotope ^3He. Since the rhodium helide ions had traveled without dissociation from the tip through the magnetic field, a lifetime of at least 3 μs and most likely complete stability is assured.

While field adsorption of neon was definitely established by the early atom-probe work [29—33], and is confirmed by direct localization at the apex through ion energy distributions to be discussed later, the existence of metal—neon molecular ions as reported by Müller et al. [29,31,69] was somewhat doubtful when other observers were unable to confirm the results [35]. The low abundance of ions $20/n$ mass units above those of the parent metal ions made a discrimination from random artifact ions difficult. The interpretation of Müller et al. [31], that the heavy mass tail in histograms of Ta^{3+} evaporating in the presence of neon indicates a dissociation in the high field near the tip according to $TaNe^{3+} \rightarrow Ta^{3+} + Ne$, may in hindsight be replaced by simply assuming artifacts due to energy deficits of Ta^{3+} in premature field evaporation. The deficits are particularly large in neon because of its evaporation promoting effect. In addition, afterpulses may also have contributed to the tail.

With this scarcity of evidence for the validity of the early observations of metal neide ions, a recent study of palladium in the magnetic atom-probe was most surprising. In a mixture of hydrogen and neon as the imaging gas, the resulting ion spectra showed about 90% $PdNe^+$, 10% $Pd\,H^+$ and 1% Pd^+ ions [42,66]. The identification of the palladium neides is unambiguous, as the characteristic isotope pattern of Pd is displayed 20 amu above the metal isotope spectrum, and the compounds with ^{22}Ne are also seen (Fig. 16). In addition, the mass scale calibration was ascertained with xenon. Although not yet definitely established, the presence of hydrogen seems to be essential in this experiment, most likely not only in order to promote evaporation of palladium with a single charge and, not unexpectedly, predominantly as a palladium hydride. Rather, the main role of hydrogen is the provision of an electron shower falling on the surface with an energy of several tens to hundreds eV as a result of the free-space ionization of hydrogen molecules. These electrons, coming in with a current density of the same order of magnitude as that of hydrogen field ionization, that is up to 1 A/cm^2 or some 10^4 electrons per surface atom every second at 10^{-4} torr H$_2$, seem to excite the surface atom—neon adsorbate complex to field evaporate subsequently as a molecular ion. In the d.c. or slow pulse evaporation of the magnetic atom-probe, the neon capped surface atom can wait for a time of the order of a millisecond until it is excited by electron impact. By contrast, in the TOF atom-probe it is highly unlikely for an excitation to occur during the nanosecond evaporation pulse. Indeed, a TOF atom-probe experiment with a palladium tip in the same hydrogen—neon ambient showed an abundance of about only one palladium neide among 1000 metal or metal hydride ions.

This experiment illustrates the complementary performance of the two types of atom-probes. Moreover, it suggests that in all slow field evaporation

364

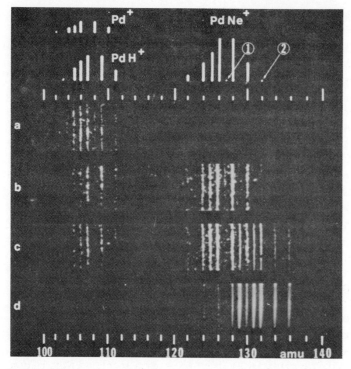

Fig. 16. Magnetic atom-probe spectra of palladium field evaporating in the presence of hydrogen (a) mostly as PdH$^+$, and (b) with neon added, mostly as PdNe$^+$. Spectrum (c) was obtained with the addition of xenon for mass calibration, and spectrum (d) is the pure Xe spectrum taken with increased gas pressure to display the rare isotopes ^{124}Xe (0.096%) and ^{126}Xe (0.090%), indicated by about 20 single ions along each spectral line.

experiments, the process must be strongly affected by electron impact excitation of the surface adsorbate complex with the electrons coming from free space ionization of the imaging gas.

From general field ion microscopy, the formation of excessive surface vacancies during the field evaporation of tungsten in the presence of neon, as discovered by Nishikawa and Müller [75], and the disturbing excess artifact vacancies encountered in the recent studies of vacancies in gold by Ast and Seidman [76] are relevant examples. As the probability of noble gas adsorption is chemically and crystallographically specific, it appears likely that the changes of field evaporation endform FIM patterns as observed by Hren et al. [77] in their experiments with varying pulse duration, are also to be explained by the effect of electronic excitation of the surface complex. The ratio of pulse width or repetition rate to the electron shower density determined by the supply rate of free space ionizing image gas, as well as the energy distribution of the electron shower affected by the ratio of pulse to holding voltage must both be essential parameters.

D. IONS FROM THE FORBIDDEN ZONE

The atom-probe discovery of image gas adsorption is supported by subsequent field ion microscopy studies of image spot brightness and flickering in He—Ne gas mixtures by Schmidt et al. [78] and by Rendulic [79], and of the recovery of image brightness following a desorption pulse by McLane et al. [80]. The location of the adsorbate at the apex of the surface atom, postulated by Tsong and Müller [73] with the suggestion of dipole—dipole binding, was directly confirmed by the appearance of ions originating within the forbidden zone at $x < x_c$. In their measurements of the energy distribution of field ionization at a single atomic site with an atom-probe apparatus modified by the inclusion of a Möllenstedt analyzer, Müller and Krishnaswamy [54] found He and Ne ions with energies of 16 and 12 eV above those of the ions from the main peak, or only 3.5—5.3 V below the tip potential. The ion potential may be considered to be determined by contributions of the external field Fx, by the attraction of the ion to the surface by its negative image, $-e/4x$, and by a repulsion term of the form $+Ax^{-n}$ of unknown coefficients. As an approximation, the repulsion may be seen as coulombic with $n = 1$, and the fractional charge $A = \epsilon e$ of the protruding surface atom should be proportional to the applied field. Assuming $\epsilon = 0.5$ at $F = 5$ V/Å, the observed potential of 3.5—5.3 V gives distances of $x = 1.3$—1.6 Å, reasonably close to the radii of the noble gas adsorbates. At this close position, the ground state of the adsorbate lies below the Fermi level, so that ionization can only occur after the adsorbate has been excited by external electron impact. In the original experiments [54], the abundance of He^{+*} ions from the forbidden zone never exceeded 2% of the normal ions originating beyond x_c. In a more detailed investigation with the magnetic atom-probe [81], helium was mixed with deuterium, the free space ionization of the latter providing the exciting electron shower. In the mass spectrum, the He^{+*} ion from the forbidden zone forms a doublet with the normal helium ion He^{+}. The higher energy ion is seen in a field range from 2.4 to 5.0 V/Å at the (011) plane of tungsten. The fading out of the He^{+*} spectral line at higher fields does not indicate the disappearance of the adsorbate, but rather a decreasing excitation probability as the average electron reaches 400 eV, far beyond the peak of the probability of helium excitation at about 30 eV. The doublet of the noble gas is seen only when an electron shower from a low ionization potential auxiliary gas like D_2, Ar or Kr is provided. With a deuterium partial pressure constant at 10^{-3} torr, the He^{+*} appears first when a He partial pressure of 10^{-7} torr allows the formation of an adsorbate layer. With increasing helium pressure, the normal He^{+} line becomes visible and equals the apex-adsorbate line He^{+*} at 5×10^{-5} torr. Inversely, at a given higher helium gas pressure the He^{+*} intensity increases with the partial pressure of the auxiliary gas producing the electron shower. When at 78 K neon is gradually admitted to a He—D_2 mixture showing the He^{+}—

He^{+*} doublet, the high energy He^{+*} line begins to fade and a neon line increasing in intensity at 20 amu turns out to be Ne^{+*}, the higher energy part of a neon doublet that finally develops at a further increased Ne partial pressure. Clearly, the He adsorbate is displaced by a neon adsorbate even at a low partial pressure of neon, as the latter gas has a higher field binding energy. The preferential adsorption of neon is the cause of the improved imaging properties of a helium—neon gas mixture as introduced in general field ion microscopy by Nishikawa and Müller [75].

Field ionization by electron tunneling through the adsorbate must be responsible for the narrow half width of about 0.5 eV of the primary peak in the field ionization energy distribution, which according to Tsong and Müller [82] localizes the source of the majority of the imaging ions to within a disc of high ionization probability which is not thicker than 0.2 Å, suggesting a hard sphere collision between the incoming gas atom to be ionized and the apex adsorbate. It is now clear that in a field ion microscope, all protruding imaged surface atoms are capped by an apex adsorbate [20], and the simple free space tunneling model of field ionization must be modified. The increase of the ionization rate by the presence of the intermediate adsorbate atom has been explained by Nolan and Herman [83] as a result of exchange effects associated with antisymmetrization of the electronic states used in the time-dependent perturbation theory of the electron transition process. The original calculation was improved [84] by including field induced distortions of the atomic orbitals, and the calculated enhancement factors now agree with Rendulic's measurements. However, Müller's earlier idea [85] that the characteristic regional brightness variations in field ion patterns of various metals may be explained by the presence of the ionization rate enhancing adsorbate at the bright regions only has to be modified with the recent atom-probe finding of the presence of adsorbed helium at all surface areas of tungsten [62]. On the other hand, atom-probe analysis [86] of ordered Pt$_3$Co, where the Co atoms in the FIM image are selectively invisible, shows He atoms are adsorbed only at the imaged Pt sites, while no He atoms are desorbed with the field evaporation of the invisible Co atoms.

There are still a good number of open questions about the noble gas adsorbate. Field desorbed helium is always singly charged, as is to be expected with its very high second ionization potential of 79 V. Neon desorbs predominantly as Ne$^+$, but at the high fields needed for desorption from the refractory metals, about 10% may come off as Ne^{2+}, which is surprising in view of the high second ionization potential of 62 V. Strangely enough, no neon ions could be desorbed from graphite, in spite of the latter's high evaporation field of above 5 V/Å. The somewhat less refractory metals Ir, Pt and Rh produce Ne$^+$ only, while the metals Ni, Fe and Co, which evaporate at fields of about 3.5 V/Å, yield only a few neon ions or none at all. Although the gas is most likely field adsorbed, its presence cannot be ascertained by the atom-probe. Upon field desorption of the substrate, the noble gas adsorbate probably comes off in neutral form and thus escapes detection. A similar sequence of ionization stages is seen with argon, which field desorbs predominantly as Ar^{2+} from the highly refractoring metals W, Ta, and Re, with only 5 to 10% Ar^+. This ratio reverses for the less refractory metals; nickel still produces a few Ar^{2+} besides the most abundant Ar^+ ions, but Co, Cu and Au surfaces yield only singly charged argon ions.

E. SURFACE INTERACTIONS WITH MOLECULAR GASES

As it seemed more urgent to clarify the behavior of the noble imaging gases which are of primary importance for general field ion microscopy, the study of the more complex interactions of the common molecular gases with the atom-probe techniques has been neglected. Early observations already established the field evaporation of various highly charged ions of metal compounds with residual gases. Hydrogen turned out to be most ubiquitous, showing up abundantly as a field desorbed H^+ ion, and occasionally also as a H_2^+ ion. Simultaneously, hydride ions seem to be formed with all metals, even those like FeH^+ for which neutral hydrides are not known. In fact, CuH_2^+ was the first field evaporating compound to be identified mass spectroscopically [9], followed by BeH^+, FeH^+, and NiH^+ in the early work by Barofsky and Müller [11]. The clear distinction of hydride ions from the metal ions requires a sufficiently high resolution which has not been attained in most of the early TOF atom-probes. Thus, the reality of some hydrides such as $TaHeH^{3+}$ by Müller et al. [31,32] or multiple hydrides of up to CuH_8^+ as recently reported by Goodman et al. [50] is still not ascertained particularly as the strong reduction of the evaporation field by hydrogen causes tails towards heavier masses by energy deficits in premature evaporation. This, of course, is no longer a problem for the new, high resolution energy focused TOF and magnetic atom-probes. For instance, with the latter, palladium was found to field evaporate in the presence of hydrogen as 90% PdH^+ and 10% Pd [42,66] (Fig. 16).

A field desorbed metal hydride ion may dissociate in the high field near the tip, and the life time of the hydride may be derived from an analysis of the apparent masses of the dissociation products [31,32], as has been shown for single isotopic Ta, which appears at $m/n = 60.3$, accompanied by a peak at 60.7 indicating the hydride ion TaH^{3+}. However, there are many signals within a mass range from 85 to 90 amu, definitely below Ta^{2+} at 90.5, as well as at 1.05–1.15 just above H^+. These odd species are interpreted as representing the products of a dissociation,

$$TaH^{3+} \rightarrow Ta^{2+} + H^+ \tag{26}$$

which takes place at some distance d from the tip where the triply charged hydride has fallen through a potential difference V_d and has acquired a velocity v_d. From there on the dissociation products are separately accelerated by the total evaporation voltage V_e to their end velocities V_{eTa} and V_{eH}, obtaining the kinetic energies

$$\tfrac{1}{2} m_H v_{eH}^2 = \tfrac{1}{2} m_H v_d^2 + e(V_e - V_d) \tag{27}$$

$$\tfrac{1}{2} m_{Ta} v_{eTa}^2 = \tfrac{1}{2} m_{Ta} v_d^2 + 2e(V_e - V_d) \tag{28}$$

At the instant of dissociation, v_d is determined by

$$\tfrac{1}{2} m_{TaH} v_d^2 = 3e V_d \tag{29}$$

while the apparent masses M as obtained from the TOF are

$$M = 2e V_e / v_e^2 \tag{30}$$

From these equations one obtains the apparent masses for the dissociation products

$$M_H^+ = \frac{m_H m_{TaH} V_e}{3 m_H V_d + m_{TaH}(V_e - V_d)} \tag{31}$$

$$M_{Ta^{2+}} = \frac{m_{Ta} m_{TaH} V_e}{3 m_{Ta} V_d + 2 m_{TaH}(V_e - V_d)} \tag{32}$$

Using V_d as a parameter for a given tip radius and approximating the field in the acceleration region by one between confocal paraboloids, the apparent masses for a tip with $r_0 = 800$ Å and $V_e = 20$ kV have been calculated to show the correlation of the two dissociation species. The peak of the Ta^{2+} distribution between 87 and 89 amu places the dissociation of TaH^{3+} at a distance 150—500 Å away from the surface, to which the TaH^{3+} ion travels within $5-12 \times 10^{-13}$ s.

A similar dissociation takes place with WH^{3+}, showing itself by two groups of signals at slightly above 1 amu and in a range from 83 to 93 amu, but the determination of a life time is not unambiguous due to the spread of the four isotopes.

Adsorbed hydrogen not only reduces the evaporation field, but also causes the peculiar effect of promoted field ionization. An addition of a few percent of hydrogen to helium as the image gas in an FIM surprisingly produces a very sharp helium ion image at 2/3 of the normal best image field [87]. This strange effect, practically useful for imaging the less refractory metals, has been explained through various mechanisms, which all seem to be superseded by some new observations by Sakurai et al. [88]. Magnetic atom-probe studies reveal no shift in the energy distribution of the promoted He^+ ions, assuring their origin at $x_c = (I - \phi)/F$. The flickering of the promoted image spots as seen on the channel plate screen supports a flip-over of a hydrogen atom from a recessed position to form a stretched H_2 molecule with another hydrogen atom already apex adsorbed. Thus, an ionization promoting electronic path is formed from the surface atom to the ionization disc at x_c. This apex adsorbed H_2 molecule has only a short life time at the promotion field of 3.3 V/Å before being field desorbed and dissociating into two H^+ ions. The predicted occurrence of these H^+ ions solely in the narrow field region of 3.3 ± 0.07 V/Å of the promoting effect as observed with the TOF atom probe strongly supports the new mechanism.

Very few definite results have been obtained so far with the other molecular gases, simply because most of the investigations carried out were of an exploratory rather than a systematic nature. Oxygen on tungsten field desorbs in the form of O^+, and mostly as various oxide molecular ions such as triply or doubly charged WO, WO_2 and WO_3. Various oxide complexes with platinum were also found at high temperatures in the early work by Vanselow and Schmidt [10]. Nitrogen, which is supposed to be only molecularly adsorbed on a noble metal such as Rh, is found to form the species N^{2+} and N^+, while N_2^+ is prevalent and equally abundant as the Rh^{2+} ion. In addition, there are some 5% RhN^{2+} ions seen. Similarly, field evaporation of Rh in the presence of CO shows not only the predominant CO^+ at 28 amu besides the abundant Rh^{2+} and some Rh^{3+} ions, but also many C^{2+}, very few C^+ and many O^+ ions. In addition, a RhC^{2+} ion is frequently seen. Dissociation of CO is also evident in field desorption experiments with W, where again C^{2+}, C^+ and O^+ are obtained besides CO^+, W^{3+} and WC^{3+}. Strangely, when neon was added to CO, mass peaks of C^{2+}, C^+, O^+, Ne^+, W^{4+}, W^{3+}, WC^{3+}, W^{2+} and WC^{2+} clearly indicate the adsorption of CO, but there was no mass 28 signal from the undissociated adsorbate. These observations of a considerable dissociation of field desorbed N_2 and CO contradict the interpretation of most experiments using thermal desorption, from which adsorption in molecular form only is concluded. The understanding of the atom-probe results as field induced effects is complicated further by the possibility of excitation and dissociation of the surface complex by the electron shower from free space ionization of the supplied gases, making the effects strongly dependent upon the ratio of d.c. holding voltage to pulse voltage. The appearance of double charges with C^{2+} and N^{2+} is not necessarily connected with the high evaporation field used in experiments on W and Rh tips. In exploratory experiments with H_2S on nickel, S^{3+} and S^{2+} were found abundantly besides H^+ and Ni^{2+}, while S^+ was completely absent. A number of the observations of the interactions of various tip metals with molecule gases are listed in the review article by Müller and Tsong [20].

Except for an occasional attempt to identify some of the bright spots appearing in an FIM image upon admission of various molecular gases [89], none of the experiments described above fully utilized the ability of the atom-probe to pinpoint the probe hole onto a selected atomic site. In most cases, the identification of all the atomic species in the immediate environment of a selected point is equally useful. The variation of surface catalytic reactions with crystallographic surface structure such as lattice steps is a wide field open for future work, but the little information obtained so far is at least of some help for the interpretation of atom-probe observations of tip surfaces in various ambients.

F. A FIELD CALIBRATION VIA FREE-SPACE IONIZATION

Field ion microscopy is used increasingly for quantitative investigations of surface phenomena which may be affected by the field. Yet only the applied voltage and not the field can be measured directly with high accuracy. So far, all voltage—field calibrations are based on a single measurement by Müller and Young [90] who used the field emission current—voltage characteristic of a tip, applied the Fowler—Nordheim equation and determined the reasonably reproducible best image field of tungsten in helium to be 4.5 V/Å. Ionization fields for other gases were derived by comparing their imaging voltages with the helium best image voltage of the same tungsten tip. Evaporation fields of all other metals [8,19] were obtained via their imaging fields in neon or argon under the unproven assumption that these are the same for all metals. A new method employing the energy distribution of free-space field ionization as measured in the magnetic atom-probe has been developed by Sakurai and Müller [91]. In the confocal hyperboloid approximation, the field at a distance x from the tip at an applied voltage V is

$$F(V,x) = V/k(r_t + 2x) \tag{33}$$

Free space field ionization depends essentially upon the field at the location of the molecule. The position x_i, where field ionization peaks at a certain field $F(V_i,x_i)$ moves outward when the applied voltage V_i is increased, so that

$$F(V_1,x_1) = F(V_2,x_2) = \dots F(V_i,x_i) \tag{34}$$

An ion formed at the tip surface would have the energy eV, when it arrives at the ground surface, while when it originated at x_i it would have an energy deficit

$$\Delta E(V_i,x_i) = e \int_0^{x_i} F(V_i,x)dx \tag{35}$$

$$\Delta E(V_i,x_i) = -\frac{eV_i}{2k}[\ln(r_t + 2x_i) - \ln r_t] \tag{36}$$

For two applied voltages V_1 and V_2 we obtain from eqns. (33) and (34)

$$\frac{V_1}{V_2} = \frac{1 + 2x_1/r_t}{1 + 2x_2/r_t}, \tag{37}$$

and eliminating x_i and r_t, the difference of relative energy deficits is

$$\frac{\Delta E(V_1,x_1)}{eV_1} - \frac{\Delta E(V_2,x_2)}{eV_2} = \frac{1}{2k}\ln\frac{V_1}{V_2} \tag{38}$$

The field calibration factor k may thus be determined experimentally by measuring energy deficits at two different applied tip voltages. Using deuterium and knowing the local tip radius from counting the rings, the peak of free-space ionization is found to be at 2.92 V/Å. For a specific tip with $r = 375$ Å, the free-space ionization peaked at $x_1 = 37$ Å with $V_1 = 10,400$ V applied, and at $x_2 = 137$ Å with $V_2 = 15,000$ V. It should be noted that the free-space ionization field is independent of the tip material. The determination of this critical field and, thereby also of the calibration of the final field is as accurate as the determination of local tip radius by counting rings [19] of a large radius tungsten tip, which is about ±3%. This compares favorably with the possible error of the established Fowler—Nordheim field emission calibration of ±15%.

When a tip radius cannot be measured by counting rings, as is the case with small tip radii or with irregularly field evaporating tips of alloys or highly imperfect lattice structures, eqns. (33) and (36) and the peak field for $D_2 = 2.92$ V/Å can be used to obtain the tip radius from

$$r_t = \frac{V_i}{2.92k \, \exp(2k\Delta E_i/eV)} \tag{39}$$

Since the new calibration technique measures the local field strength at distances from the surface that are large compared to lattice step heights, but only a fraction of the tip radius, regional field strengths over various crystallographic tip areas, which are known to have quite different local radii, can now be determined via magnetic atom-probe energy deficit data of free-space field ionization peaks. Table 1 gives some results obtained by Sakurai [81].

TABLE 1

Evaporation fields (in V/Å) of various metals in He (or He—Ne) measured by free-space field ionization energy distribution

	(011)	(111)	(112)	(113)	(001)	Temp. (K)	Gas
W	5.5	6.2	5.7			78	He
	5.7	6.3	5.9			21	
Mo	4.6	4.7	4.7	4.7		78	He
	5.0	5.0	5.0			21	
Ir		5.1		5.4	5.2	78	He
		5.4		5.6	5.4	21	
Rh		4.5		4.8	4.5	78	He
		4.8		4.9	4.8	21	
Ni		3.2		3.5	3.2	78	He—Ne
		3.6		3.8	3.5	21	
Pt					4.8	78	He

References pp. 375—378

VIII. Metallurgical applications

While the emphasis of this report, reflecting the preference of the author, is on the design of atom-probes and the basic phenomena of surface physics encountered with their operation, the main applications, just as of the field ion microscope itself, have been in physical metallurgy. The potential for studies of short range order of alloys, for the identification and localization of impurities, interstitials, small scale segregation and precipitates was recognized in the first publications [12,13]. The location of two of the three active atom-probes at U.S. Steel, where Brenner pioneered the applications, and at the Department of Metallurgy and Materials Science at Cambridge University, where Turner et al. made their important contributions, testifies to the practicality of the new device.

So far, very little work has been done towards the identification of individual atom spots in the field ion image of a specimen. Brenner and McKinney [89] looked at the bright additional spots appearing on tungsten after exposure to carbon monoxide and nitrogen, finding that most of them are displaced tungsten atoms while invisible adsorbate molecules are still present, as had been suspected earlier [29]. Much work still needs to be done to determine the nature of individual bright spots caused by interstitials [8,20], as those shown by Nakamura and Müller [92] in the FIM study of oxygen diffusion in tantalum, or the interstitial spots diffusing to the surface during or subsequent to in situ particle irradiation [93,94]. The extra bright spots in a Pt—8% W alloy, thought to represent tungsten atoms, yielded only Pt atoms when the top layer was pulsed in an atom-probe (Müller 1974, unpublished) but a subsequent pulse removed the next atomic layer and produced a tungsten signal. This result suggests that the bright spot is due to a higher ionization probability above a Pt surface atom, when its substrate lattice is disturbed by the local effect of a foreign atom on the band structure and its related reduced reflection of the ionizing electron attempting to tunnel into a band gap. This explanation will also apply to the local disturbance of the lattice and concomitantly of the band structure for self-interstitials from radiation damage.

Most metallurgical applications are based on the unique FIM capability of atom-layer by atom-layer dissection of the tip specimen with a sequence of controlled field evaporation pulses. The atom-probe hole may be aimed at easily discernable precipitation clusters or at a grain boundary [32] (Fig. 17). Even when locally varying evaporation fields in solute alloys and around precipitates prevent the attainment of clearly interpretable FIM images, the composition profile into the depth of the specimen obtained by long evaporation sequences in the atom-probe gives useful information. Such profiles were first used by Brenner and McKinney [35].

A particular careful analysis is an atom-probe study of the precipitation of Cu in an Fe—1.4% Cu alloy by Goodman et al. [50]. Small precipitates, with

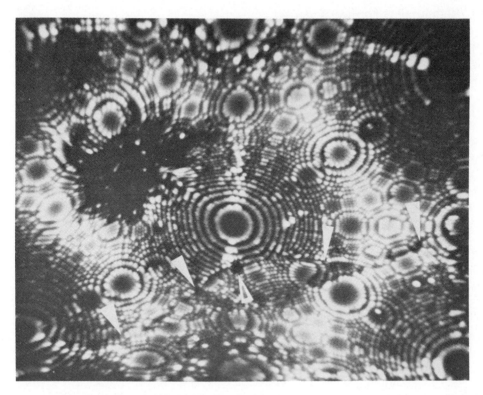

Fig. 17. Probe hole (hollow arrow) pointed at a grain boundary in a tungsten specimen. The large void at the upper left is caused by impurities doped into the incandescent lamp wire.

diameters down to 8 Å and thus well below the 50 Å limit of visibility in a transmission electron microscope could be analyzed. Following the change in composition with size, which determines the interface energy and thereby the activation barrier, clearly demonstrated the unique analytical capability of the atom-probe, although the instrument used had only a limited resolution by a TOF determination of ±25 ns and suffered from up to 15% after-pulses. Some aiming errors may arise because, in the early stages, precipitate particles are not much larger than the area covered by the probe hole and they appear dark against the matrix, but positioning was no problem for 50 Å diameter precipitates. Depth profiles of composition were obtained by plotting the cumulative count of Cu ions versus the count of Fe ions. The overall slope of this stepped curve gives the composition. When a particle has been penetrated, the slope is zero, indicating the pure Fe matrix. The result of this investigation was that at the earliest stages of nucleation with an 8–15 Å precipitate diameter the composition was about 50% Cu, after prolonged aging to 50 Å diameter it was 69% Cu, while thermodynamically the equilibrium of the precipitate should be nearly pure copper.

In contrast to the Fe—Cu system just described, the precipitates in Cu—1.7% Fe and Cu—2% Co appear as bright and often fully atomically resolved particles. Brenner [95] found the copper-containing nuclei in the alloy with iron to be coherent with the Cu matrix up to a size of 50 Å, while those in the Co-containing alloy are coherent to their full size of 100 Å. The deformation in the first case may be due to the field stress. An interesting side observation by Brenner is that neon image gas atoms could be field desorbed from the bright iron-containing precipitate below the evaporation field of the metal substrate, which is unusual. A similar situation was later noticed by Krishnaswamy in the case of stainless steel.

Brenner and Goodman [96] have also studied the nitride platelets appearing in a Fe—3% Mo alloy in the process of ammonia hardening the surface of steel. Again, an in situ analysis is exclusively an atom-probe task. Specimen tips were made of a fully nitrided alloy wire, and the precipitates appear as platelets not more than 2—3 atom layers thick lying on {100} matrix planes. The effective size of the probe hole was adjusted to the thickness of the platelets seen edge-on by increasing the tip-to-probe hole distance. The composition of the platelets was found to be $Fe_3Mo_3N_2$ rather than the previously assumed compound with one nitrogen atom only. Some uncertainty arose from the inability to resolve FeN^{2+} ($m/n = 34—36$) from MoN^{3+} (35.3—38) and also Fe^+ (54—59) from MoN^{3+} (53—57), but it appears that almost all of the Mo is converted to the nitride. In addition, one extra large precipitate cluster was nearly pure Mo_3N_2 with some of the Mo atoms substituted by Fe.

Other alloys forming precipitates were studied with the Cambridge atom-probe. Turner et al. [37] report small regions of enhanced Ti concentrations prior to the precipitation of Ni_3Ti, and of carbide and nitride particles in ferrous alloy. Somewhat limited by the resolution and strong artifact background, the authors conclude that from molybdenum carbide precipitates the metal appears as Mo^{2+} and Mo^{3+} as it does from the pure metal, carbon comes off as C^+ and C^{2+}, while molybdenum—carbon complex ions occur infrequently, if at all. By contrast, in the case of a molybdenum nitride precipitate, the presence of a nitride ion is confirmed.

One of the least utilized capabilities of the atom-probe is the analysis of segregation to grain boundaries. In field ion micrographs, the boundaries are often found to be decorated by bright spots, the identification of which has not yet been attempted. Other impurity atoms at grain boundaries may be simply invisible in the FIM, but they still can be identified with the atom-probe. Turner and Papazian [97] analyzed an alloy of zone refined iron which was carburized to a level of 0.05% C. The specimen wire was recrystallized in hydrogen to produce a grain size of 10 μm. With a tip having a grain boundary at a large distance down the shank, as seen by preceding electron microscopical inspection, no carbon could be detected by the atom-probe.

With a tip from the same alloy containing a grain boundary in the field of view of the tip cap, 0.6—0.4% carbon ions were obtained with the probe hole 500 Å away from as well as right across the boundary. In another sample, the carbon content at the grain boundary was 0.7%. This result is surprising in that it indicates the concentration of carbon at the boundary to be far from a monolayer, and rather diffuse over a distance of at least 500 Å. With such a wide zone of a high carbon content, the cause of the driving force for the segregation is mysterious. It is obvious that this important work should be resumed with a new atom-probe in which the artifact level is much lower than the level of the few carbon signals available in such an experiment.

Atom-probe work with the common transition metals and their alloys, particularly ferrous alloys such as stainless steels, definitely requires a good resolution to distinguish safely the possibly overlapping isotopes of the metals and their compound ions with the ubiquitous hydrogen, with carbon, nitrogen, and oxygen. A two-compound alloy such as stainless steel 410 can still be handled reliably with a straight atom-probe, as has been shown by Krishnaswamy et al. [98] in their analysis of chromium depletion by annealing. However, real progress in the reliability and practicality of atom-probe analysis of a great variety of metallurgical specimens in a range from atomic dimensions to sizes that overlap with the lower limit of electron microscopy and some related microanalytical techniques can now be foreseen with the use of the new high-resolution, artifact-free atom-probe. We have shown that technical difficulties still encountered with most of the 15—20 atom-probes that have been constructed so far in as many laboratories may be overcome with the addition of energy-deficit compensation or energy focusing features. Thus, an exciting new research tool of unique microanalytical sensitivity is at hand.

Acknowledgments

The author wishes to acknowledge his appreciation for the many contributions of his former and present co-workers J.A. Panitz, T.T. Tsong, S.V. Krishnaswamy, S.B. McLane, and T. Sakurai. The entire work on the atom-probe at The Pennsylvania State University was supported by grants from the National Science Foundation.

References

1 E.W. Müller, Z. Phys., 106 (1937) 541; 108 (1938) 668.
2 E.W. Müller, Naturwissenschaften, 29 (1941) 533.
3 E.W. Müller, Z. Phys., 131 (1951) 136.

376

4 E.W. Müller, ACS-Lecture, Univ. of Chicago, Nov. 1952, unpublished.
5 M.G. Inghram and R. Gomer, J. Chem. Phys., 22 (1954) 1274; Z. Naturforsch. A, 10 (1955) 863.
6 H.D. Beckey, Field Ionization Mass Spectrometry, Pergamon Press, Oxford, 1971.
7 J. Block, Z. Phys. Chem. (NF) 39 (1963) 169.
8 E.W. Müller, Advan. Electron. Electron Phys., 13 (1960) 83.
9 R. Thomson, Master's Thesis, The Pennsylvania State University, 1963, unpublished.
10 R. Vanselow and W.A. Schmidt, Z. Naturforsch. A, 21 (1966) 1960.
11 D.F. Barofsky and E.W. Müller, Surface Sci., 10 (1968) 177.
12 E.W. Müller and J.A. Panitz, Abstr. 14th Field Emission Symp., National Bureau of Standards, Washington, D.C., June 1967.
13 E.W. Müller, J.A. Panitz and S.B. McLane, Rev. Sci. Instrum., 39 (1968) 83.
14 E.W. Müller, Lab. Pract., 22 (1973) 408.
15 E.W. Müller and K. Bahadur, Phys. Rev., 102 (1956) 624.
16 R.J. Walko and E.W. Müller, Phys. Status Solidi (a), 9 (1972) K9.
17 J.A. Panitz, Rev. Sci. Instrum., 44 (1973) 1034.
18 J.A. Panitz, J. Vac. Sci. Technol., 11 (1974) 206.
19 E.W. Müller and T.T. Tsong, Field Ion Microscopy, Principles and Applications, Elsevier, Amsterdam, 1969.
20 E.W. Müller and T.T. Tsong, in S.G. Davison (Ed.), Progress in Surface Science, Vol. 4, Part 1, Pergamon Press, Oxford, 1973.
21 J.R. Oppenheimer, Phys. Rev., 31 (1928) 67.
22 E.W. Müller, Phys. Rev., 102 (1956) 618.
23 D.G. Brandon, Surface Sci., 3 (1965) 1.
24 R. Gomer, J. Chem. Phys., 31 (1959) 341.
25 R. Gomer and L.W. Swanson, J. Chem. Phys., 38 (1963) 1613; 39 (1963) 2813.
26 M. Vesely and G. Ehrlich, Surface Sci., 34 (1973) 547.
27 W.A. Schmidt, O. Frank and J.H. Block, Abstr. 20th Field Emission Symposium, The Pennsylvania State University, 1973, p. 49.
28 E.W. Müller, S.B. McLane and J.A. Panitz, 4th Europ. Reg. Conf. Electron Microsc., Rome, 1968, Vol. 1, p. 135.
29 E.W. Müller, Quart. Rev., 23 (1969) 177.
30 E.W. Müller, S.B. McLane and J.A. Panitz, Surface Sci., 17 (1969) 430.
31 E.W. Müller, S.V. Krishnaswamy and S.B. McLane, Surface Sci., 23 (1970) 112.
32 E.W. Müller, Naturwissenschaften, 57 (1970) 222.
33 E.W. Müller, Ber. Bunsenges. Phys. Chem., 75 (1971) 979.
34 S.S. Brenner and J.T. McKinney, Appl. Phys. Lett., 13 (1968) 29.
35 S.S. Brenner and J.T. McKinney, Surface Sci., 23 (1970) 88.
36 P.J. Turner, B.J. Regan and M.J. Southon, Proc. 25th Anniv. Meet. E.M.A.G., London, 1971, p. 252.
37 P.J. Turner, B.J. Regan and M.J. Southon, Vacuum, 22 (1972) 443.
38 P.J. Turner, B.J. Regan and M.J. Southon, Surface Sci., 35 (1973) 336.
39 S.S. Brenner and J.T. McKinney, Rev. Sci. Instrum., 43 (1972) 1264.
40 E.W. Müller, S.V. Krishnaswamy and S.B. McLane, Rev. Sci. Instrum., 44 (1973) 84.
41 E.W. Müller, J. Microscopy (Oxford), 100 (1974) 121.
42 E.W. Müller, Proc. 2nd Int. Conf. Solid Surfaces, Kyoto, Japan, March 1974, Jap. J. Appl. Phys., Suppl. 2 Part 2 (1974) 1.
43 E.W. Müller and S.B. Krishnaswamy, Rev. Sci. Instrum., 45 (1974) 1053.
44 J.A. Panitz, S.B. McLane and E.W. Müller, Rev. Sci. Instrum., 40 (1969) 1321.
45 S.B. McLane, Abstr. 16th Field Emission Symp., Pittsburgh, Pa., 1969, p. 22.
46 C.A. Johnson, Abstr. 16th Field Emission Symp., Pittsburgh, Pa., 1969, p. 23.
47 C.A. Johnson, Rev. Sci. Instrum., 41 (1970) 1812.
48 A.S. Berger, Rev. Sci. Instrum., 44 (1973) 592.

49 S.B. McLane, Abstr. 20th Field Emission Symp., The Pennsylvania State University, 1973, p. 55.
50 S.R. Goodman, S.S. Brenner and J.R. Low, Jr., Met. Trans., 4 (1973) 2371.
51 A.A. Lucas, Phys. Rev. B, 4 (1971) 2939.
52 A.A. Lucas and M. Sunic, Surface Sci., 32 (1972) 439.
53 A.J. Jason, Phys. Rev., 156 (1967) 266.
54 E.W. Müller and S.V. Krishnaswamy, Surface Sci., 36 (1973) 29.
55 S.V. Krishnaswamy and E.W. Müller, Abstr. 20th Field Emission Symposium, The Pennsylvania State University, 1973, p. 50.
56 S.V. Krishnaswamy and E.W. Müller, Rev. Sci. Instrum., 45 (1974) 1049.
57 E.W. Müller and S.V. Krishnaswamy, Abstr. 20th Field Emission Symp., The Pennsylvania State University, 1973, p. 52.
58 E.M. Purcell, Phys. Rev., 54 (1938) 818.
59 H. Ewald and H. Liebl, Z. Naturforsch. A, 10 (1955) 872.
60 H. Wollnik, in A. Septier (Ed.), Focusing of Charged Particles, Vol. II, Academic Press, 1967, p. 163.
61 W.P. Poschenrieder, Int. J. Mass. Spectrom. Ion Phys., 9 (1972) 357.
62 E.W. Müller, S.V. Krishnaswamy and S.B. McLane, Phys. Rev. Lett., 31 (1973) 1282.
63 W.P. Poschenrieder and G.H. Oetjen, J. Vac. Sci. Technol., 9 (1972) 212.
64 R.J. Walko and E.W. Müller, Phys. Status Solidi (a), 9 (1972) K9.
65 E.W. Müller, S.V. Krishnaswamy, S.B. McLane, T. Sakurai and R.J. Walko, Abstr. 19th Field Emission Symp., University of Illinois, 1972, p. 61.
66 E.W. Müller and T. Sakurai, J. Vac. Sci. Technol., 11 (1974) 878.
67 E.W. Müller, in E. Drauglis, R.D. Gretz and R.I. Jaffee (Eds.), Molecular Processes on Solid Surfaces, Battelle Kronberg Conference, May 1968, McGraw-Hill, New York, 1969, p. 67.
68 E.W. Müller, in R.F. Hochman, E.W. Müller and B. Ralph (Eds.), Applications of Field Ion Microscopy, Georgia Institute of Technology Conference, May 19, 1968, Georgia Technical Press, Atlanta, Ga., 1969, p. 59.
69 E.W. Müller, Structure et Propriétés des Surfaces des Solides, CNRS, Conf., July 1969, Paris, 1970, p. 81.
70 E.W. Müller and S.V. Krishnaswamy, Phys. Status Solidi (a), 3 (1970) 27.
71 G. Ehrlich and C.F. Kirk, J. Chem. Phys., 48 (1968) 1465.
72 E.W. Plummer and T.N. Rhodin, J. Chem. Phys., 49 (1968) 3479.
73 T.T. Tsong and E.W. Müller, Phys. Rev. Lett., 25 (1970) 911.
74 T.T. Tsong and E.W. Müller, J. Chem. Phys., 55 (1971) 2884.
75 O. Nishikawa and E.W. Müller, J. Appl. Phys., 35 (1964) 2806.
76 D.G. Ast and D.N. Seidman, Surface Sci., 28 (1971) 19.
77 J.J. Hren, A.J.W. Moore and J.A. Spink, Surface Sci., 29 (1972) 331.
78 W. Schmidt, Th. Reisner and E. Krautz, Surface Sci., 26 (1971) 293.
79 K.D. Rendulic, Surface Sci., 28 (1971) 285.
80 S.B. McLane, E.W. Müller and S.V. Krishnaswamy, Surface Sci., 27 (1971) 367.
81 T. Sakurai, Ph.D. Thesis, The Pennsylvania State University, 1974.
82 T.T. Tsong and E.W. Müller, J. Chem. Phys., 41 (1964) 3279.
83 D.A. Nolan and R.M. Herman, Phys. Rev. B, 8 (1973) 4099.
84 D.A. Nolan and R.M. Herman, Phys. Rev. B, 10 (1974) 50.
85 E.W. Müller, J. Less Common Metals, 28 (1972) 37.
86 T.T. Tsong, S.V. Krishnaswamy, S.B. McLane, and E.W. Müller, Appl. Phys. Lett., 23 (1973) 1.
87 E.W. Müller, S. Nakamura, O. Nishikawa and S.B. McLane, J. Appl. Phys., 36 (1965) 2496.
88 T. Sakurai, T.T. Tsong and E.W. Müller, Phys. Rev. Lett., 30 (1974) 532.
89 S.S. Brenner and J.T. McKinney, Surface Sci., 20 (1970) 411.

378

90 E.W. Müller and R.D. Young, J. Appl. Phys., 32 (1961) 2525.
91 T. Sakurai and E.W. Müller, Phys. Rev. Lett., 30 (1973) 532.
92 S. Nakamura and E.W. Müller, J. Appl. Phys., 36 (1965) 3634.
93 E.W. Müller, Proc. 4th Int. Conf. Reactivity of Solids, Elsevier, Amsterdam, 1960, p. 682.
94 M.K. Sinha and E.W. Müller, J. Appl. Phys., 35 (1964) 1256.
95 S.S. Brenner, Abstr. 19th Field Emission Symp. University of Illinois, 1972, p. 58.
96 S.S. Brenner and S.R. Goodman, Scr. Met., 5 (1971) 865.
97 P.J. Turner and J.M. Papazian, Met. Sci. J., 7 (1973) 81.
98 S.V. Krishnaswamy, S.B. McLane and E.W. Müller, J. Vac. Sci. Technol., 11 (1974) 899.

Chapter 9

FIELD ION MASS SPECTROMETRY APPLIED TO SURFACE INVESTIGATIONS

J.H. BLOCK in cooperation with A.W. CZANDERNA

I. Introduction

The field ionization phenomenon, predicted by Oppenheimer [1] and Lanczos [2] was experimentally verified in the field ion microscope (FIM) by Müller [3]. The ionization of positive ions, which are formed from neutral particles by an electric field exceeding 10^8 V/cm at a field emitter surface, was investigated first by mass spectrometric methods in 1954, when Inghram and Gomer [4,5] analyzed the field ionization of positive ions. They found that gaseous molecules could be detected in the form of ions, independent of the state of adsorption, as partly charged negative or positive surface layers and independent of the strength of the adsorption bond. Since these early observations, basic contributions have been made by Beckey [6] and a few other groups of scientists. Indeed, all kinds of gaseous compounds can be ionized by fields at solid surfaces. However, considerable detailed knowledge is now available for explaining the behavior of molecules in high electric fields and at surfaces.

The fundamentals of field ionization have been described in the excellent publications by Müller and Tsong [7], Ehrlich [8], and Gomer [9]. Recent survey articles by Beckey [10], Robertson [11,12], and Block [13] are concerned with the mass spectrometry of field ions. The purpose of this contribution is to describe recent developments where the mass spectrometric analysis of field ions gives more detailed information about the molecular behavior on solid surfaces. Although the method of field ion mass spectrometry (FIMS) has been known for over 20 years, only a few laboratories have utilized it. While various interesting problems can be studied effectively with FIMS, very few field ion mass spectrometers are being used for surface chemical research.

When the first experiments with LEED were performed in 1929, nearly 40 years elapsed before routine analysis of surface structure was achieved and this was not attained without the help of commercially available equipment. Compared with LEED, the experimental requirements for carrying out FIMS are much more severe. The ultra high vacuum requirements for maintaining

the clean surfaces are more difficult to achieve and the electronic requirements for identifying single ions or measuring currents of the order of
10^{-22} A are extremely demanding. These and other technical difficulties
have been solved so it can now be expected that this advanced technique will
be applied by more investigators to various problems of surface science.

When the first experiments were performed with FIMS, the principal goal
was to analyze atoms or molecules at or near a solid surface. Much impetus
for this direction of research was given by the results of field ion microscopy, where the identification of image spots of an atomically enlarged
picture of a surface was desired. In the previous chapter, the identification of
atoms on solid surfaces was described by Müller.

The extremely high electric fields, which are applied in this analytical
method, have a fundamental influence on the properties of matter, and
disturb the usual properties of atoms and molecules. At much lower fields,
the diversity of electric field effects on molecular systems was studied and
reported in the classic work of Debye [14]. The concept of molecular dipoles and of polarizabilities, which evolved from these investigations at fields
of less than 10^4 V/cm, has contributed fundamentally to the understanding
of molecular structure and arrangements. From the interaction of molecules
in electric fields, various properties of matter could be evaluated such as the
configuration of isomers, aspects of the charge distribution in molecules, the
electrical permittivity and dielectric constants. Variable electric fields are
closely related to the whole discipline of optical properties of molecular
systems. The energy introduced into molecules by stationary electric fields
of less than 10^4 V/cm is small compared with the energies of chemical
bonds. All the experimental methods for determining molecular structures
under these conditions are related to specific molecular orientations or to
small amounts of energy absorption by vibrational or electronic levels.

Molecular properties will be quite different in the experimental conditions
of FIMS. At external electric fields of 10^7 V/cm and greater, internal molecular potentials are severely disturbed. The energy absorbed by molecules
is comparable with the energy of chemical bonds. From a thermodynamic
point of view, we can regard the electric field as a new variable. Like temperature or pressure, the field alters the properties of molecular ensembles.
Therefore, as a parallel with high-temperature or high-pressure chemistry, we
now have to consider a new discipline, "high-field chemistry".

For the primary task of FIMS, the analysis of atomic or molecular species
at or near surfaces, these field induced alterations of the molecular properties make straightforward conclusions intricate. However, compared with
other methods of ion formation, FIMS is still a very favorable procedure. In
contrast to electron impact or photoionization, which are widely used in
mass spectrometric applications, the energies involved are low and the fragmentation of the resulting molecular ions is relatively limited. Therefore, one
of the important developments of FIMS is the possibility for analyzing

rather unstable compounds. Since Beckey [15] has already demonstrated the advantage of FIMS compared with other ionization methods for the identification of organic substances of biological and medical significance, this important domain of FIMS will not be considered further in this contribution.

At surfaces investigated by FIMS, high electric fields are not necessarily an artifact. Fields of the same order of magnitude as used in this method occur naturally at surfaces of ionic crystals, in the Helmholtz double layer at electrodes, in intermolecular interactions of particles, etc. It is these types of electric field interactions at surfaces that will be discussed in this chapter.

II. Experimental methods

The field ion mass spectrometer consists of three major parts: the ion source and beam focusing unit, the mass separator, and the ion detector. In principle, field ion sources can be combined with all types of mass spectrometers. The particular properties of field ion sources, however, have advantages or disadvantages for the two different combinations which will be described.

A. THE FIELD EMITTER

One of the great restrictions for applying FIMS to surface problems is the preparation of a field emitting tip from the material chosen for study. To produce an electric field exceeding 10^8 V/cm, the needle-shaped emitter has to have a tip with a radius of curvature of about 1000 Å. Fields of 10^8 V/cm are easily reached with a potential difference, ΔU, of several kV. From the laws of electrostatics, the field is given by $F = \Delta U/kr$, where k (≈ 5) is a correction factor for the emitter shank.

The preparation of field emitters is usually accomplished by electrolytic etching, which has not been elaborated in detail for most cases. For the conditions normally used, bulk crystal domains etch at a slower rate than grain boundaries, which simplifies the preparation of single crystal end forms on emitter tips. Many different materials have been tested so far in FIM and the preparation techniques for these have been published by Müller [16], and Bowkett and Smith [17]. Some new methods developed in the author's laboratory are compiled in Table 1. For investigations using other emitter materials, thin layers of a material have been deposited in situ onto a stable field emitter. For example, Ag or Cu have been deposited on W-emitters. An oxide surface, like ZnO, can also be prepared by depositing Zn onto a W-field emitter and then oxidizing to ZnO. Complicated crystallographic structures, like zeolites on Pt-emitters, have been deposited mechanically and used in special cases, as will be discussed in Section VI.

TABLE 1

Some techniques for emitter preparation*

Material	Etchant and/or conditions	
Ag	Conc. aq. sol. of KCN.	2—5 V a.c.
Al	$(HClO_4$ in $H_2O) : (C_2H_5OH) = 1 : 1.$	< 10 V a.c.
Au	Conc. aq. sol. of KCN.	2—6 V a.c.
Ge	Aq. NaOH conc. <0.1 wt. % in H_2O. (The concentration critically depends on the conductivity of Ge, such that Ge cannot shunt the electrolyte.) 30—120 V a.c., such that the current density >1 A cm^{-2}.	
Se	Vapor deposited with field induced whisker formation; the evaporation furnace was at $470°K$ and the Pt surface at $\approx300°K$ [19].	
ZnO	H_2SO_4 16 wt. % in H_2O.	
Zeolites	Slurry of sub-micron grains deposited on Pt, heated and fused to Pt support under microscope at $T \approx 900°K$.	

* In addition of those cited in refs. 7 and 17.

Useful field emitters can also be produced by whisker growth on Pt, W or other surfaces. As a result of field induced chemical reactions, various organic compounds decompose and form whiskers of a carbonacious material. As shown by Fig. 1, Beckey et al. [18], obtained a multitude of emitters, but with varying radii. Examples which have been investigated intensively are the field polymerization of acetone and benzonitrile. These preparation methods have gained considerable importance in the analytical applications of FIMS. Typically, a 5 μm tungsten wire is heated to $900°C$ in 10^{-3} torr of benzonitrile, while a potential of several kV is applied. After a few hours, semiconducting microneedles, several μm in length, are obtained.

The formation of selenium whisker field emitters is of particular interest. Usually, the radius of a field emitter is determined from the rigid shape of a whisker. After forming a whisker, field evaporation of the deposited material only can increase the emitter radius. Indeed, in most of the experimental observations, the emitter radius does increase with the time of experiment. An exception has been observed with Se [19]. Under the proper conditions of Se vapor supply, temperature, and rate of field evaporation, the radius of Se whiskers adapts quickly to new potential differences in such a way that a constant field strength is preserved at all times at the surface of the whiskers.

The precautions which have to be considered in preparing field emitters depend on whether FIMS will be used to analyze field ionized gas molecules or to relate surface interactions to surface structure. For the analysis of field ionized gas molecules, which may be formed without any surface interaction, any kind of emitter material may be employed provided the surface

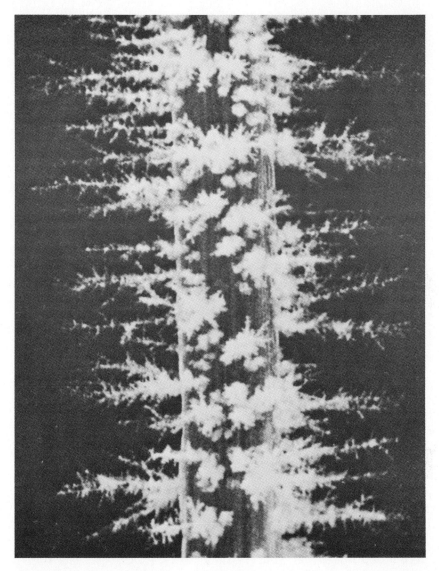

Fig. 1. Needle-shaped field ion emitter, produced by the field induced polymerization of benzonitrile. (After Beckey et al. [18].)

curvature yields the necessary field strength. Even razor blades and the edges of evaporated films have been used in addition to a variety of whiskers. However, for study of the dependence of surface interactions on surface structures, there are extreme requirements for the preparation of a suitable emitter. For example, it has been reported that a single foreign molecule on an emitter surface, investigated mass spectrometrically, disturbed consider-

384

ably the ionization behavior of interacting particles. The purity of the emitter material has to be extraordinary because ppb impurities may seriously influence the results. Diffusion processes at elevated temperatures do permit surface segregation of these trace impurities but caution must be observed. Furthermore, surface contamination from adsorption of residual gases can only be minimized with ultra high vacuum conditions. Investigations on surfaces by FIMS are best begun with residual gas pressures of 10^{-10} torr or less. The vacuum requirements are much more severe in FIMS than in other methods of surface analysis because of the dipole attraction of the residual gases into the electric field.

Other limitations to the application of field emitters result from the mechanical strength and chemical bonding energy of the emitter material. Since a force of the order of 400 N/m^2 results at a field strength of 10^8 V/cm, the creep limit of some materials may be exceeded. Field ionization of gases like He or H_2 can only be achieved at field strengths where field evaporation of many metal tips occurs before field ionization of the gas. For example, F_{evap} is 2×10^8 V/cm for Ag at 80°K so the field ionization of H_2 can be studied; Ag$^+$ ions at 300°K preclude a field ionization study of hydrogen because of reduced field evaporation fields.

B. THE FIELD ION SOURCE

The essential features of an emitter tip, T, mounted in a field ion source are shown in Fig. 2. The emitter tip may be reproducibly and reversibly placed in position 1 for FIM observation, or in position 2 for FIMS analysis about an axis of rotation in the UHV chamber, V. The optimum location of the tip is controlled with the adjustment, A, on one of the flexible bellows, B. In addition to being able to align the tip on the optical axis of the spectrometer, there is a device that permits moving the tip for analysis of the different crystal surfaces on the surface of the tip. Observation of FI images, which are made using the channel plate, C, luminescent screen, S, mirror, M, window, W, and an appropriate imaging gas, provides the necessary elucidation of the surface structure being investigated.

In position 1, field ion currents of the order of 10^{-8} A, which are obtained as a maximum only when single emitters are used, are focused by the lens, L, onto the entrance slit of a 60° magnetic sector mass spectrometer. The ion current is emitted over a solid angle of ≈120°. Therefore, in constructing the ion source, the beam focusing devices are necessary to improve the transmission coefficient for better sensitivity.

The temperature of the tip is controlled by using the cold finger, F, or the resistance of the filament loop, T. Automatic temperature control is obtained with an electronic bridge; the temperature is measured with a thermocouple or an optical pyrometer.

This device has been found to be reliable in both FIM and FIMS analysis.

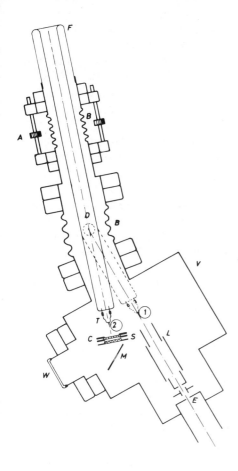

Fig. 2. The field ion source of a magnetic sector type mass spectrometer combined with a field ion microscope. A, tip position adjustment; B, bellows; C, channel plate; D, rotation axis; E, entrance slit to spectrometer; F, cold finger; L, ion focusing lens; M, mirror; S, luminescent screen; T, emitter tip; V, vacuum chamber; and W, window. Position 1 is for FIMS analysis and position 2 is for FIM observation. (After Schmidt et al. [20].)

An easy and quick control of the surface emissivity may be made at any time during the mass spectrometric investigations. Of course, this simple technique does not compete with the atom probe for analyzing single surface sites, since ion currents from a solid angle of approximately $2°$ of the emitter are focused onto E. However, this ion source provides immediate FIM information about the surface, which is a significant advantage compared with other mass spectrometric techniques.

C. MASS SEPARATORS

1. Magnetic sector instruments

In the initial FIMS studies, magnetic sector instruments were used for the mass separation as described [6]. From a geometrical point of view, this is not the most suitable combination, since a point source with a 120° solid angle of emission has to be adapted with an ion optical arrangement and the accompanying slit geometry. A further disadvantage of this combination is a rather slow scanning speed. The latter is a severe restriction because only a small selected part of the mass scale can be focused onto the exit slit of the spectrometer at any given time. Nevertheless, magnetic sector instruments are still used most frequently in FIMS studies. Auxiliary electronic devices for the measurement and control of the focusing magnetic deflection fields are used for the selection of different parts of a mass scale, which have to be resolved, and for precise determination of the mass.

Occasionally, ion intensities are so small that only few ions of the proper M/n ratio, (M = mass, n = charge number), will be formed at the instant when the focusing conditions are fulfilled. The large statistical fluctuations

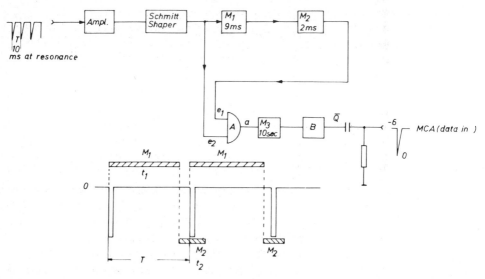

Fig. 3. Circuit diagram for automatic and repeated scanning of a specific mass region. The resonance signal (upper left) consists of peaks with 10 msec time difference. This signal is amplified and correctly shaped. The delay units M_1 and M_2 cause two subsequent peaks with 10 ± 1 msec time distance to arrive simultaneously (through e_1 and e_2) at the switch A. The current, a, then charges a capacity ($\overline{Q}$) which starts the multichannel MCA. During reset of the resonance signal the delay M_3 automatically and B manually prevents the start of MCA.

in intensities can be averaged by repeatedly scanning the mass range, storing the signal intensities in a multichannel analyzer, and then obtaining the average with a small computer. One way for automatically scanning a desired mass range is illustrated by the circuit diagram of Fig. 3. Here, the magnetic field is measured by the Lamor frequency of the nuclear magnetic resonance of protons or ^{7}Li nuclei. The resonance signal, which determines a selected point on the mass scale, (usually, to better than 10^{-2} amu) is used to start a multichannel analyzer of typically 100 channels. Different masses of interest are stored successively into different channels. The magnetic field is then automatically reset to a value beneath the starting point on the mass scale and the scanning repeated. Ion intensities of less than 10^{-22} A can be measured without disturbing intensity fluctuations, provided the emitter surface remains stable during the prolonged time of measurement. The accuracy of the measurement can be controlled easily by comparing naturally occurring isotopic distributions such as ^{13}C in carbon. Thus, mass intensities from ion currents of less than 10^{-22} A could be measured to within a few per cent accuracy after several scans of the mass range without disturbing the background noise.

For analytical purposes, double focusing magnetic mass spectrometers have been used in FIMS [21]. The high resolution, using mass defects for chemical structure analysis, has not yet been applied for surface studies, since more fundamental questions still need to be resolved.

2. Resonance mass separators

Quadrupole mass filters, which have been applied recently for the analysis of field ions [22,23] have several promising advantages over present methods. First, the principal advantages of this instrument are a compact design, which makes it easier to obtain the necessary UHV conditions, and a capability to scan masses up to 1000 amu/sec. However, the mass filter like the magnetic sector but not like the time-of-flight mass spectrometer, can only analyze a small selected part of the mass scale at a time.

Second, the mass analysis is independent of the energy of the ions. Compared with the other methods described, this feature is unique and permits this instrument to be used for measurements on dielectric surface layers or for problems where a disturbed energy distribution is introduced during the ion formation. The transmission coefficient for ions, however, depends strongly on the ion energy. Thus, calibrations of the measured ion intensities are required and quantitative intensity measurements are still difficult to obtain.

Third, the total ion transmission is more favorable in the mass filter than for the other mass separators because the geometry of the ion trajectory is much more adaptable to the "point source" of an emitter. The allowed entrance aperture of the quadrupole field is large; consequently, ions entering at an angle of $10°$ or even more off the axis will be focused. The permissible off-axis angle depends on the radial energy of the ions and varies with the focusing properties of the lens system.

Finally, the operational mode of the instrument can be changed rapidly from a high resolution to a high transmission mode. Since the product of mass resolution and ion transmission is nearly constant for mass filters, either small ion intensities can be detected at low mass resolution or vice versa.

There are two principal disadvantages to the quadrupole [24]. First, there is a relatively small mass resolution, which permits a mass separation of about 0.5 amu. This mass resolution is determined by the ion energy, the position and angle of incidence of an ion relative to the optical axis and the parameters which determine the ion oscillations in the quadrupole field, e.g. the position of the scan-line in the stability diagram for the quadrupole field. Second, there is an uncertainty in the relative ion intensities detected at different positions of the mass scale. The transmission coefficients are a complicated function of the mass-to-charge ratio of particles to be analyzed because of the resonance conditions which permit an ion of a certain mass to reach the detector during its oscillating trajectory.

From an experimental point of view, the energy of ions which can be separated successfully by the quadrupole field is a major problem. In the quadrupole field, a residence time of about 10^{-5} sec is required to obtain separation of the ions. Since the required precision of positioning the quadrupole rods [24] can only be established for short rods, (the length of quadrupole rods in commercial devices is less than 20 cm), ion energies have to be less than 100 eV and it is better if they are only 10 eV.

The required low ion energies, which are usually obtained from electron impact ionization, are not directly compatible with FI, where high potential differences are needed for the ionizing electric field. A field ion source combined with a quadrupole mass filter is shown in Fig. 4. The emitter (T) is mounted on a cold finger (F), with an adjustable bellows (A,B) for aligning the source onto the optical axis of the quadrupole. The emitter is heated using current leads; the temperature can be controlled through additional connections by regulating a constant voltage drop. A series of electrodes (L) in front of the emitter is used for creating the high electric field and to retard the high energy ions. The emitter potential is usually less than 100 V with respect to the field at the axis of the quadrupole rods (Q). A retarding field electrode (R) is necessary for suppressing low energy ions and for measuring energy distributions of field ions and must be installed in front of the quadrupole rods. An energy filter located behind the mass separator is useless because the quadrupole field alters the original energy distributions. The ion detectors consist of a Faraday cage (FC) on the optical axis of the instrument, an electron multiplier (EM) and an electron channeltron (C). The latter two are perpendicular to the center axis. Deflection electrodes (D)

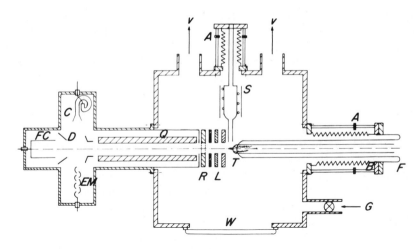

Fig. 4. A field ion source combined with a quadrupole mass filter. The sensitive ion detectors are off the axis. For the evaporation of solids onto the tip the additional source, S, was introduced. A, tip or source position adjustment; B, bellows; C, electron channel-tron; D, deflection electrodes; EM, electron multiplier; F, cold finger; FC, Faraday cage; G, gas inlet valve; Q, quadrupole rods; R, retarding field electrode; S, evaporation source for other solids; T, emitter tip; V, vacuum; W, window.

are used to direct the ions to one of the detectors. The off-axis placement of sensitive detectors is necessary to avoid background noise from electrons or Bremsstrahlung, which are created by high energy ion impact at electrodes [25]. Heinen et al. [26] tried to suppress this noise by tilting the ion source with respect to the quadrupole axis.

The quadrupole mass filter cannot compete in mass resolution with magnetic sector field instruments. However, the identification of mass numbers up to 400 amu has been demonstrated. The energy of ions cannot be ana-

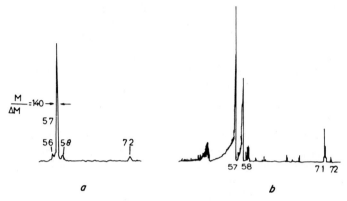

Fig. 5. Comparison of field ion mass spectra of neo-pentane using (a) the quadrupole mass filter [23] and (b) the magnetic sector instrument [28].

lyzed using the quadrupole field. For example, an energy loss resulting from a fast molecular decomposition [27] will not be detected, as shown in Fig. 5, for the field ion mass spectra of *n*-pentane. Decomposition of the parent molecular ion according to

$$
\begin{array}{ccc}
\mathrm{CH_3} & & \mathrm{CH_3} \\
| & & | \\
\mathrm{CH_3 - \overset{+}{C} - CH_3} \rightarrow & \mathrm{CH_3 - \overset{+}{C} - CH_3 + CH_3^*} \\
| & & | \\
\mathrm{CH_3} & & \mathrm{CH_3} \\
\mathrm{M72} & & \mathrm{M57} \qquad \mathrm{M15}
\end{array}
$$

requires times in excess of 10^{-12} sec. Fast $(t > 10^{-11}$ sec) decomposition products, that have lost energy, exhibit broadening on the low mass side in the magnetic sector instruments (Fig. 5(b)), but in the quadrupole mass spectrum, (Fig. 5(a)), the signal at M57 is symmetric. Figure 5(a) shows that lower energy ions are not revealed by the mass spectra.

For reactions on oxides or other compounds, where the energy loss results from non-equipotential emitter surfaces, the quadrupole mass filter is advantageous for determining the real masses of surface reaction products, independent of the energy of ions.

3. Time-of-flight analysis of field ions

In the atom probe described by Müller et al. [24] focusing of the ion trajectories is not required for time-of-flight (TOF) analysis of ions. The combination of a probe hole field ion microscope, with subsequent TOF mass analysis of ions entering the probe hole, is a fascinating invention for the identification of single surface atoms within the atomic resolution of a surface.

By combining pulsed fields at the emitter with TOF mass analysis, further important studies of time-dependent and field-dependent processes can be carried out. The temporal change of the field strength, which is proportional to U in Fig. 6, permits reactions to occur without disturbance for the time, t_R. At the desorption field strength, U_D, surface particles are field desorbed during the time t_D. The reaction time, t_R, can be adjusted to lower limits of 10^{-5} sec and the desorption time, t_D, can range from 10^{-8} to 10^{-2} sec. With t_R, kinetic processes can be monitored in the μsec time range. The variation of the reaction field, $\sim U_R$, is of particular interest for the chemist because it offers the possibility for studying the field dependence of chemical reactions in electric field ranges of 10^5 to 10^7 V/cm and, to date, this was not possible with other experimental methods.

The device which has been used extensively for TOF measurements is shown in Fig. 7 [30]. The emitter tip, T, mounted on the isolating support, I, can be tilted by a mechanical device within the optical axis of the TOF

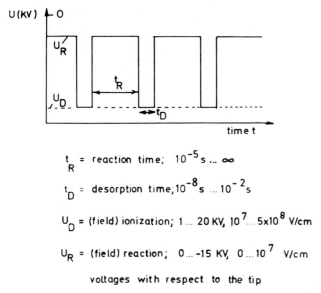

t_R = reaction time; 10^{-5} s ... ∞

t_D = desorption time; 10^{-8} s ... 10^{-2} s

U_D = (field) ionization; 1 ... 20 KV, 10^7 ... 5×10^8 V/cm

U_R = (field) reaction; 0 ... -15 KV, 0 ... 10^7 V/cm

voltages with respect to the tip

Fig. 6. Schematic representation of field pulses for time-of-flight measurements of field ions.

instrument. The field emission image can be observed through the window, W, on the luminescent screen, S, for orienting the emitter surface. The high electrostatic field for desorption is created by the pulsed electrode, E_p; the constant potential electrode, E_s, shields the ion trajectory from time-dependent potentials, which cause time-focusing effects [29—32] with improper mass resolution. The positive tip potential, U_T, of 2—4 kV determines the ion energy with respect to the field-free flight tube, F, which is stabilized by U_F at a potential of −2.7 kV. The lenses L and A serve to focus the ion beam onto M_c, which is part of a Bendix type glass strip multiplier with stabilized potentials M_D and M_F.

The mass signals are received at M_A and analyzed in the following way: a master oscillator, G, supervises the different time regulating installations which are required for TOF measurements. After the high voltage pulse, H_p on E_p, is started for field desorption, signals arriving at M_A can be registered on eight different channel correlations of the multichannel analyzer, MCA. This is achieved by the time delay units D_1 to D_8, which start the appropriate time-to-pulse-height converters, TPC, for receiving signals from M_A at different flight times or mass scale regions. The circuitous use of the TPC's solved the problems caused by the simultaneous registration of different ions from a single desorption pulse; this method was established when a time resolution of about 10 nsec had not been verified directly. With this design, eight different masses from a single desorption pulse can be registered at different positions on the mass scale. In addition, coincidence measurements

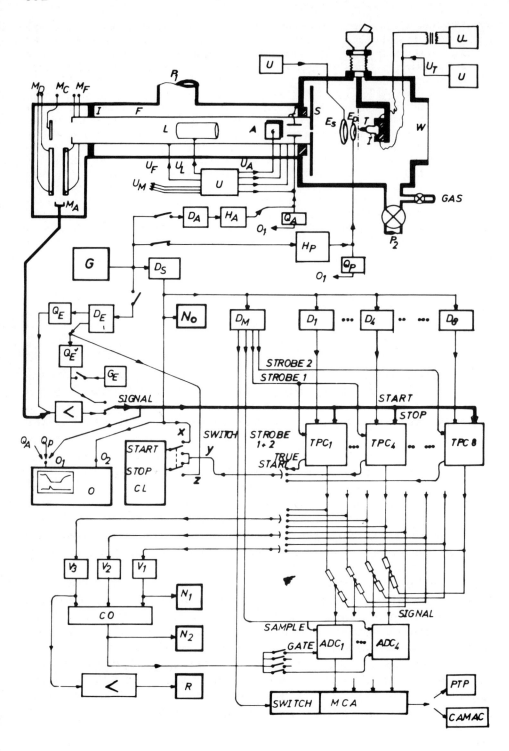

Fig. 7. The time-of-flight mass spectrometer with electronic circuits for data handling and sampling [30]. In addition to the explanation in the text, A, deflection plates, horizontal and vertical ≈ 100 V for $1°$ deflection at 5 kV; ADC, analog-to-digital converter, Laben type 8213 positive (bipolar) input, 8 mV/channel, 128/156 channels sample input optimal, gate coincidence or anticoincidence, delayed 0—8 sec after signal (adjustable); CL, clock, Systron type donnor, 10 nsec resolution; CO, coincidence, anticoincidence unit; CAMAC, on-line data transmission to computer; D, delay and gate generators, D_A for driving the pulse generator H_A, D_E for test pulses, D_M for stroke pulses and D_S for synchronization with HV-pulses; G_E, gate pulse generator for test pulses at low repetition rates for checking possible drift of delays; N, scaler, Ortec type 484, for counts up to 10^{10}, mechanically reset and used for counting the cycles (No) and ion currents in some regions (N_1 and N_2), where N_2 is only for selected coincidence; O, dual trace oscilloscope (Tetronix 454); P, mercury pumps; PTP, auxiliary computer unit using paper tape punch; Q, attenuators and shapers for the required pulse shapes; R, strip chart recorder for controlling total ion currents; V, discriminators, producing standard output pulses suitable for the ADC's; x,y,z, time signals where x is the start of cycle, synchronized by D_S to the leading edge of the HV pulse, y is the true start output of a selected TPC, and z is the adjusted test pulses.

with high pulse repetition rates ($\lessgtr 70$ kC) are possible. Further details are explained in Fig. 7.

D. ENERGY ANALYSIS OF FIELD IONS

The energy distribution of field ions yields reliable information about the electric potential and distance from the emitter surface where the ionization process actually occurs. Our knowledge about different field ionization mechanisms has been improved considerably by measurements of energy distributions.

In the magnetic sector type mass spectrometer, retarding field electrodes have been applied for this purpose. The energy analysis performed by Jason [33] and by Heinen et al. [34] used retarding electrodes in back of the magnetic mass separator as shown in Fig. 8. Goldenfeld et al. [35], mounted the retarding grid between the ion source and magnetic analyzer. In this case, a spherical grid has to be used for the divergent ion beam, otherwise the distribution of momenta would be measured instead of the energy distribution.

In the design and operation of the energy analyzer in Fig. 8, the retarding electrode consists of two gold meshes, with about 20 strips per linear mm and 70% transmission. The two meshes are 2 mm apart. With this construction, a minimum potential penetration is achieved. Since the work function of the retarding electrode is part of the measured potential difference ($U_E - U_R - \Phi_R$), gold is a convenient material which is least affected by adsorption processes. The potential difference between the emitter electrode and retarding grid is supplied by an additional variable low voltage source of 0—100 V. The ion intensity at a given mass number is measured as a function

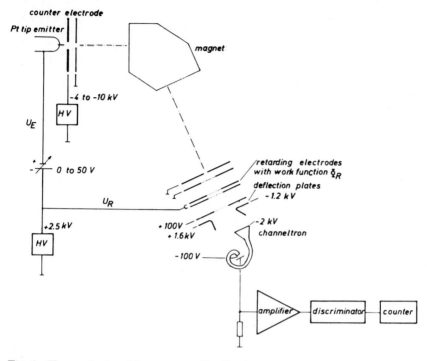

Fig. 8. The analysis of ion energy distributions using a magnetic sector tape mass spectrometer. Although the focusing of the magnetic field leaves an energy spread of 20 eV or more, the retarding electrode can analyze 0.1 eV or less.

of the potential difference. Ions, which are created several angstroms in front of the emitter tip at a potential $U_E - \Delta U$, arrive with an energy loss and can pass the retarding electrode only if the potential difference is sufficiently high. The ion detector, which could be a channeltron, is placed off the optical axis of the instrument. Deflection plates focus the ion beams onto the entrance aperture of the detector whereas secondary electrons and Bremsstrahlung are not detected.

In TOF mass analysis of field ions produced by field pulses, time-dependent potentials affect the trajectory and energy of ions. In order to prevent energy uncertainties, Müller and Krishnaswamy [36] modified the atom probe by adding a Möllenstedt energy analyzer. This technique takes advantage of the strong chromatic aberration which a slightly off-axis beam suffers in an electrostatic saddle-field lens. This device offers an energy resolution of 5×10^{-5}, i.e. field ions with an initial energy of 10 kV can be resolved to less than an electron volt. In another development, an energy focusing device (according to Poschenrieder) is used at the exit of the TOF tube as described by Müller in the previous chapter.

E. ION DETECTORS

For the identification of ion currents of 10^{-22} A (1 ion/min) and discrimination against background noise, a special analytical technique has been applied which is based on the fact that the pulse height distribution of a secondary electron multiplier is mass dependent [37].

The pulses, which are created by single impinging ions onto a Cu/Be multiplier, are stored in a multi-channel analyzer. An impedance transformer distributes the different pulse signals according to their pulse height into different channels. As is seen in Fig. 9, the maximum in the pulse height distribution depends on the mass of the ion. This experimental observation depends on the inverse correlation between the momentum of ions and the intensity of secondary electrons at the first dynode. Thus, this distribution function gives an additional statistical indication of ions with different masses. The distribution statistics usually follow the Poisson equation [38]. The maximum for electrons or γ-rays is different from that of ions. Back-

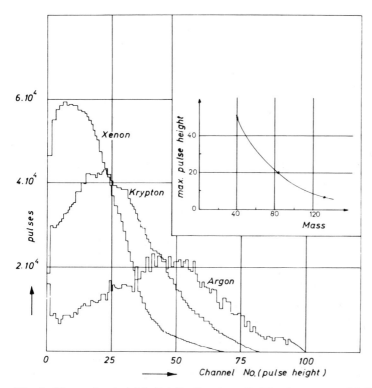

Fig. 9. The pulse height distribution in a Cu/Be electron multiplier as measured for the field ions Xe$^+$, Kr$^+$, and Ar$^+$ (after Goldenfeld et al. [35]) and the dependence of the maximum pulse height on the mass of the impinging ion (inset).

ground signals can be measured independently and subtracted from ion intensities.

III. Mechanisms of ion formation

A. FIELD IONIZATION

In the early investigations, the field ionization process was attributed to a quantum mechanical tunneling effect, which resulted from the deformation of molecular potentials in the extreme external field. For this model of FI, the tunneling probability D depends exclusively on the ionization potential I of a molecule, the work function Φ of the surface, and the field F. By applying the one-dimensional WKB approximation, the penetration probability is

$$D \approx \exp[-A(I - \Phi)I^{1/2}F^{-1}]\,f(F,I) \tag{1}$$

where A is a constant and $f(F,I)$ is a small correction term. This equation explains the usually observed exponential increase of the ionization probability with the field. However, the dependence on the term $(I - \Phi)I^{1/2}$ for different gas molecules and surfaces is an exception. The validity of this correlation is only fulfilled for rare gases, but even there, field induced adsorption phenomena have been found to interfere.

The reason that the theoretical potential energy model for electron tunneling cannot be applied to calculate ionization probabilities in electric fields is that intermolecular forces are affecting the ionization process simultaneously. Therefore, a particular kind of chemical ionization at surfaces is predominant in these mechanisms.

The main reaction types, which result from field interaction at emitter surfaces, can be classified as proton or cation transfer, charge transfer and donor—acceptor complex formation, ion—molecule reactions, and heterolytic bond cleavage. The latter includes the bond rupture of surface atoms that occurs during the field desorption or field evaporation processes. Since all these reactions proceed under unusual field conditions and with the interaction of surface sites, the products cannot be compared directly with those of ordinary homogeneous chemical reactions at zero field.

B. PROTON TRANSFER

Field ion mass spectra of molecules with sufficient proton affinities yield proton attachment ions rather than molecular ions, if intermolecular interactions are involed in the ionization process. This type of behavior indicates clearly that field ionization cannot be described solely by the tunneling of electrons. The following two examples illustrate the proton attachment.

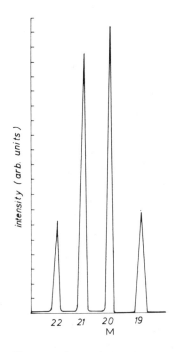

Fig. 10. One of the first mass spectra indicating proton (deuteron) attachment. (After Beckey [39].) The mass signals of a H_2O-D_2O mixture are attributed to M19(H_3O^+), M20(H_2DO^+), M21(HD_2O^+) and M22(D_3O^+).

In one of the first investigations, Beckey [39] studied the behavior of water in fields. Since the accurate mass scale determination was still a problem at that time, H_2O-D_2O mixtures were used to determine if H_2O^+ or H_3O^+ is the principal intensity of the field ion current. For the H_2O^+ intensity in the isotope mixture, three mass signals are expected, namely, M18(H_2O), M19(HDO) and M20(D_2O), whereas four mass signals will be formed for the proton attachment, H_3O^+. As shown in Fig. 10, four mass signals were obtained, clearly indicating the proton (deuteron) attachment process. Beckey also found [39] that many other gases like N_2, CO, CO_2, and Ar also form proton attachment ions if traces of water are present. The proton transfer process occurs because of the energetics of these reactions and the fast reaction rates which are involved in proton transfer. The intensity of proton attachment ions depends strongly on the applied field strength as is shown for water in Fig. 11. Here Schmidt's [40] experiments show that the ratio, $I_{H_2O^+}/I_{H_3O^+}$, increases with increasing field strength, which can be explained by considering the kinetics of the process. The sharp onset of ion emission above 6×10^7 V/cm is due to the presence of a condensed multilayer, which is removed during the ionization process. At elevated fields, the field ionization probability of impinging water molecules increases

398

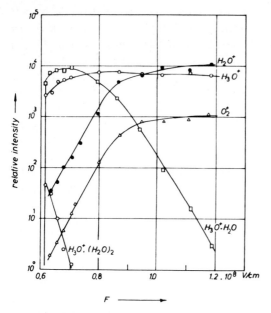

Fig. 11. The intensity of field ions during the interaction of H_2O with a Pt emitter with increasing field strengths.

and, consequently, bimolecular processes are less frequent. Accordingly, the intensity of molecular ions of associated molecules decreases with increasing field strength.

Field ion mass spectra of water demonstrate another phenomenon of surface interactions. Higher molecular weight associates, $H_3O^+(H_2O)_n$, are formed where n, which depends on the temperature and field strength [41], has a maximum value of ten at low temperatures. These ions are formed in multilayers of adsorbed H_2O molecules, which are adsorbed or condensed by field compression even though the field free partial pressure of H_2O is less than 10^{-6} torr. A remarkable property of these ions is that the molecular structures, $H_3O^+(H_2O)_n$, are held together by hydrogen bonding of only about 5 kcal/bond/mole. The molecular excitation during the ion formation process is minor and does not affect these weak chemical bonds.

C. CHARGE-TRANSFER AND INTERMOLECULAR INTERACTIONS

One of the important quantities in field ionization is the minimum field strength which is necessary for ion formation. Since an exponential increase of ion intensities with field strength is usually observed, the threshold can be measured reasonably. Threshold fields for various molecules are compared for a definite small ion current, typically 1 ion/sec, measured under equal conditions of tip radius, gas pressure, and temperature.

In FIM, the promotion of field ionization or field desorption is a well established phenomenon. The best voltage for imaging a surface with an image gas, such as He, is drastically lowered if another gas, such as H_2, is admixed. The promotion of field ionization results from a shift of the threshold field, usually from intermolecular interactions. Other mechanisms have been proposed, in particular, the excitation of electronic or vibronic levels in the molecule. In early work, Ehrlich and Hudda [42], while studying the He-promoted field desorption of H_2, tried to account for the phenomenon by an electron impact mechanism. Electrons from a field ion formed several angstroms in front of the emitter would have sufficient impact energy to cause desorption. The extremely sharp energy distribution of field ions, which was measured by Tsong and Müller [43] to a half width of about 0.6 Å, seems to exclude this possibility. However, a conceivable possibility for an electron-assisted desorption mechanism exists, since Jason [33] found resonance peaks in the field ion energy distribution spectrum of H_2, CO, and Ne. At resonance peaks, electrons are released with energies exceeding the values obtained from electron impact desorption. Another possible excitation mechanism was proposed by Nishikawa and Müller [44], in which part of the polarization energy of an impinging gas molecule is transferred to an adsorbed particle. The importance of this mechanism has been discounted by Bassett [45]. Field pulse experiments clearly indicate that the promotion effect can be observed without impinging molecules or electrons. Therefore, it is necessary to conclude that this phenomenon has a much more general nature.

There are numerous experimental results showing that threshold fields depend sensitively on intermolecular interactions, in particular, when charge transfer complexes may be formed between different molecules at the emitter surface. For example, remarkable differences in threshold field, which are measured by the potential difference U_E between the emitter and counter-

TABLE 2

Threshold fields of some gases on cleaned Pt without and with chemisorbed layers (after Block [46]).

Molecule	U_E clean Pt (V)	Adsorption layer	U_E with an adsorption layer (V)
Hydrogen	8500	Naphthalene	3500
Argon	8380	Naphthalene	4250
Argon	8380	Benzene	4000
Benzene	1500	Naphthalene	700
Naphthalene	920	Chinone	495

electrode, have been found if hydrogen or argon is ionized on a cleaned Pt-emitter or on an adsorbed layer of a hydrocarbon on Pt (Table 2). There have been three approaches to explain the observed decrease in threshold fields. If we consider [46] a molecule with a dipole μ, then $U(r)$ varies according to μ/r^3. Between two molecular dipoles, the interaction energy E_W, caused by dispersion forces, is given by

$$E_W = \frac{1}{r^6}\left(\frac{2\mu_1^2\mu_2^2}{3kT} + \mu_1^2\alpha_2 + \mu_2^2\alpha_1 + \frac{3}{4}h\alpha_1\alpha_2\,\frac{\gamma_1\gamma_2}{\gamma_1+\gamma_2}\right) \qquad (2)$$

where α is the polarizability, γ the frequency of molecular oscillators, and the indices 1 and 2 identify the two molecular species. For non-polar molecules, dispersion forces are described exclusively by the last term of eqn. (2). According to Debye [47], the magnitude of the dispersion forces may be estimated from the enthalpy of evaporation Λ by $\Lambda \approx \alpha F^2$. For naphthene, Λ is 46.38 kJ/mole and with $\alpha = 17.5 \times 10^{-24}$ cm^3, F is 6.2×10^7 V/cm. With large dispersion forces, the electrical field strength reaches values at the molecular interaction level scheme which are sufficient in a macroscopic sense for field ionization. While the calculation given is merely qualitative, it clearly indicates that molecular fields are of importance and will influence ionization processes if external macroscopic fields are superimposed upon local molecular fields.

The field strength for desorption of several molecules adsorbed on tungsten surfaces is considerably reduced if an imaging gas is introduced. When He was introduced, Bell et al. [48] reported the reduction in the threshold field for desorption from W was 55% for CH_4, 64% for H_2, and 83% for Xe. In addition to the gas- or electron-impact mechanism, a gas-complex-promoted mechanism was considered. If quasi-stable heteronuclear ions are formed, the ionization can occur more readily because the ionization potential of theoretical compound gases (XeHe, $CH_4 \cdot H_2$, etc.) would be considerably smaller than the ionization potentials of the separated atom or molecule. Thus, the formation of intermediate complex ions seems to offer another possible explanation. Since field ions like HHe or $CH_4 He^+$ have not been detected by mass spectrometric methods, intermediate complexes of this kind must have a life-time of less than 10^{-12} sec.

Intermolecular forces can retard as well as promote ion desorption. This was shown by Schmidt et al. [49], who reported threshold fields for ion formation increased for some molecules. A comparison for different gases at Fe- and Pt-emitter surfaces is given in Fig. 12. The values of U_E are related to ion intensities of one ion/sec and are plotted according to eqn. (1), e.g. the electron tunneling mechanism without surface interaction. The expected behavior, which is shown by the straight lines, is followed only by the rare gases (the deviation in Xe on Fe was caused by H_2O impurities in the gas supply). Here, O_2 on Fe, CO on Pt, and N_2 on Pt all exhibit a promotion

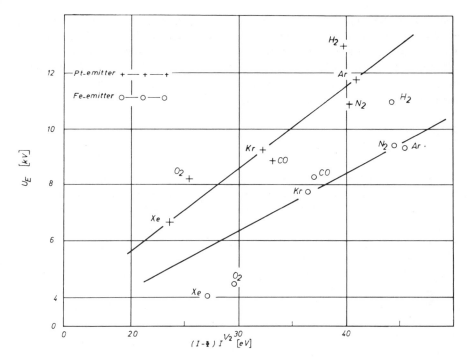

Fig. 12. Threshold fields for field ionization of different gases on platinum and iron. (After Schmidt et al. [49].) The straight lines are in agreement with an electron tunneling mechanism without surface interaction.

effect but H_2 on Fe and Pt, O_2 on Pt, and CO on Fe require higher values of U_E than expected. The change in the work function Φ due to adsorption cannot explain this experimental result because a decrease in Φ would increase U_E an equivalent amount.

The decrease or increase in threshold fields is closely connected to chemical bond formation and distribution of orbitals at surfaces, as described by Knor and Müller [50] and Knor [51]. The problems of interpreting FIM images, where certain adsorbed species are visible and others are not [52], have a direct relation to intermolecular forces and threshold fields in FIMS. Since these threshold fields in FI are as important as values of ionization potentials in photo- or electron-impact ionization, the application of FIMS for quantitative surface analysis will be complicated as long as these correlations are unknown.

D. HETEROLYTIC BOND CLEAVAGE

The ion formation at surfaces can proceed either by electron transfer, as discussed so far, or by the dissociation of chemical bonds. Ion pairs are formed in a heterolytic dissociation process, as is usually observed in electrolyte solutions, where positive and negative molecular ions are easily separated in an environment with a high dielectric constant. This process is also possible in high electric fields at surfaces. However, kinetic processes severely limit the extent of heterolytic dissociation compared with electron transfer mechanisms. In electron transfer, times of $>10^{-16}$ sec are likely but heterolytic dissociation requires at least the time of one vibrational mode of $>10^{-12}$ sec. Therefore, a molecule entering a high electric field will react first by an electron transfer process.

Heterolytic cleavage will proceed only if the environment of an adsorption layer or a condensed film at a surface favors the bond cleavage for kinetic or energetic reasons. The field evaporation of metals and the field desorption of surface layers are essentially processes of this kind. In both field evaporation and in field desorption, surface lattice atoms or molecules have to break chemical bonds to form positive ions. The problems associated with this mechanism have been discussed in the previous chapter by Müller.

Many reactions in condensed layers at emitter surfaces involve dissociation processes of chemical bonds. Examples will be given for arsenic oxide, selenium and some substances which are known to dissociate spontaneously at zero-field conditions.

E. ION—MOLECULE REACTIONS

The ions which are formed primarily in the adsorption layer, either by electron transfer or by heterolytic bond cleavage may react further, so that secondary products of ion–molecule reactions are desorbed. The energetics of these reactions can frequently be compared with those of processes between ions and molecules in the gas phase. There are, however, two additional influences of the surface. First, it acts as a third body and highly exothermic adiabatic reactions, which are impossible in bimolecular gas collisions, can occur. In other cases, the specific chemical interaction of surface radical sites has been observed. The energy of these sites can be involved in certain endothermic reactions, which for thermodynamic reasons will not occur in the gas phase. As an example, Röllgen found very intense H_3O^+ signals in benzene—water mixtures, which under identical conditions of pure H_2O pressure and at the same temperature could not be observed on the needle emitters under investigation. Since evidence for benzene cations is found in the mass spectrum (Fig. 13), where $C_6H_6^{2+}$ ions are observed, it is evident that the formation of H_3O^+ ions is catalyzed by benzene cations chemisorbed on the surface. The field ionization process has to be excluded for the formation of these ions because of the high ionization potential of doubly charged ions. It has to be concluded that a benzene ion, which is

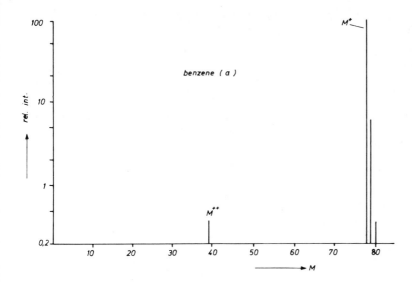

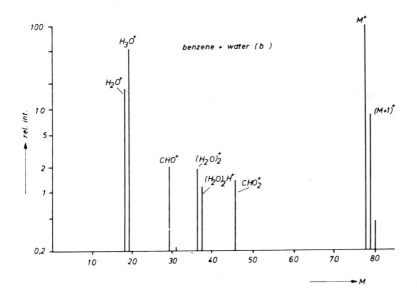

Fig. 13. Field ion mass spectra of (a) benzene, and (b) benzene—water mixtures on needle emitters. The doubly charged benzene ion at M39 is evidence for chemisorbed benzene ions, which also cause decomposition products in (b). (After Röllgen and Beckey [53].)

chemisorbed on the carbonaceous needle surface is field desorbed.

The mass spectrum of a benzene—water mixture (Fig. 13(b)) contains various proton attachment products, which are also explained by ion—molecule reactions. The formation of H_3O^+ can be described by the mechanism

The addition of the OH radical results in the cleavage of the benzene ring and leads to the subsequent desorption of numerous secondary products. From experiments with isotopes, it follows that the H atom of the CHO^+ ion originates from benzene and the H atom of CHO_2^+ from water.

An unusual molecular ion reaction is found in the mass spectrum of acetonitrile [54]. In experiments with pulsed fields, parent molecular ions of CH_3CN^+ could not be observed. The mass spectrum consists of $(M + H)^+$, $(2M - H)^+$ and $(2M + H)^+$ ions. This unusual behavior can be explained as follows: acetonitrile has a rather high ionization potential of 12.2 eV. On the other hand, there is an energy release of more than 4 eV in the ion—molecule reaction

$$CH_3CN^+ + CH_3CN \rightarrow CH_2CN^+ + CH_3CNH^+$$

This reaction may occur nearly adiabatically at fields where it is not necessary to supply the ionization energy of the proton donor. For these ion—molecule reactions, the highly polarized configuration of chemisorbed molecules or even ions may supply additional field energy, thus, favoring certain types of reactions.

IV. The identification of surface interactions

Under the usual FI conditions of a steady electric field and dynamic gas supply, the origin of field ions may be different. First, they may be formed in the gas phase several angstroms in front of the emitter surface. This will be the normal case at elevated field strengths, where the ionization probabilities of impinging gas molecules is high enough to form ions before a surface interaction occurs. Under these circumstances, FIMS is a tool for gas analysis.

Second, parent molecular ions of the homogeneous phase FI are often unstable. Although the energy gained during FI is usually small, molecular ions will decay, particularly if they have lost a bonding orbital electron. Typical examples of this are the positive ions of n-pentane, $C(CH_3)_4^+$, or CCl_4^+.

Finally, ions may be formed by molecular surface interactions either directly with the metal emitter atoms or with a chemisorbed layer on top of the metal surface. This mechanism will occur only if the process described in the previous paragraphs can be excluded.

There are different diagnostic methods used to distinguish between these processes. However, in several cases, the correct interpretation of mass spectra is still doubtful.

A. SURFACE SELECTIVITY OF FIELD IONS

The most straightforward way is to compare FIMS of one gaseous substance at different surfaces. This requires the knowledge of the actual surface

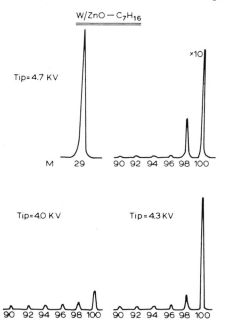

Fig. 14. Field ion mass spectra of n-heptane from a ZnO surface. The ions desorbed from the tungsten field emitter coated with a ZnO film, differ conspicuously from desorption from tungsten. On ZnO, the absolute intensities of C_7H_{16} (M100) increase with increasing field strength (tip at 4.0—4.7 kV), but the relative intensities of dehydrogenation products decrease. (After Block and Bozdech [56].)

state during FI. In the first measurements of field ions of hydrocarbons on different metal surfaces, no difference in the intensity distribution was observed for the different ions [55]. The reason, of course, is that the clean metal surface is covered with a carbonaceous deposit, which becomes the actual site of FI. Thus, an independent knowledge of the surface cleanliness, for example by FEM or FIM, is needed for these experiments. Under properly controlled conditions, mass spectra do display characteristic surface-dependent mass signals as shown by the FIMS of n-heptane on W and on ZnO in Fig. 14 [56]. On the oxide surface, the hydrocarbon is partially dehydrogenated but on the W surface only the parent molecular ion and the dissociatively chemisorbed structure $C_5H_{11}^+$ can be observed. The signal at $M29(C_2H_5^+)$ is the well-known field fragment of $C_5H_{12}^+$. From these experiments with n-heptane, another argument for surface interactions can be derived directly. The usual field dependence of surface ions is shown by Fig.

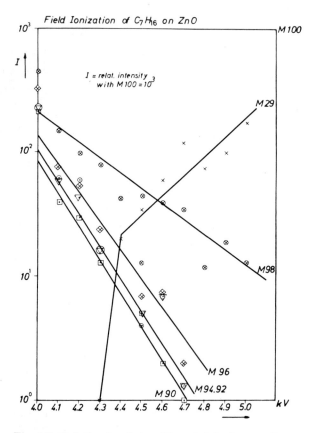

Fig. 15. Relative ion intensities of dehydrogenation products (M98 to M92) and the field fragment (M29) as a function of the field strength (in kV). (After Block and Bozdech [56].)

15 [56]. With increasing field strength, the relative ion intensities from surface compounds will decrease since the kinetic competition between field ionization and chemical surface reaction favors field ionization with increasing field. However, the interpretation of field dependencies often remains ambiguous, since in some cases a field-retarded homogeneous ion decomposition has been observed.

B. APPEARANCE POTENTIALS

A reliable method for identifying molecular species on a surface involves the measurement of the energy distribution of field ions as described in Section II.D. On a metal emitter tip, an energy deficit is good evidence that ions are formed in the homogeneous phase before reaching the tip or on top of a condensed surface layer. On the other hand, Tsong and Müller [43] have shown by measuring energy distributions in a FIM that rare gases are ionized over a range of only 0.6 Å relative to the critical distance for field ionization. With the same type of experiment, Jason [33] discovered resonance peaks in field ionization, which required a quantum mechanical explanation.

Furthermore, the measurement of appearance potentials in field ion formation provides information about the energies involved in surface processes. As in other ionization methods, the appearance potential is defined as the minimum energy which is supplied by the electric field to generate an ion. The absolute value of the appearance potential, AP, can be obtained directly from plots of the ion intensity versus energy deficit, as shown in Fig. 16. The AP value is given by AP $= U_E - U_R - \phi_R$, where ϕ_R is the work function of the retarding electrode and, with U_R, determines the potential barrier. It should be noted that the AP value is independent of the work function of the emitter. For an ion, the AP value determines the energy uptake from the field, since the AP is related to the ionization potential I and the polarization energy E_p of the neutral molecule by AP $= I + E_p$, the energy of polarization of the ion has no effect on the measured AP since it vanishes from the energy balance after removal of the ion from the high field region. Considerable information can be obtained from energy deficit measurements, as illustrated in Fig. 16. During the field ionization of n-heptane, masses from different fragments are observed which may be partially surface products and partially fragments of a homogeneous ion decomposition. Appearance potentials have to be measured at a relatively low field strength, where essentially no ionization occurs at a considerable distance from the emitter surface. The experimental curve for the parent ion (M100) appears at a different position and has a different shape from the fragments (M43 and M29). The intensity threshold for the $C_7H_{16}^+$ ion is 5.4 ± 0.1 eV. When this value is corrected for the work function of the retarding electrode (Au, ϕ_R = 4.9 eV), an AP of 10.3 eV is obtained, which is slightly above the value for I

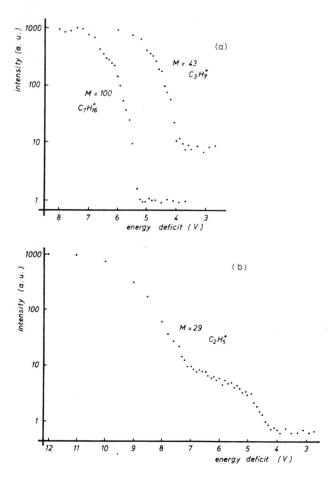

Fig. 16. The ion intensity of n-heptane and the M43 and M29 fragments versus energy deficit during field ionization of n-heptane on Pt. (After Heinen et al. [34].)

of 9.9 eV for n-heptane. In a similar way, the AP value for $C_3H_7^+$ is 7.9 ± 0.4 eV; for $C_2H_5^+$, two values are obtained at 9.1 ± 0.5 eV and 12.1 ± 0.3 eV.

The AP values of the fragments of n-heptane are all below the values obtained for the molecular ion. For a molecular decomposition, AP values for the fragments should be higher than for the parent ion, $C_7H_{16}^+$. Thus, $C_3H_7^+$ and $C_2H_5^+$ must have gained energy due to a surface process. For $C_2H_5^+$, the greater energy loss of 12.1 eV exceeds the value of 10.3 eV for $C_7H_{16}^+$ and must be attributed to a monomolecular decay process

$$C_7H_{16}^+ \rightarrow C_2H_5^+ + C_5H_{11}$$

which occurs in times $>10^{-12}$ sec and is field-dependent. Since most of the intensity of the $C_2H_5^+$ ion in Fig. 16(b) results from this homogeneous decomposition process, the determination of the field strength from the fragment intensities can still be used as a good approximation [57].

It is still not completely understood why the AP of 10.3 eV of $C_7H_{16}^+$ exceeds its adiabatic ionization potential of 9.9 eV, although an internal excitation by Franck—Condon transitions seems likely. These transitions are not possible with rare gas atoms and appearance potentials in agreement with adiabatic ionization potentials have been measured in these cases.

The determination of appearance potentials, as described above, has only been applied to a few cases. The technique looks promising and should become a routine method for the identification of surface molecules.

C. PULSED FIELDS

A third reliable method for the investigation of surface reactions is to use the pulsed field methods, described in Section II.C,3. Here, the ionizing field is applied only for short times t_D. Chemical surface reactions will occur during t_R, e.g. between desorption pulses. Different ion intensities, which depend on the rate of the reaction, will be found if the times t_R are altered. With a dynamic gas supply, a pressure of 10^{-5} torr, and a high sticking coefficient, nearly 100 msec are required to form an adsorption layer. Such an adsorption or chemisorption process can be studied by pulsed field techniques. The dependence of ion intensities can be used to deduce that ions are formed at surfaces.

As an example, cyclohexane may physically adsorb on platinum or chemisorb according to

$$C_6H_{12\,(gas)} \xrightarrow{Pt} C_6H_{11}-Pt + H-Pt$$

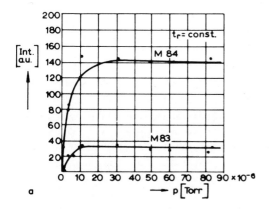 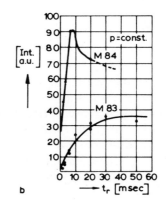

Fig. 17. Pulsed field desorption of cyclohexane from Pt. (a) Normalized intensities for physically adsorbed (M84) and chemisorbed (M83) cyclohexane at a constant pulse repetition rate as a function of the gas pressure; (b) ion intensities as a function of the pulse repetition rate.

As shown in Fig. 17 [58], physically adsorbed and gaseous cyclohexane appear at M84, but the chemisorption product appears at M83. The pressure dependence of the mass signals M84 and M83, at constant reaction time t_R (Fig. 17(a)), indicates that these adsorption phenomena are being monitored. As is seen from the pressure dependence, both physical and chemical adsorption occur simultaneously. The kinetics of this process were then studied by varying the pulse repetition rate (Fig. 17(b)). With increasing reaction times t_R, the concentration of physically adsorbed molecules reaches a maximum and then decreases. The increase in M83 shows that chemisorption occurs by an activated process where the species is at least partially formed from the physically adsorbed state.

The time dependence of mass signals, as observed in Fig. 17, can only be attributed to surface species. A gas phase ionization in front of the emitter would always yield ion intensities which are proportional to the length of desorption pulses t_D and independent of t_R.

Finally, the correlation of field ions with surface processes is given by numerous ionic structures, which cannot be formed in the gas phase. In d.c. fields, bimolecular reactions of gas molecules can be excluded, even at the highest possible gas pressures, because characteristic energy losses due to ion—molecule collisions would have been observed in the mass spectra. Therefore, bimolecular reactions, like proton transfer or association processes, proceed on surfaces or in layers condensed onto surfaces.

V. Field induced surface reactions

A. FIELD INDUCED ADSORPTION

The adsorption or chemisorption of gas molecules on solid surfaces is usually influenced by external electric fields. Two kinds of interactions can be distinguished as described briefly below.

First, polarization forces will be involved, giving rise to a "field bonding energy." In this case, the van der Waals forces of a physically adsorbed particle will be increased by the electric field. The conclusions, first drawn by Drechsler [59], were treated more quantitatively later by Tsong and Müller [60]. They showed that short range dipole—dipole forces lead to adsorption energies in excess of 0.1 eV for rare gases on an emitter surface. The existence of rare gas adsorption under field conditions is an important point for the interpretation of field ion images and the field evaporation process.

Second, the field can cause the formation of regular chemical bonds to the surface due to the shift of electronic levels. This mechanism is, in particular, to be expected on semiconductor surfaces; Röllgen and Beckey [61] demonstrated that even ionic surface structues may be stabilized by field interac-

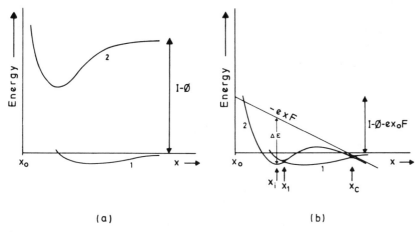

Fig. 18. The potential energy for a neutral atom (1) and for an ion (2) as a function of the distance x; (a) without, and (b) with an external field. (After Röllgen and Beckey [61].)

tion, as is depicted in the potential diagram of Fig. 18. Here, it is assumed that a metal surface is covered with an organic deposit up to the distance x_0 from the metal surface. Under certain conditions, only ionic structures will be chemisorbed at the surface but neutral molecules will not. At zero field, the potential energy of the ion (curve 2 in Fig. 18(a)) exceeds the potential energy of the neutral (curve 1) by the difference $I - \Phi$. At smaller values of x, curve 2 has a deeper minimum because the interaction forces are greater for ions. At zero-field conditions, the neutral molecule is the stable one and may only be physically adsorbed. In a field (Fig. 18(b)), the potential energy of the ion is decreased according to $-exF$ and a minimum is produced at x_i, which is at a lower potential energy than the neutral. There are now two intersections (x_1 and x_c) between the potential curves for the ion and the neutral molecule. A transition from the neutral state to the ionic one is possible at distances below x_1. This behavior can also be interpreted by the fact that the ionization potential of the molecule is reduced by ΔE, the bond energy of the ion to the emitter surface. The field induced chemisorption of ions has severe consequences for various reaction mechanisms, such as ion—molecule reactions, formation of doubly charged ions, etc.

B. FIELD INDUCED DESORPTION

The usual field desorption of positive ions, can be understood directly from Fig. 18(b). If the electric field is increased further, the activation barrier at x_1 will disappear and the ion desorbs. Attention has to be drawn at this point to field desorption of neutral molecules, which severely restricts the possibility of using FIMS for quantitative surface analysis. Neutral mole-

412

cules are not revealed in the mass spectrum, unless additional electron impact ionization is applied.

There are some recent experimental observations [20,62] which indicate that chemisorption bonds of negatively charged moleules may be unstable in an electric field. One example is the adsorption of oxygen on silver. Because of the unique behavior of Ag to oxidize ethylene to ethylene oxide, the oxygen adsorption on this metal has been studied by different techniques [63]. From these experiments, it was concluded that three different adsorption states exist at different surface coverages: (1) an almost non-activated $(E_a < 3$ kcal/mole) dissociative adsorption which covers nearly half a monolayer at room temperature, (2) an activated $(E_a \approx 8$ kcal/mole) non-dissociative adsorption, and (3) an activated $(E_a \cong 14$ [63b] or 22 [63a] kcal/mole) dissociative adsorption. The oxygen adsorption on a silver surface has been studied selectively by FEM [64,65].

In FIMS studies from temperatures between 80 and 425°K, none of the expected oxygen—silver compounds could be detected [62]. The absence of

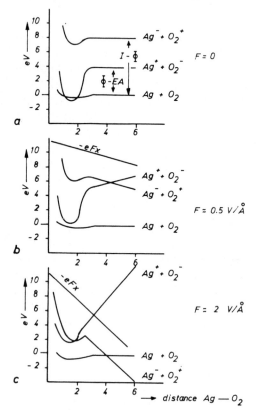

Fig. 19. Potential energy diagrams of Ag and O_2 for three different regions of the electric field.

these compounds results from a field induced desorption process of neutral oxygen, which can be described by potential energy diagrams. For constructing the diagrams, it is assumed first that oxygen is negatively charged, as $O^{\delta-}$ or $O_2^{\delta-}$, in the chemisorption bond, which is in agreement with several experimental investigations [62]. For the molecular form of oxygen, the potential diagram can be drawn from data available. The activation energy for desorption is of the order of 20 kcal/mole [66], which is slightly higher than the electron affinity of oxygen of about 10 kcal/mole. The ionization potential of O_2 is 12.2 eV and the work function of Ag is 4.3 eV. The potential diagram for Ag + O_2, $Ag^+ + O_2^-$ and $Ag^- + O_2^+$ can then be constructed, where it is necessary to assume that the interaction energies of Ag + O_2 and $Ag^- + O_2^+$ are smaller than that for $Ag^+ + O_2^-$.

In Fig. 19, the potential energies for ions are given as a function of the Ag—O_2 distance. At $F = 0$ (Fig. 19(a)), in accord with experimental results, the negative oxygen is the most stable adsorption state. With increasing field, there are two regions of different behavior. At medium fields ($F \approx 5 \times 10^7$ V/cm, Fig. 19(b)), the potential energy curve of $Ag^+ + O_2^-$ is situated above the Ag—O_2 state. Since the interaction energy is small, O_2 will desorb as a neutral molecule. At elevated fields ($F \approx 2 \times 10^8$ V/cm, Fig. 19(c)), oxygen will be field ionized and O_2^+ will be formed, as confirmed experimentally.

The absence of Ag—O compounds in field desorption products can also be qualitatively explained by a particular kind of surface bond. In some systems, the chemical bond will be stable only if molecular orbitals of the chemisorbed particle can be filled by metal electrons. This produces semipolar surface bonds and there is an observed increase in work function. The charge distribution will be disturbed by external fields and occasionally this may lead to a non-bonding orbital configuration. As a consequence, neutral molecules can desorb.

C. THERMODYNAMIC EQUILIBRIA

From a macroscopic point of view, the electric field represents an additional state variable, which, like pressure or temperature, can alter chemical equilibria. The thermodynamic relationship was established by Bergmann et al. [67] who studied the dielectric absorption in liquid solutions at higher fields ($>10^5$ V/cm). The equilibrium constant of a chemical reaction will depend on the field strength if the reaction has an electric momentum. This momentum ΔM is defined by the dipole moments μ and the polarizabilities α of the reactants and products of a chemical reaction

$$\Delta M = [\Sigma \mu_p - \Sigma \mu_r] F + \tfrac{1}{2} [\Sigma \alpha_p - \Sigma \alpha_r] F^2 \qquad (3)$$

where p and r indicate products and reactants, respectively. Thus, the electrical momentum is determined by the field energies $E_F = \mu F + \tfrac{1}{2} \alpha F^2$ on either side of an equilibrium. Using Van't Hoff's law, the approximate field dependence of the equilibrium constant K is given by $\partial \ln K/\partial F = \Delta M/kT$, where kT is the statistical energy.

In FIMS, thermodynamic data are not available directly since the methods of ion desorption deviate from the required closed thermodynamic system. However, the shift of an equilibrium constant or steady-state concentration can be measured if pulsed field methods are applied. A system is brought to equilibrium at a field strength which is too low for ion desorption. With a rapid field pulse, this state may be ionized suddenly and is analyzed mass spectrometrically. The shift of the equilibrium constant is measured from the variation of the "reaction field" at repeated desorption pulses with uni-

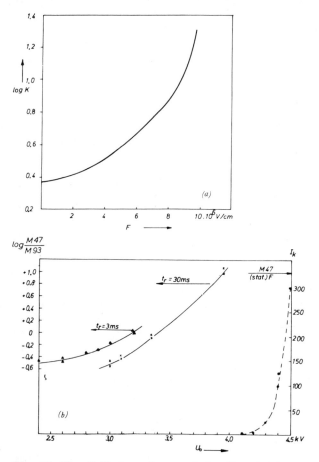

Fig. 20. The field dependence of the dimerization equilibrium of formic acid. (a) The calculated shift of the equilibrium constant K with the field strength F. (b) Experimental values for concentration ratios of monomers and dimers; the right hand scale describes intensities I_k at constant (stat F) fields, which are measured above $U_0 = 4.1$ kV.

form pulse amplitudes. The variation of the pulse repetition rate, in addition, indicates whether equilibrium was reached before field desorption occurred.

Under conditions of dynamic gas supply, it seems more reasonable to consider steady-state conditions which are altered by external fields. The dimerization reaction of formic acid, $2\,HCOOH \rightleftharpoons (HCOOH)_2$, has been investigated [68,69]. The monomeric form has a dipole moment μ_m of 1.35 Debye. Since the dimer consists of a planar structure with two hydrogen bonds, the dipoles are compenstated and μ_d is 0. Except for small contributions by the hydrogen bonds, the polarizabilities of the reactants and product are roughly the same. From the known equilibrium constant of $K = 2.40$ at $298°K$, the variation of $\log K$ with F can be calculated (Fig. 20(a)). It is seen that as more field energy is accepted by the system, K increases and larger monomeric concentrations will result.

Experimental studies of this reaction are presented in Fig. 20(b), where the dependence of the ratios of M47/M93 with the potential difference U_0, which is proportional to the reaction field, are plotted. Measurements were made for reaction times of 3 and 30 msec. The concentration ratio increased by more than an order of magnitude, which agrees with the theoretically expected change. In comparing quantitative values, however, the experimental slope is somewhat less than the calculated one. This deviation results because steady state adsorbed multilayers are formed so ideal thermodynamic relations obviously are not applicable. The differences in the values for 3 and 30 msec also indicate that thermodynamic equilibrium is not reached. The ions actually detected are the proton-attachment products $HCOOH \cdot H^+$ and $(HCOOH)_2 \cdot H^+$. These are formed by bimolecular proton transfer process during the desorption pulse. In Fig. 20(b), the right-hand scale indicates the threshold of ion formation, which is the upper limit for the reaction field.

For reasons given above, quantitative correlations between equilibrium constants and electric fields cannot be expected in FIMS experiments. Nevertheless, the field dependence of thermodynamic values strongly changes the chemical expectations and has to be considered carefully. Some chemical reactions may even be observed under field influences, which are thermodynamically impossible at zero field conditions.

The reaction coordinates of surface reactions will also be altered in the presence of electric fields. Changes in the activation energies for surface reactions would also be anticipated. The Polanyi energy surface of a heterogeneous reaction will be field-dependent, since Coulomb interactions are involved in the calculation of energy barriers. To date, no detailed experiments on this problem have been carried out.

D. FIELD POLYMERIZATION AND FRAGMENTATION

Polymerization reactions may be initiated by radicals, ion-radicals or ionic

416

states of a molecule. The last ones may be produced by field interaction at surfaces and be the origin of chemical reaction products which are absent without fields. The field polymerization of acetone is one of these reactions which has been studied thoroughly [61]. A radical site at the surface will bind tightly an acetone molecule by a field induced chemisorption. Thus

Chemisorption   Proton transfer

A surface molecular cation is formed during this chemisorption. The field induced electron transfer is denoted as FI. Subsequent reactions are proton transfer, dimerization, loss of oxygen, etc.

Dimerization   Loss of oxygen

The C=CH$_2$ double bond, which is formed during dimerization may react further to form higher molecular weight compounds.

In similar reactions, the masses M115, M113 and M71 are formed besides molecular ions of much higher mass numbers. Actually, these reactions lead to the formation of carbonaceous whiskers, which can then be used as needle emitters as described in Section II.A.

Molecular ions smaller than the parent ion can also be formed by these surface interactions. Typical examples include the masses M50, M31 and M29. This is an important type of reaction since fragments are usually regarded as products of homogeneous decay processes. The identification of surface reaction products with lower masses was made using the pulsed field methods and by the energy analysis of field ions.

In the present case, the origin of ions at the surface has been convincingly

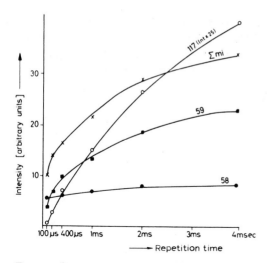

Fig. 21. Pulsed field experiments with acetone. The time dependence of M58[(CH$_3$)$_2$CO$^+$], M59[(CH$_3$)$_2$CO$^+$H] and M117[(CH$_3$)$_2$CO$^+$]$_2$H corresponds with first, second and third order kinetics, respectively. (After Röllgen and Beckey [61].)

demonstrated using kinetic arguments. By applying pulsed field techniques, Röllgen and Beckey [61] measured different ion intensities as a function of the pulse repetition rate (or reaction time), as shown in Fig. 21. At short repetition times, the $(P + 1)^+$ ion current, where P is the intensity of the mass of the parent ion, increases much faster than the P^+ ion, while at long repetition times, the slope of the $(2P - 1)^+$ ions is greatest. The reason for these different time dependencies is that P^+ is formed in a monomolecular ionization process, whereas $(P + 1)^+$ and $(2P - 1)$ must be preceded by a bimolecular or trimolecular reaction, respectively. All these can only be possible at a surface. Further confirmation of a surface mechanism is obtained from the pressure dependence of the signal intensities, which obey the expected kinetic laws, accordingly.

E. FIELD DESORPTION OF SURFACE COMPLEXES

The phenomenon of field induced chemisorption has already shown that several surface compounds may be formed by field interferences which do not exist without electric fields. In a recent investigation on the field desorption of surface layers [70], another fundamental property of products formed in FIMS, was elaborated. As described in Section V.A, field induced chemisorption could be explained from the field dependence of potential energy curves, which permit ionic structures to be stable only in high electric fields. Field desorption experiments now indicate clearly that such ionic structures can be desorbed in connection with metal atoms of the surface.

The stability of these metal-complex ions very much resembles the properties of metal ion ligands, which are known in solution.

Some experimental examples illustrate the general rules governing the
formation of ions. In experiments with silver emitters, singly charged Ag^+
ions were formed, exclusively. In the temperature range between $80°$ and
$375°K$, the activation energy was found to depend linearly on the square
root of the field strength. The threshold field for field evaporation is $2.2 \times$
10^8 V/cm at $80°K$. The field evaporation behavior is changed completely if a
layer of H_2O molecules interacts with the Ag surface. In this case,
$Ag^+(H_2O)_n$ complexes predominate. As is shown in Fig. 22, the intensities
are temperature dependent and $Ag^+(H_2O)_n$ complexes are formed at lower
temperatures and lower threshold fields than Ag^+ ions without ligands. The
Ag^+ ions were only detected above $250°K$. At lower temperatures, maxima
occur for the $Ag^+(H_2O)_n$ intensities at $120°K$ and at $220°K$. The low temperature maximum is of particular interest. The highest molecular weight
complex is $Ag^+(H_2O)_{12}$, which occurs in low concentrations. The intensities
here decrease with increasing values of n. In the complex chemistry of Ag^+,
the structure $Ag^+(H_2O)_4 (H_2O)_8$ represents a metal ion with a completely
saturated first and second ligand sphere.

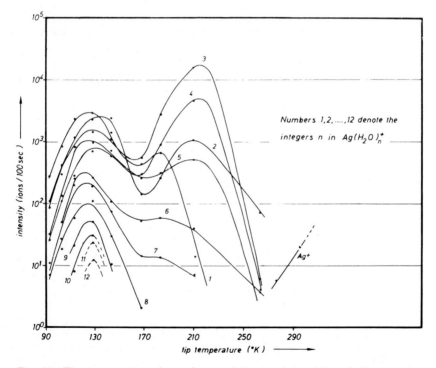

Fig. 22. The temperature dependence of the ion intensities of silver—water complex ions
formed at a Ag field emitter under steady field conditions and constant field strength.

The results of Fig. 22 can be interpreted in the following way. A silver atom will be preferentially field evaporated from a kink site or a position where it is already partly separated from the solid surface. These selected Ag atoms will be highly polarized on top of the metal surface and be chemisorbed in an ion-like state, which is unsaturated with respect to the coordination number of Ag^+. The surface complexes, $Ag^+(H_2O)_n$, stabilize the surface ion and remove the potential barrier for ion desorption. The intensity maxima are not correlated directly with the stability of the surface complexes, since the supply of H_2O molecules may be the rate-determining step in complex formation. Thus, at low temperatures where the mobility of H_2O is low, the chance that 12 H_2O molecules collide with a Ag^+ ion simultaneously is rather low. At higher temperatures, the stability of the second ligand sphere, on the other hand, must be rather small.

Similar results have been found with NH_3. In this case, the maximum number for n is eight at low temperatures, again from the complex chemistry of Ag^+, where $[Ag^+(NH_3)_2]\,(NH_3)_6$ can be formed at most. At the higher temperatures of about $200°K$, $Ag^+(NH_3)_2$ is the most abundant intensity; silver diamino complexes are known to be very stable. The metals Cu, Ag, and Au of the first group of the periodic table, which have been studied recently, unequivocally demonstrate that the stability of complex ions determines the structures of field desorbed positively charged molecular species. They differ from adsorption states of molecules under field-free conditions. The gases, H_2, CO, CH_4, N_2, etc., which are not adsorbed on Ag at room temperature, or Au, gave the most abundant ion intensities, whereas oxygen, as discussed before, neither forms ion-compounds with Ag nor with Au or Cu.

TABLE 3

Complex ions formed between Cu, Ag or Au and different gases and values of n observed

Gas	Cu	n	Ag	n	Au	n
H_2	$Cu(H_2)_n^+$	1–3	$Ag(H_2)_n^+$	1, 2	$Au(H_2)_n^+$	1
N_2	$Cu(N_2)_n^+$	1, 2	$Ag(N_2)_n^+$	1	$Au(N_2)_n^+$	1
CO	$Cu(CO)_n^+$	1, 2	$Ag(CO)_n^+$	1, 2	$Au(CO)_n^+$	1, 2
H_2O	$Cu(H_2O)_n^+$	1	$Ag(H_2O)_n^+$	1–12	No complex	
NH_3	$Cu(NH_3)_n^+$	1	$Ag(NH_3)_n^+$	1–8	$Au(NH_3)_n^+$	1
C_2H_4	$Cu(C_2H_4)_n^+$	1–3	$Ag(C_2H_4)_n^+$	1–3		
Kr, Xe			No complex			
Pyridine			$Ag(Py)_n^+$	1–4		

Experimental results for various gases on Cu, Ag and Au are compiled in Table 3. Complex ions are found below the field strength of the field evaporation of pure metals and at different temperatures, $Ag(H_2)$ at $80°K$, $Ag(Py)_4^+$ at $300°K$, etc. The values in Table 3 provide only a qualitative survey on the nature of complex compounds which are formed. There are also characteristic intensity distributions for the different values of n. Higher values of n than those indicated in Table 3 may be missing in some cases because of insufficient gas supply or surface mobility. Dissociation products of H_2 or CO were not bound in any case, whereas NH_3 yielded the catalytic decomposition product N_2 at Cu surfaces.

Normally, pyridine forms a doubly charged silver complex $Ag(Py)_4^{2+}$. During extended investigations at different temperatures and fields, only singly charged ions could be observed in our experiments. In particular with Ag, a constant emission of complex ions was observed for many hours without any change in the tip radius. This can only be explained by a high surface mobility of silver, even at $80°K$.

To date, field evaporation and field desorption processes of metals and chemisorbed layers on metals have been treated in accordance with theoretical models as described by Müller [71], Brandon [72], Tsong [73], and McKinstry [74]. The experimental results presented in this chapter suggest other models should be considered. For example, the transition of a neutral metal lattice atom into a ligand-stabilized ionized state is exactly the same process as that which occurs in electrochemistry when metal electrodes are dissolved in an electrolyte. The solvation energy of ions is a decisive factor which makes this reaction energetically possible. In comparison, field desorption products will be stabilized and activation barriers will be reduced if part of the "solvation energy" is gained by formation of a complex. The reaction conditions in FIMS certainly differ from the nearly equilibrium conditions of electrode reactions. Thus, in all the cases of Cu (and for Ag and pyridine), only singly charged ions were identified whereas in electrolytic solutions, Cu^{2+} ions prevail. As discussed before, the kinetic competition between the successive steps of ion formation, $Cu \rightarrow Cu^+ + e^-_{(Me)} \rightarrow Cu^{2+} + 2 e^-_{(Me)}$, and field desorption may greatly reduce the number of Cu^{2+} complex ions formed in FIMS.

VI. Surface reactions without field perturbance

Three types of surface reactions or surface structures, which are not influenced by external electric fields and can be investigated by FIMS, will be discussed. First, ions can be formed at a solid surface according to the Langmuir—Kingdon [75] mechanism. A second type of field-independent reaction exists if the electrical momentum, as defined before, is negligible. As an example, the condensation of sulfur molecules will be discussed. Finally, the pulsed field desorption method may be used, with appropriate pre-cautions, to investigate surface reactions in an unperturbed mode.

If the ionization potential I of a molecule is comparable with the work function Φ of the surface, the thermally desorbed ion current j_+ is given by [76]

$$j_+ = j_0 \left(\frac{q_+}{q_0}\right) \exp[\Phi - I/kT] \tag{4}$$

where j_0 is the current of desorbing neutrals; q_+ and q_0 are electronic partition functions. There are chemical reactions known to form ionic sur-face compounds even though, in these cases, the ionization potential of the isolated molecule is larger than the work function of the surface. For this, the interaction energy of the ionic states at the surface must exceed the difference between ionization potential and work function. These ions can also be desorbed thermally, but sometimes only at elevated temperatures where organic molecules will decompose [77]. These surface ionization processes, which can be observed without external electric fields, may be studied using FIMS.

A. CARBONIUM IONS ON SURFACES

Because of various experimental observations, carbonium ions have been assumed to be intermediates in many surface chemical reaction mechanisms. The isomerization of hydrocarbons or the dehydration of alcohols on oxide surfaces are typical surface reactions with ionic intermediates. There are a few cases, however, where carbonium ions are chemisorbed as a stable surface compound. One such case was described by Block and Zei [78], who used zeolite surfaces as shown in Fig. 23. Zeolites are active catalysts in isomeriza-tion and dehydration reactions. For adsorption and desorption measure-ments, a group of hydrocarbons was chosen which easily form carbonium ions, e.g. triphenylmethyl compounds $Ph_3 CX$, where $Ph = C_6 H_5$ and $X = H$, Cl, Br, COOH, or OH. The experimental conditions for desorption of $Ph_3 CCl$ ions are compared in Fig. 24. On a CaY-zeolite surface, ion emission can be observed at electric fields of less than 10^5 V/cm which were just sufficient to focus ion beams in the mass spectrometer. Ions were not de-tected until a temperature of 450°K was reached; great intensities were detected with increasing temperatures. The ion current was composed exclu-sively of $Ph_3 C^+$ fragments but no parent molecular ions were obtained. However, both the parent ion and fragments were obtained from $Ph_3 CCl$ on Pt surfaces (right-hand side of Fig. 24). The temperature dependence of ion intensities was, as usual in field ionization, negative; threshold field strengths in the 10^7 V/cm region were necessary for ion desorption from Pt.

Obviously, the mechanisms which are involved in ion formation and de-sorption are completely different on the CaY-zeolite and on the Pt surface.

a. Zeolite coatings on Pt-wollaston-wire

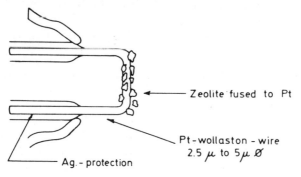

b. Zeolite grains on Pt-tips

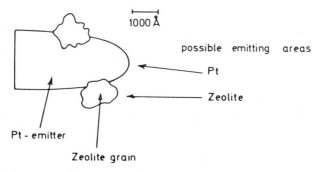

Fig. 23. For FIMS investigations of carbonium ions on zeolite surfaces, the emitter is either (a) a Pt—wollaston wire, coated partially with zeolite, or (b) a Pt-emitter with zeolite grains. (After Block and Zei [78].)

In the first case, a chemical surface ionization (reactive surface ionization) with the formation of carbonium ions is involved but in the second case, the usual electron transfer process occurs. The desorption of Ph_3C^+ from the CaY—zeolite surface is a thermally activated process with an activation energy of only 8 kcal/mole. Coulombic interaction energies have to be overcome during the desorption. This conclusion can be drawn from the observed linear dependence of ion intensity with the exp $F^{1/2}$ of external field [78]. Thus, at constant temperature, the ion intensity i_+ increases in accordance with a Schottky barrier assumption: $i_+ \propto \exp(eF^{1/2}/kT)$.

The location of the active sites for forming carbonium ions on the zeolite surface is answered partially by FIMS. The active sites may be a high local electric field at the cations [79], at Brönsted OH groups [80], or at acid surface sites of the Lewis type [81]. The FIMS results on ion desorption excluded the possibility of reactive surface ionization occurring at the local

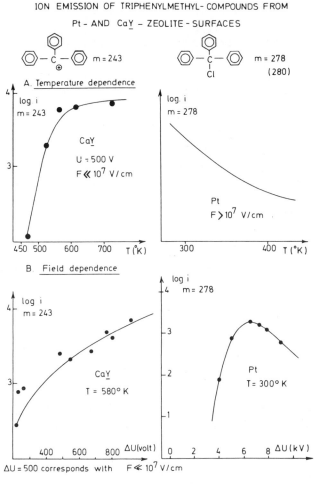

Fig. 24. A comparison of ion intensities of triphenylmethyl compounds from CaY-zeolite (left) and Pt surfaces (right) at various temperatures and fields. (After Block and Zei [78].)

electrostatic fields of cations. The question of whether Brönsted or Lewis sites are involved was solved by Karge [82] with infrared absorption spectroscopy. He used exactly the same materials and confirmed the existence of Ph_3C^+ carbonium ions on CaY and several other zeolites, in particular by the intense absorption band at 1357 cm^{-1}, which belongs to the $\nu_5(Ph-C)$ vibration. Furthermore, it could be concluded that Brönsted OH groups are not altered by the chemisorption process and Lewis sites *are* involved in the formation of surface carbonium ions.

The identification of surface carbonium ions is an example of one of the powerful possible applications of FIMS. For the elucidation of mechanisms of heterogeneous surface reactions, the structure of adsorbed molecules is one important requirement. For the ionic structures discussed above, the ion desorption represents a selective tool for analysis. Of the numerous other structures which might be present on a surface, only the surface ions present will be detected. The interesting question of whether it will be possible to use preferential desorption to identify the short lifetime ionic intermediates of catalytic reactions now arises. A first attempt was made, recently, by Zei [83]. He compared the behavior of t-butanol on Pt, Au, and MY-zeolite surfaces. The principal intensities at the zeolite surface were the possible ionic intermediates $(CH_3) \cdot C^+$, $(CH_3)_2 CH_2 \cdot C^+$, or $(CH_3)_3 COH \cdot M^+$. However, the conclusions are complicated by other fragmentation processes due to the instability of the t-butanol parent ion.

B. CHEMICAL REACTIONS WITHOUT ELECTRIC MOMENTUM

One of the chemical reactions without electrical momentum is the condensation of S_2-molecules onto the adlayer of a sulphurized tungsten surface. The reaction scheme

describes molecular sulphur, which forms the intermediates of S_4 to S_7 and the stable S_8 configuration. None of these molecules is known to have a permanent electric dipole. The different polarizabilities cancel in the computation of the reaction momentum.

In recent experiments by Davis et al. [84], an electrochemical source for evaporating S_2 molecules onto a W emitter was used to study the reaction $x/2\ S_2 > S_x$ $(4 < x < 8)$ on a sulphurized tungsten emitter surface. Since the first sulphur layer is strongly chemisorbed and needs rather high fields for field desorption, condensation reactions of sulphur can be investigated independently at medium electric fields of $\leqslant 5 \times 10^7$ V/cm. During these experiments, impingement rates of 3×10^{12} molecules cm^{-2} sec^{-1} were used up to the levels where the formation of multilayers is expected. Field desorbed ions were analyzed in a quadrupole mass filter at temperatures from 150 up to $500°$K and various electric fields.

Under steady field conditions, the ions, S_2^+, S_4^+, S_5^+, S_6^+, S_7^+ and S_8^+, were observed but S_4^+ was monitored only under certain conditions and at the limit of sensitivity. The field dependence of these intensities is given in Fig. 25. At higher fields, S_2^+ is ionized in the gas phase before the molecule arrives at the surface and the S_2^+ intensity is proportional to the impingement rate. Under steady field conditions, surface interactions and condensation processes can be observed only in a small region of field, where the ionization

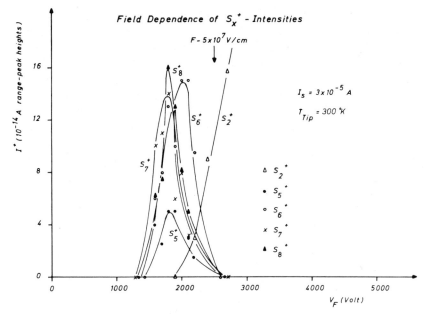

Fig. 25. The field dependence of S_x intensities in the physisorbed surface layer of sulphur on tungsten. Impinging S_2 molecules react quickly to form intermediates S_5, S_6, S_7, but the reaction to form S_8 is relatively slow.

probability is less than one. In this field region, the S_x^+ intensities display maxima.

To understand the reactions of S_2 molecules on the emitter surface, it is convenient to follow their behavior as the field is gradually decreased from steady state conditions. The ionization probability of S_2, which is unity at elevated fields because only S_2^+ is observed, decreases exponentially near the threshold field strength. At or slightly below this field (1900 V in Fig. 25), S_2 molecules may reach the surface before being ionized. During this interaction, the rate of chemical reaction to form S_5, S_6, S_7, and S_8 is equal to or faster than the net rate of adsorption and probably faster than the rate of thermal accomodation of S_2. Otherwise, a maximum in S_2^+ intensities at or below ≈ 1900 V would be expected. Therefore, it can be concluded that the chemical reaction of S_2 is a relatively fast process under low field conditions where S_2 reaches the surface. This conclusion is not influenced by the ionization probabilities of the different S_x species, since their ionization potentials differ only slightly. The minimum time for this reaction is greater than 10^{-12} sec.

Next, the question must be considered of whether S_5, S_6, S_7, and S_8 are transient species from the reaction of S_2 towards solid sulphur or whether these concentrations are "equilibrium values" for the temperature, pressure, and field under consideration. In the first case, the concentrations will be

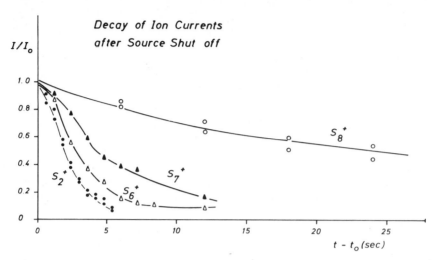

Fig. 26. The decay of field ion currents after shutting off the source. The S_2^+ intensities agree with the times for having the S_2 beam off. The S_8^+ intensities at $(t - t_0) > 15$ sec indicate that back reactions $S_8 \to S_x$ $(x < 8)$ are negligible.

given by the relative rates to form S_x or S_8. In the second case, the back reaction of S_8 to S_x to S_2 will contribute to the concentrations of S_5, S_6, and S_7. This question can be resolved by kinetic experiments in which the S_x^+ intensities are measured with pulsed field methods or at non-steady conditions. The behavior when the S_2 supply is discontinued defines a kinetic regime.

During the source-shut-off experiments (Fig. 26), S_8 was the only detectable molecule after about 15—20 sec. If we assume that sulphur is condensed onto the shank and diffuses into the ionization region, the sulphur activity under steady ionization conditions will probably be less than one. Even then, S_8 is the only stable molecule in the weakly adsorbed layer. A reaction towards S_5—S_7 is not observed. The intensities S_6 and S_7 at less than 12 sec indicate that the reaction $x/2\ S_2 \to S_x$ $(x < 8)$ is faster than the reaction to form S_8 out of the condensed phase. A preferred surface mobility of S_8, which could also explain this behavior, could be excluded because of other experimental results. A fast reaction of S_2 towards S_x $(x < 8)$ and a slow formation of S_8 is in agreement with other reported results. Condensed S_2 is stable only for short times in an isolated matrix and at low temperatures; S_8 is not formed from S_6 with a measurable rate until about $330°K$ is reached. The physisorbed layer of sulphur on tungsten is S_8, but the formation of these molecules is a relatively slow process, which permits the intermediates to be observed.

VII. Applications of FIMS

A. REACTIONS OF WATER

The results of FI show that straightforward conclusions between field desorbed ions and surface structure in a zero-field regime, as present in heterogeneous catalysis, are limited by the presence of field induced reactions. Investigations of structure sensitive reactions, which are related to catalytic processes, cannot ignore these field induced alterations [85].

As discussed in Section V.E, water usually forms ions like H_3O^+, H_2O^+, and $(H_2O)_n H^+$ where n varies from 1 to 10. Details of the mechanisms involved in these reactions have been studied by Beckey [39,41], Schmidt [40], Goldenfeld et al. [86], Anway [87], and Röllgen and Beckey [88].

The formation of associated ions occurs in a multilayer of water molecules. The field induced force, which is acting on each gas molecule, is the gradient of the polarization energy $\partial E_p / \partial r = (\mu + \alpha F) \, \partial F / \partial r$. For H_2O, μ is 1.84×10^{-18} esu and α is 1.57×10^{-24} cm^3. At a field strength of 3×10^7 V/cm, the field induced force acting on H_2O molecules produces a field enhanced pressure of 16 atm (compared with a field free pressure of 10^{-5} torr). This value was not corrected for reduction of the field in the dielectric layer. This rough calculation shows that it is necessary to consider multilayer formation of H_2O molecules. Anway [87] actually measured the layer thickness from the energy deficit and reached the conclusion that up to 100 monolayers of water will be condensed on the tip surface at low fields. In general, condensed layers will form easier when the values of μ, α, and I are high and the condensation pressures are low.

Several mechanisms have been formulated to explain the formation of H_3O^+ ions in the liquid-like surface layer. According to Onsager's theory [89], the self-dissociation of water has to be field dependent. At a field strength of 3×10^7 V/cm, the conductivity of water does not exhibit ohmic behavior but has a parabolic dependence according to the second Wien effect. The dissociation equilibrium

$$2H_2O \rightleftharpoons H_3O^+ + OH^-$$

is field dependent. The immediate neutralization of OH^- ions is considered the source of secondary reactions.

The field ionization of one H_2O molecule, which subsequently reacts to form H_3O^+ by

$$2H_2O \xrightarrow[F]{-e} H_2O^+ + H_2O \rightarrow H_3O^+ + OH^*$$

is considered to be another possible mechanism. A third mechanism involves surface sites of the emitter, e.g.

$$Me + nH_2O \xrightarrow[F]{-e} MeOH + (H_2O)_{n-1} \cdot H^+$$

It is necessary to consider this mechanism because the threshold fields for ion formation depend on the tip material (2×10^7 V/cm for W, 5.1×10^7 V/cm for Ir, and 6.1×10^7 V/cm for Pt) [40]. The formation of different reaction products at different metal surfaces also favors this mechanism. With tungsten, the intensity ratio $[H_2O^+]/[H_3O^+]$ is 10^{-3} to 10^{-4} but with Pt and Ir, it is about 0.5. Tungsten oxide ions were found but with noble metals, no compounds were observed. However, under certain conditions, O_2^+ ions could be detected with Pt and Ir, which presumably form according to the reaction

$$6H_2O_{ads} \xrightarrow{F} 4H_3O^+ + O_2^+ + 5e^- \,(Me)$$

Furthermore, the recombination of OH radicals leads to H_2O_2 and O by

$$2OH \rightarrow H_2O_2 \rightarrow H_2O + O$$

Only OH could not be measured in the form of a positive ion.

Some definite conclusions have been reached about the different mechanisms from measurements of the energy deficit and appearance potential of field ions. Anway [87] measured the energy distributions on a tungsten field emitter, as shown in Fig. 27, for $H_3O^+ \cdot 3H_2O$. The broad signals with a large energy deficit originate from ionization processes occurring at the surface of the condensed layer. At higher field strength, the maximum in the distribution shifts to lower values since the thickness of the layer is decreased. This type of behavior is also found for other associated ions of water. In addition, there is a "fixed peak" at a constant low energy loss, which is produced from ionization at a minimum distance and is independent of the field strength. The appearance potential, which is different for various ionic species, is obtained from the low energy threshold of this peak. The appearance potentials of the associated ions of $H_3O^+ \cdot nH_2O$ on W, are as follows: H_3O^+, 8.83 ± 0.05 eV; $H_3O^+ \cdot H_2O$, 10.03 ± 0.07 eV; $H_3O^+ \cdot 2H_2O$, $11.6 \leqslant 11.5 \leqslant 11.2$ eV; $H_3O^+ \cdot 3H_2O$, $11.7 \leqslant 11.6 \leqslant 11.1$ eV. The values obtained for $H_3O^+ \cdot 2H_2O$ and $H_3O^+ \cdot 3H_2O$ are considered to be the upper limit. A comparison of these experimental values with data for different reactions given in the literature can be used to exclude several of the proposed mechanisms. The heats of formation, ΔH_r, of H_3O^+ in the following reactions are

$$2\,H_2O_{lig} \rightarrow H_3O^+ + OH^- \qquad\qquad \Delta H_r = 10.6 \text{ eV} \qquad\qquad (5a)$$

$$H_2O(v) + H_2O_{lig} \rightarrow H_3O^+ + OH + e^- \qquad \Delta H_r = 12.0 \text{ eV} \qquad\qquad (5b)$$

$$6\,H_2O_{lig} \rightarrow 4\,H_3O^+ + O_2 + 4\,e^- \qquad\qquad \Delta H_r = 13.6 \text{ eV}/H_3O^+ \qquad (5c)$$

$$2\,OH^- + 2\,H_2O \rightarrow 2\,H_3O^+ + 2\,O_2 + 4\,e^- \quad \Delta H_r = 8.7 \text{ eV}/H_2O^+ \qquad (5d)$$

The ΔH_r values given by Anway [87] are based on the value of 6.17 eV for

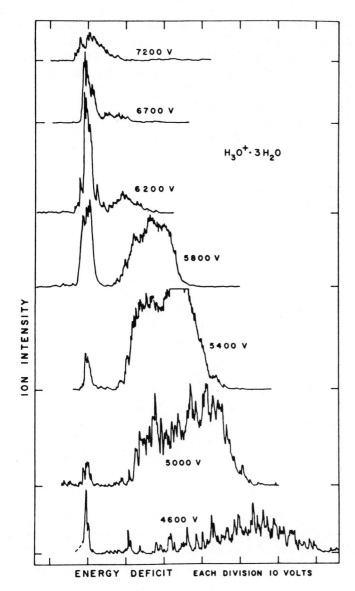

Fig. 27. Energy distribution of the H_3O ion at ΔU values between 4600 V and 7200 V. (After Anway [87].)

the proton affinity for H_2O. His values differ slightly from those given by Heinen et al. [34] who used the more reliable value of 7.1 eV for the reaction

$$H_2O(g) + H^+ \rightarrow H_3O^+(g)$$

References pp. 443—446

In any event, the formation of H_3O^+ by reaction (5a) is unlikely since the experimentally measured appearance potential (8.83 eV) is considerably lower than for this reaction. Reactions (5b) and (5c) are also excluded because the energy values are even higher; reaction (5d) occurs only as a side reaction of the OH radicals from (5a) and (5b). Based on the energy of the reactions considered, it has to be concluded that surface sites are involved in the formation of H_3O^+ ions. This conclusion was also drawn by Heinen et al. [34] who found an appearance potential of 10.5 eV for forming H_3O^+ on Pt surfaces. The most likely reaction is

$$2\ H_2O \underset{F}{\overset{-e}{\rightleftarrows}} H_3O^+ + OH$$

where ΔH_r = 11.7 eV. This reaction is energetically possible if the binding energy of OH radicals to the Pt surface is assumed to be 1.2 eV. The assumed binding energy of OH to the surface is not necessarily the bond energy of Pt–OH; it is more likely that the OH radicals are attached to free radical sites on a contaminated Pt surface. Furthermore, it is suggested that the proton affinities are not too different for H_2O and $(H_2O)_n$ and that the mechanisms for the formation of $H_3O^+ \cdot (H_2O)_{n-1}$ will be comparable. A quantitative explanation for the experimental values given by Anway [87] is still not available.

There are other observations in the field ionization of water which are still not completely understood. It is not likely that the ion $H_2O^+ \cdot H_2O$, which can be obtained on noble metals, is an ionized dimer. The appearance potential of about 11.7 eV is consistent with the suggested reaction [61]

$$2\ H_2O \overset{-e}{\longrightarrow} (OH \cdot H_3O)^+ + e^-$$

where the dimer, a product of a proton transfer reaction, gains the additional bonding energy of an OH radical.

Small amounts of H_4O^+ have been identified [87], but the energy of the reaction is unknown. Some considerations have been made on the structure of these ions. Using the LCAO MO method, the structure of H_3O^+ was ascertained to be a flat pyramid H_3O^+, which is isoelectronic with NH_3 and is assumed to have the same electronic ground state configuration $^1A_1$ of the C_3 group. The addition of $^2S_{1/2}$-state hydrogen atom would result in a triangular pyramid structure for $H_3O^+ \cdot H$.

The geometrical and chemical structure of the condensed water layer is still unresolved. In the energy deficit measurements (Fig. 27), "fixed peaks" and high loss intensities have been obtained simultaneously. This means that ions are emitted from the top of the condensed layer simultaneously with ions formed closest to or on the surface. Therefore, it seems unlikely that the emitter surface is homogeneously covered by the condensed phase. Although a whisker model has been suggested for polymeric water molecules [86], it is more likely, from observations in the FIM, that water forms small drop-shaped islands on different regions of the emitter because of the extreme surface tension on the highly charged surface.

B. THE ANALYSIS OF EVAPORATION PRODUCTS OF SOLIDS

1. Evaporation of solid selenium

Investigations on selenium were stimulated by observations in the thin film preparation of this material. Semiconducting properties of evaporated films of pure selenium showed a "memory" of the original evaporated sample. This information must have been carried by the molecular structures of possible gaseous species.

In the analysis of Se vapor, Langmuir-type evaporation was compared with Knudsen equilibrium composition of Se vapor [19]. Evaporation temperatures were in the range 150—200°C, where Se exists only in the trigonal configuration. Field ionization was achieved by a Pt emitter of approximately 1000 Å radius of curvature. The intensities in the mass spectra of Se are represented in Fig. 28 and indicate quantitative differences between Langmuir and Knudsen evaporation but no fundamental qualitative alterations in the evaporation process. The mean molecular weight is $Se_{5.2}$ in molecular evaporation and $Se_{6.2}$ in equilibrium evaporation, which is in contrast to

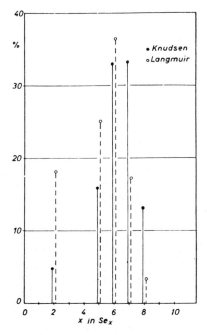

Fig. 28. Molecular distribution of evaporated Se_x, $x \leqslant 10$. (After Saure and Block [19].)

432

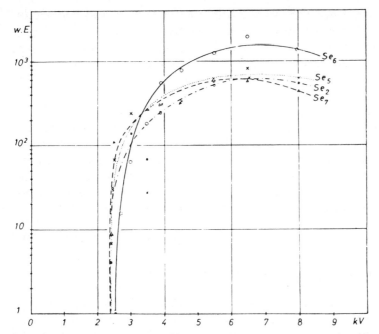

Fig. 29. Field ion intensities (W.E., arbitrary units) of Se_x^+ as a function of the field strength ($\Delta U \leqslant 9$ kV).

earlier literature [90], but in agreement with results of Detry et al. [91] that evaporating molecules are mainly Se_2, Se_5, Se_6, and Se_8. Only occasionally have Se_4 and Se_9 been observed at the limit of detectability. The intensity of Se_3 was beneath the sensitivity in these experiments.

The difficulty in determining molecular composition by electron impact mass spectroscopy is caused by the uncertainty in fragmentation probabilities. Detry et al. [91] have used thermodynamic relations of an electrochemical source, which produced the Se molecular beam to circumvent this difficulty. Also, in FI, the possibility of fragmentation of parent molecular ions has to be examined. The dependence of ion intensities on field strength is given in Fig. 29. From these experiments, the occurrence of a possible misleading fragmentation of Se molecular ions can be excluded. The onset field of all the species Se_2^+ to Se_8^+ is nearly the same because the ionization potentials are only slightly different. At elevated fields, none of the smaller molecular ions grow at the expense of larger ones which would be expected from the usual field dependent fragmentation probability (The slight decrease of all of the intensities is caused by an ion optical property of the $60°$ magnetic sector field instrument.)

From these experiments, arguments could be derived that with the exception of Se_2, the major evaporating molecules must have ring structures.

There are analogies between structural properties of gaseous Se_x molecules and unbranched hydrocarbons. Although the bonding orbitals in Se are (sp^2) hybrids, with predominating p-contribution, both groups of compounds can form ring structures in addition to chains, if according to the bond angles a closed structure can be established without, or with only minor, strain. Assuming the bond angle in solid Se ($103-105°$), the Se_5 molecule could form a ring with little strain (the Se_6 has none). As with the behavior of hydrocarbons, the probability of ring formation decreases for medium sized molecules with 8 atoms because of the diminished probability that chain ends touch. Therefore, ring structures of Se should have the main intensities with the molecules Se_5, Se_6, and Se_7. The Se_5 ring is kinetically favored, the Se_6 ring thermodynamically favored. This also agrees with the observed differences between Knudsen and Langmuir evaporation.

The field enhanced condensation of selenium vapor onto a field emitter surface causes a complete change in the structure of field ions. On a Pt emitter, the nucleation and growth of Se whiskers in the field direction was observed, when selenium vapor condensed on the surface at a field strength which was still insufficient for field evaporation.

The formation of the whiskers can be observed by optical microscopy, as well as by field ionization properties. After this whisker formation, the necessary potential differences ΔU for ion formation of a test gas (i.e. cyclohexane) is much less than on the initially cleaned Pt emitter. This is due to the reduced radius of curvature at the surface of the conducting Se whiskers. Under the experimental conditions used, these whiskers thermally evaporated around $360°K$. Above this temperature, test gases were only ionized at potential differences ΔU of the clean field emitter.

The ions field evaporated from a Se whisker are completely different from ions obtained from a molecular beam. Se_5^+ ions prevail and furthermore, the mass spectrum contains molecular ions Se_x^{n+} with $x/n = 7\frac{1}{2}$, $8\frac{1}{2}$, 9, $9\frac{1}{2}$.... 13, which indicate doubly charged species of Se_{15}^{2+} to at least Se_{26}^{2+}, (the upper mass limit of the ion detection system). Furthermore, there are mass signals at $x/n = 4\frac{1}{4}$, $4\frac{3}{4}$, etc. These ions have been observed between Se_{17}^{4+} and Se_{33}^{4+}, but not all of them are present. In Fig. 30, a part of the original spectrum around $x/n = 4\frac{3}{4}$, 5, $5\frac{1}{4}$ is illustrated, which shows that the mass signals around M375 and M412 are multiply charged whereas the signal around M396 is mainly composed of singly charged molecular ions. The isotopes of Se (M74 to M84) require a theoretical intensity distribution which depends on the charge of the molecular ions in the following way. Although the center of ion intensities in the mass scale is nearly the same for Se_5^{1+}, Se_{10}^{2+}, and Se_{20}^{4+}, the breadth of mass signals of these differently charged particles diminshes with increasing charge. Computer calculations have shown that four-fold charged molecular ions should have roughly half the breadth in mass scale as singly charged ones. In Fig. 30, the resolved mass signal of Se_5^+ (around M396) has satellites for $x/n = 4\frac{3}{4}$ and $5\frac{1}{4}$, which have

References pp. 443—446

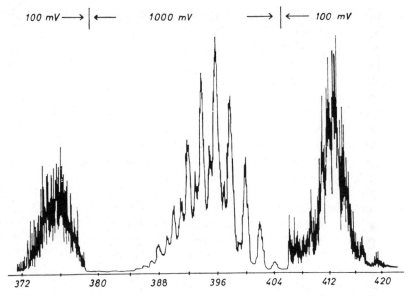

$100\ mV \longrightarrow$ | $\longleftarrow$ $1000\ mV$ $\longrightarrow$ | $\longleftarrow$ $100\ mV$

372   380   388   396   404   412   420

Fig. 30. Part of the original mass spectrum of Se_x for $x = 4\frac{3}{4}$, $x = 5$, and $x = 5\frac{1}{4}$.

about half the breadth and which are not resolved, as is to be expected for the mass resolution used.

These observations are consistent with the following explanation. With the Pt field emitter at room temperature, the selenium vapor at this surface is oversaturated (unless removed by field ionization). Therefore, Se vapor will condense to a solid phase. There is, however, a particular behavior. Although Se vapor is usually deposited as an amorphous phase, the presence of an electric field causes immediate whisker formation. According to well-known theories of nucleation [92], the initiation of a nucleus is favored by the presence of an electric charge (or field) and its growth, probably due to surface migration, is in the field direction. At higher fields, surface layers are removed in the form of field desorption or field evaporation products. According to the theory of field evaporation [7], which only describes the evaporation of atomic species, singly charged Se_1^+ ions should be expected. In the case of solid Se, the particular properties of the solid are represented by the evaporation products. Whole parts of the lattice are evaporated which are conglomerated at the surface or formed by lattice defects of the Se lattice.

The most prominent evaporation product under these conditions (dynamic Se gas supply to the whisker surface) is Se_5^+. During field evaporation, the Se_5^+ signal is more intense than the Se_6^+ signal during field ionization. During the impingement of gas molecules, an immediate rearrangement of molecular size must occur, presumably with the temporary incorporation of these molecules into the whisker lattice. The high intensity of Se_5^+ signals indicates that Se_5 surface groups have a high preference in surface mobilities. The higher molecular weight compounds are regarded as parts of the Se lattice, which will have a chain structure. The reason for the bond cleavage in the Se lattice can be found in the high mechanical tension which these whiskers will have under field conditions (about 400 N/m^2 mm^{-2} at 1 V/Å). Mechanical stress is known [93] to create lattice defects in selenium by the formation of p-defects or disturbed covalent Se—Se bonds. The piezo-electric properties of solid Se can create lattice defects due to electrostriction.

The fact that doubly and quadruply charged ions, but no triply charged ones, can be measured, supports the suggestion that these ions are formed by heterolytic bond cleavage of an Se—Se bond. In this case, electrons are removed pairwise from occupied orbitals.

2. Evaporation of arsenic oxide modifications

The thermal evaporation of solids has been investigated extensively by Stranski and Wolff [94] who used the modifications of As_4O_6 as a model substance. With substances which have metallic or molecular lattices, the evaporation coefficient is usually near to one and the activation energy of this process equals the enthalpy of evaporation. The situation is completely different if evaporating molecules have to be formed by cleavage of chemical bonds within the lattice prior to the thermal evaporation. Then, the activation energy of the process may be determined by this bond fission and the thermal evaporation coefficient is much less than one.

These principles of a "normal" and "abnormal" evaporation have been demonstrated with the example of the modifications of arsenic oxide. Both the cubic arsenolite and the monoclinic claudetite are composed of $AsO_{3/2}$ polyhedra. In the first case, four of these polyhedra form a saturated As_4O_6 molecule which crystallizes in the diamond lattice type. In the second case (claudetite), no such saturated molecular units are found, the $AsO_{3/2}$ polyhedra instead form a two-dimensional network which crystallizes in a monoclinic crystal lattice. The evaporation coefficient of arsenolite is one and of claudetite is 10^{-4} to 10^{-5} since As—O bonds have to be broken prior to evaporation.

To understand the evaporation mechanism, it was necessary to analyze the molecular structure of desorbing particles above these different crystal modifications. This had already been tried by integral methods applying microbalance techniques [95]. FIMS gave a direct answer to this problem [96]. With an experimental technique as used before with Se, the Langmuir evaporation of As oxides was investigated at different evaporation temperatures and different electric fields on a Pt-field emitter.

The arsenolite modification, which we investigated, evaporated under all conditions exclusively in the form of As_4O_6 molecules (M = 396) with the proper isotope contribution of 1.2% at M397. At the field strength used, no fragmentation of $As_4O_6^+$ ions could be observed (within 0.1% intensity). However, a very complicated mass spectrum occurred as soon as As_4O_6 molecules were allowed to condense on the Pt-emitter surface.

The evaporation of both forms of claudetite (I: crystal platelets, II: crystal needles) also gave mainly As_4O_6 molecules when the evaporation process was investigated at comparatively high temperatures in order to have sufficient intensities. However, particularly in claudetite II, products were observed which differ from the thermodynamically stable As_4O_6 molecule. In claudetite I, this was mainly AsO (M91) and As_4 (M300), in claudetite II, As_2O_3 was one of the observable evaporation products. These results are in accordance with observations of integral microbalance techniques and indicate in detail, that during the Langmuir evaporation of claudetite, bond cleavage may occur statistically in such a way that products are formed that differ from the thermodynamically stable As_4O_6 gas molecules. With claudetite again, mass spectra are completely different if molecules are allowed to condense on the Pt-emitter.

The alterations of the mass spectra after the formation of condensed As oxide layers are demonstrated for arsenolite in Fig. 31 and Table 4. The following observations are made. First, the original sharp mass signal (Fig. 31(a)) of the parent molecular ion broadens towards higher masses and acquires two new maxima at M396.5 and M397 (Fig. 31(b)). The intensity of

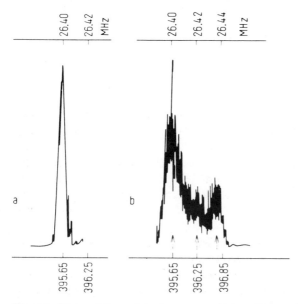

Fig. 31. Parts of the original mass spectrum of arsenolite, (a) at a cleaned emitter, and (b) with a condensed layer. The resonance frequencies are indicated at the top and the mass scale at the bottom. (After Becker et al. [96].)

TABLE 4

Mass spectrum of arsenolite vapor with a condensed arsenic oxide layer on a Pt-field emitter. (Temperature of the evaporation furnace $350°K$, temperature of the Pt-field emitter $298°K$.)

Mass number	Structure	Relative intensity
396	$As_4O_6^+$, $As_8O_{12}^{2+}$	100
388	$As_8O_{11}^{2+}$	100
190	$As_4O_5^{2+}$	50
198	$As_4O_6^{2+}$	20
289	$As_3O_4^+$	10
487	$As_5O_7^+$, $As_{10}O_{14}^{2+}$	7
412	$As_4O_7^+$, $As_8O_{14}^{2+}$	5
342.5	$As_7O_{10}^{2+}$	4
495	$As_{10}O_{15}^{2+}$	4
91	AsO^+, $As_2O_2^{2+}$	3
281	$As_6O_7^{2+}$	3
334.5	$As_7O_9^{2+}$	3
380	$As_4O_5^+$, $As_8O_{10}^{2+}$	3
404	$As_8O_{13}^{2+}$	3
503	$As_5O_8^+$, $As_{10}O_{16}^{2+}$	3

these additional maxima increases with the background pressure of H_2O, which acts as a proton donor.

Second, in a mixture with a calibrating gas (*n*-heptane was normally used) unusual reactions occurred. *n*-Heptane acts as a proton donor. As long as arsenic oxide is deposited, the $C_7H_{15}^+$ ion (M99) is formed, but after removal of the deposit, the usual $C_7H_{16}^+$ ion (M100) occurs exclusively. Another phenomenon of the arsenic oxide deposit concerns the detectability of mercury (from the background pressure). The singly charged mercury isotopes could be detected easily on the deposit and not on the Pt surface. The reason for this promotion effect is still unknown.

Third, the condensed layer of arsenic oxides creates numerous new mass signals compiled in Table 4, which represent ionic products of the dielectric surface phase. Many of the structures compiled in Table 4 have additional signals which indicate proton attachment. This is the reason for the broadening of mass signals in Fig. 31(b). The field induced self-dissociation in the $(As_4O_6)_x$ layer also creates $As_8O_{12}^{2+}$ ions. The proton attachment product is responsible for the maximum at M396.5 ($As_8O_{12}H^{2+}$). The doubly charged

References pp. 443—446

ions are preferably formed by heterolytic bond cleavage, for example

$$(As_4O_6)_x \rightarrow (As_4O_6)_{x-2} + O^{2-} + As_8O_{11}^{2+}$$

which is one of the major positive ions in Table 4.

Field ions from condensed claudetite are somewhat different. In particular, misfits of the As_4O_6 polyhedra are observed, where one, two, or three oxygen atoms are added to the M = 396 parent molecular ion. This is understandable, since the As—O— bond cleavage which is necessary in this case may statistically add the oxygen to those As atoms, which then form a thermodynamically unfavored compound.

Investigations of the evaporation of solids show that in many cases, the molecular weight and structure of gaseous evaporation products can be determined. However, in FIMS, the condensation of compounds easily creates new mechanisms of ion formation. Processes which occur in solid electrolytes at emitter surfaces are still rather unknown and present a new interesting area of research.

C. NITROGEN COMPOUNDS AT METAL SURFACES

The evaluation of the mechanism of the ammonia synthesis was one of the major research projects at the Fritz-Haber-Institut when FIMS became applicable to surface science. Although the important ammonia synthesis has been in industrial operation nearly 50 years, questions about details of the reaction mechanism between N_2 and H_2 still have to be explained. This was the motivation to apply FIMS for the identification of nitrogen compounds on metal surfaces and, in particular, on iron.

This question was investigated by Schmidt [97] who used an Fe-field emitter for analyzing interactions of NH_3 or N_2—H_2 mixtures with the metal surface. These experiments in FIMS could not be performed under high pressure and temperature conditions which have to be applied in the ammonia synthesis. The question of interest, however, was whether nitrogen or hydrogenated forms thereof are chemisorbed in a molecular form $N_2H_x (x \leqslant 6)$ or in atomic form $NH_x (x \leqslant 3)$, as claimed in the literature [98,99].

Under the experimental conditions of FIMS, with a maximum total pressure of 10^{-4} torr and at 300°K, several species could be observed. With NH_3 interacting, besides NH_3^+, the ions NH_4^+ and $NH_3 \cdot NH_4^+$ were most prominent and showed field-dependent intensities. This behavior is comparable with results of the field ion mass spectra of water and can be explained by molecular association and proton transfer reactions as described before. In view of the mechanism of the ammonia synthesis [100], the absence of NH_2^+, NH^+ and N^+ ions in these experiments was emphasized. Particularly, NH and NH_2 should increase their concentration in surface layers as a residue during the NH_4^+ formation. In fact, however, the reactions in the surface

TABLE 5

Ions formed at a Fe-field emitter during the interaction of NH_3

Composition	Relative intensity	Composition	Relative intensity	Composition	Relative intensity
N_2^+	1.0	N_3^+	3.5	NH_3^+	2.3×10^4
N_2H^+	5.0×10^3	N_3H^+	2.5	NH_4^+	1.0×10^6
$N_2H_2^+$	3.5×10^1	$N_3H_2^+$	6.5	NH_4^+ NH_3	2.85×10^3
$N_2H_3^+$	1.0×10^2				
$N_2H_4^+$	1.0×10^1	$N_3H_4^+$	1.5	N_4^+	1.2
$N_2H_5^+$	3.3×10^1	$N_3H_5^+$	1.0	N_4H^+	1.0
$FeN_2H_n^+$		$FeN_3H_n^+$		$FeN_4H_n^+$	
$n = 0, 1, ..., 6$	1.0—10.0	$n = 0, 1, ..., 9$	1.0—10.0	$n = 0, 1, ..., 12$	1.0—10.0

layer lead to products which only contain molecular nitrogen or associates with 3 or 4 nitrogen atoms. On the other hand, estimates of ionization probabilities indicate that, for instance, NH should be ionized, if present. The observed ions are compiled in Table 5. According to this result, chemisorption structures with atomic nitrogen are absent. The desorption of surface compounds involves small quantities of ions which contain Fe. Again, field ions with atomic nitrogen are absent. The occurrence of $FeN_3H_n^+$ indicates that unusual surface structures with (at zero field) non-stable neutral species are formed. On the other hand, N_2H^+ ions have been found during field ionization in mixtures of N_2 and $3H_2$. Most probably, this is a consequence of a proton transfer reaction which occurs most easily with traces of water. As the conditions of the FIMS experiments are far away from measurable equilibrium concentrations of ammonia, it would be unlikely to detect reaction products or precursor states thereof.

Although the formation of field ions on iron may be governed by the laws of complex chemistry of these ions, there are other arguments in favor of a molecular form of nitrogen at the surface. If FIMS of platinum is compared with iron during the interaction of ammonia, structures with atomic nitrogen such as PtN^+, are very abundant. This indicates a dissociative chemisorption of nitrogen on platinum which could not be observed with iron. One of the arguments in the dissociative or non-dissociative mechanism of the ammonia synthesis is the high bond energy in the nitrogen molecule ($N \equiv N$ bond = 226 kcal/mole). A partial hydrogenation of the nitrogen molecular structure $N_2 + x\, H_2 \rightarrow N_2-H_{2x}$ would circumvent this energy barrier, since in hydrazine, for instance, the H_2N-NH_2 bond amounts only to about 60 kcal/mole and should be dissociated much easier at a metal surface.

In this connection, the behavior of hydrazine is of interest as a possible

TABLE 6

Comparison of electron impact (EI) mass spectra of N_2H_4 with FIMS on a Pt-emitter at 298°K and $P_{N_2H_4}$ at 5×10^{-6} torr

Mass number	Structure	Rel. intensities		ΔH (kcal/mole)
		EI	FI	
34	$(^{15}N^{14}NH_5^+)$	0	0.7	
33	$N_2H_5^+(^{15}N^{14}NH_4)$	0.7	100	
32	$N_2H_4^+$	100	20	224
31	$N_2H_3^+$	43	10	231
30	$N_2H_2^+$	31	0.2	276
29	N_2H^+	40	0.4	312
28	N_2^+	21	0.3	396 (359 from N_2)
19	H_3O^+		0.1	
18	$NH_4^+H_2O^+$		4.0	
17	NH_3^+	26	1.0	223 (from NH_3)
16	NH_2^+	29	0.1	302
15	NH^+	8		381 (from NH)
14	N^+	5		
2	H_2^+		2.5	

The enthalpy of formation of ions ΔH is related to monomolecular reactions $N_2H_4 \rightleftarrows N_aH_b^+ + N_cH_d + H_e$, with $a = 2$, $b \leqslant 4$, $c = 1$, $d \leqslant 2$, $e = 2$.

intermediate in the mechanism in the ammonia synthesis. The field ionization of N_2H_4 on a Pt-field emitter has recently been investigated [101]. As usual, the products of field ionization have to be compared with ions of electron impact ionization to evaluate the relative stability of ionic species. This comparison is given in Table 6. With electron impact (70 eV), all possible fragments of N_2H_4 with the exception of H^+ occur with an intensity of >1% (M32 = 100%). In contrast, the field ion mass spectrum shows ions which cannot be explained by fragmentation of the parent molecular ion. Under the conditions where the field ion mass spectrum has been taken, proton attachment to form $N_2H_5^+$ prevails. Only parent molecular ions and the species $N_2H_3^+$ and NH_4^+ have an intensity of >1%. These intensities are field- and temperature-dependent.

The formation of these field ions has to be explained by the following mechanisms.

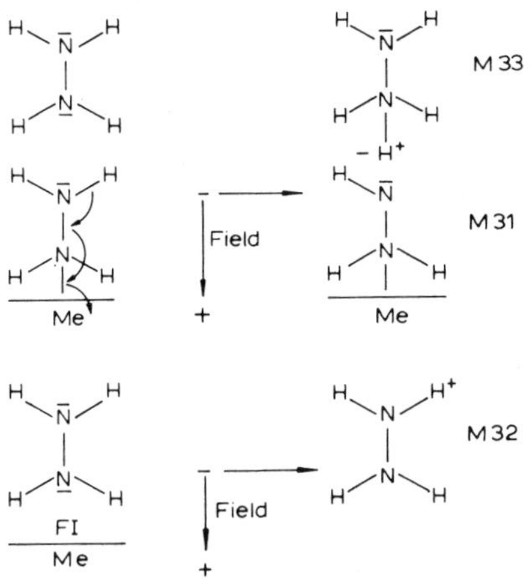

M 33

M 31

| Field

Me

+

Me

FI

| Field

M 32

Me

+

The adsorption of hydrazine is stabilized by the additional field bonding energy, due to the large dipole moment of 1.85 Debye. For an N_2H_4 molecule chemisorbed with one of the lone electron pairs and with a charge center 2.5 Å above the surface, a field bond energy of 35 kcal/mole is achieved.

The ion M33, which is the main intensity at room temperature, is produced by a bimolecular field induced reaction. The remaining radical M31 can be detected as well. At an uncovered Pt surface, the formation of $N_2H_4^+$ ions prevails, as can be deduced from the temperature dependence of the ion intensities (see Fig. 32). Additional subordinate bimolecular reactions

$$N_2H_4 + N_2H_3 \xrightarrow{\text{field}} N_2H_5^+ + N_2H_2 + e^- \text{ (Me)}$$

and

$$N_2H_4 + N_2H_2 \xrightarrow{\text{field}} N_2H_5^+ + N_2H + e^- \text{ (Me)}$$

may proceed which give rise to small intensities of M30, M29, and M28. However, the latter ions may also be produced by molecular decay of excited molecular ions. Slow decomposition processes (10^{-11} sec) could not be observed.

The temperature dependence of ion intensities yields further information about the processes occurring at the Pt surface. Due to the exothermic heat of adsorption of hydrazine, the $N_2H_5^+$ intensities decrease by two orders of magnitude if the temperature is raised from 298 to 448°K. The decrease of $N_2H_4^+$ in the lower temperature region was explained by the same reason, whereas the increase above 373°K was explained by the cessation of bimolecular reactions. Mass 32 increases at the expense of M33. With exceptional surface diffusion, N_2H_4 will reach the ionization zone from other parts of the emitter tip and contribute to the ion current, particularly at elevated

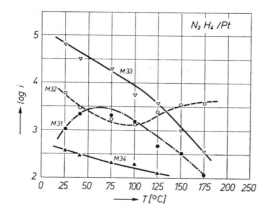

Fig. 32. Temperature dependence of ion intensities during FI of N_2H_4 on Pt. (After Block [101].)

temperatures. The radical N_2H_3 is probably more tightly bound to the surface and does not display additional contributions at elevated temperatures.

At elevated temperatures, products of the catalytic hydrazine decomposition have been expected successively. In Table 7 and Fig. 32, possible inter-

TABLE 7

Identification of FI products at $473°K$ during the interaction of N_2H_4 on a Pt-field emitter

Mass number	Structure	Rel. intensity
34	$N_2H_6^+$	1.75
33	$N_2H_5^+$	100.0
32	$N_2H_4^+$	14.4
31	$N_2H_3^+$	2.73
30	$N_2H_2^+$	0.70
29	N_2H^+	0.58
28	N_2^+	0.35
18	NH_4^+	12.2
17	NH_3^+	1.9
16.5	$N_2H_5^{2+}$	0.12
16	$N_2H_4^{2+}, (NH_2^+)$	0.09
15.5	$N_2H_3^{2+}$	0.09
15	$N_2H_2^{2+} (NH^+)$	0.07
14.5	N_2H^{2+}	0.05
14	N^+	0.07

mediates and products amount to less than 1% (NH_3^+) or 0.5% ($N_2H_2^+$, N_2H^+, N_2^+) and one of the postulated intermediates to less than 0.1% (NH_2^+). From separate tests of the ionization probabilities at different field strengths, it had to be concluded that all these substances should be ionized if present at the surface.

With a sensitive sampling technique, reaction products in the mass regions of N_2 to N_2H_6 had been compared with those in the region N to NH_4. Association products $N_2H_4 \cdot NH_4^+$, $N_2H_4 \cdot N_2H_5^+$ or $N_2H_4 \cdot H_2O^+$ (from possible traces of H_2O) could not be detected. During these experiments, ammonia is formed and ionized as NH_3^+ and NH_4^+. Nitrogen (M28) has only a small intensity, since it probably desorbs immediately. Small ion intensities in the mass range M14 to M16 are mainly doubly charged ions as indicated by ion signals at half interger mass units. At higher hydrazine pressures, the additional ion $N_2H_2^+$ could be detected.

The suspicion that possible surface amides, imides or nitrides could be field desorbed only in the form of metal complex ions was refuted by experiments at still higher field strength. The field evaporation of Pt is not involved with the structures $Pt \cdot NH_2$, $Pt-NH$ or $Pt-N$.

The conclusions drawn from these experiments were that surface amides, imides or nitrides could not be intermediates of the hydrazine decomposition. On Fe surfaces, the absence of nitrogen isotope mixing in the reaction of a mixture of $^{14}NH_2 \cdot {}^{14}NH_2$ and $^{15}NH_2 \cdot {}^{15}NH_2$ required the same explanation [102]. Similarly, as on palladium, Ertl and Tornau [103] concluded that the decisive part of the reaction is $N_2H_4 \rightarrow N_2H_3 + H$.

Acknowledgements

Valuable discussions with G. Abend, D. Cocke, W.A. Schmidt and H. Thimm are acknowledged. Support of the work, performed in the Fritz-Haber-Institut der Max-Planck-Gesellschaft, by the Deutsche Forschungsgemeinschaft and the Senator für Wirtschaft, Berlin-West, ERB Fonds is gratefully acknowledged.

References

1   J.R. Oppenheimer, Phys. Rev., 31 (1928) 66.
2   C.Z. Lanczos, Z. Phys., 62 (1930) 28; 65 (1930) 518; 68 (1931) 204.
3   E.W. Müller, Z. Phys., 131 (1951) 136.
4   M.G. Inghram and R. Gomer, J. Chem. Phys. 22 (1954) 1279.
5   M.G. Inghram and R. Gomer, Z. Naturforsch. A, 10 (1955) 863.
6   H.D. Beckey, Field Ionization Mass Spectrometry, Pergamon Press, Oxford, 1971.
7   E.W. Müller and T.T. Tsong, Field Ion Microscopy, Principles and Application, American Elsevier, New York, 1969.

444

8 G. Ehrlich, Advan. Catal., 14 (1963) 255.
9 R. Gomer, Field Emission and Field Ionization, Harvard University Press, Boston, 1961.
10 H.D. Beckey, Angew. Chem., 81 (1969) 662.
11 A.J.B. Robertson, J. Sci. Instrum., 7 (1974) 321.
12 A.J.B. Robertson, in A. Maccoll (Ed.), International Review of Science, Vol. 5, Butterworth, London, 1972, p. 103.
13 J.H. Block, Advan. Mass Spectrom., 6 (1974) 109.
14 P. Debye, Polar Molecules, Reinhold, New York, 1929.
15 H.D. Beckey, Proc. Int. Congr. Anal. Chem., San Francisco, August 1973.
16 E.W. Müller, Advan. Electron. Electron Phys., 13 (1960) 83.
17 K.M. Bowkett and D.A. Smith, Field Ion Microscopy, North-Holland, Amsterdam, 1970.
18 H.D. Beckey, E. Hiet, A. Maas, M.D. Migahed and E. Ochterbeck, Int. J. Mass Spectrom. Ion Phys., 3 (1969) 161.
19 H. Saure and J.H. Block, Int. J. Mass Spectrom. Ion Phys., 7 (1971) 145, 157.
20 W.A. Schmidt, O. Frank and J.H. Block, Surface Sci., 44 (1974) 185.
21 H.D. Beckey and H.R. Schulten, Angew. Chem., (1975) in press.
22 T. Utsumi and O. Nishikawa, J. Vac. Sci. Technol., 9 (1972) 477.
23 A. Martin and J.H. Block, Messtechnik, 5 (1973) 149.
24 W. Paul, H.P. Reinhard and U. von Zahn, Z. Phys., 152 (1958) 163.
25 J.H. Block and R. Abitz, unpublished.
26 H.J. Heinen, Ch. Hötzel and H.D. Beckey, Int. J. Mass Spectrom. Ion Phys., 13 (1974) 55.
27 H.D. Beckey, H. Knöppel, G. Metzinger and P. Schulze, Advan. Mass Spectrom., 3 (1966) 35.
28 P. Hindennack and J.H. Block, Ber. Bunsenges. Phys. Chem., 75 (1971) 993.
29 E.W. Müller, J.A. Panitz and S.B. McLane, Rev. Sci. Instrum., 39 (1968) 83.
30 G. Abend, H. Thimm and J.H. Block, unpublished.
31 H. Thimm, Diplomarbeit, Frie Universität, Berlin, 1969.
32 F.W. Röllgen and H.D. Beckey, Messtechnik, 5 (1972) 115.
33 A.J. Jason, Phys. Rev., 156 (1967) 266.
34 H.J. Heinen, F.W. Röllgen and H.D. Beckey, to be published.
35 I.V. Goldenfeld, I.Z. Korostyshevsky and B.G. Mischanchuk, Int. J. Mass Spectrom. Ion Phys., 13 (1974) 297.
36 E.W. Müller and S.V. Krishnaswamy, Surface Sci., 36 (1973) 29.
37 W.A. Schmidt and O. Frank, Int. J. Mass Spectrom. Ion Phys., 2 (1969) 399.
38 W. Quitzow and J.H. Block, Z. Angew. Phys., 31 (1971) 193.
39 H.D. Beckey, Z. Naturforsch. A, 14 (1959) 712.
40 W.A. Schmidt, Z. Naturforsch. A, 19 (1964) 318.
41 H.D. Beckey, Z. Naturforsch. A, 15 (1960) 822.
42 G. Ehrlich and F. Hudda, Phil. Mag., 8 (1963) 1587.
43 T.T. Tsong and E.W. Müller, J. Appl. Phys., 37 (1966) 3065.
44 O. Nishikawa and E.W. Müller, J. Appl. Phys., 35 (1964) 2806.
45 D.W. Bassett, Brit. J. Appl. Phys., 18 (1967) 1753.
46 J.H. Block, Z. Phys. Chem. (N.F.), 64 (1969) 199.
47 A. Prock and G. McConkey, Topics in Chemical Physics, (based on the Harvard Lecture of P. Debye), Elsevier, Amsterdam, 1962.
48 A.E. Bell, L.W. Swanson and D. Reed, Surface Sci., 17 (1969) 418.
49 W.A. Schmidt, O. Frank and J.H. Block, Ber. Bunsenges. Phys. Chem., 75 (1971) 1240.
50 Z. Knor and E.W. Müller, Surface Sci., 10 (1968) 21.
51 Z. Knor, Advan. Catal., 22 (1972) 51.

52 R.L. Lewis and R. Gomer, Surface Sci., 26 (1971) 197.
53 F.W. Röllgen and H.D. Beckey, Ber. Bunsenges. Phys. Chem., 75 (1971) 988.
54 F.W. Röllgen and H.D. Beckey, Z. Naturforsch. A, 29 (1974) 230.
55 H.D. Beckey, Z. Naturforsch. A, 17 (1962) 1103.
56 J.H. Block and G. Bozdech, unpublished.
57 F. Speier, H.J. Heinen and H.D. Beckey, Messtechnik, 4 (1972) 147.
58 J.H. Block, Z. Phys. Chem. (N.F.), 39 (1963) 169.
59 M. Drechsler, Angew. Chem., 79 (1967) 987.
60 T.T. Tsong and E.W. Müller, Phys. Rev. Lett., 25 (1970) 911; J. Chem. Phys., 55 (1971) 2884.
61 F.W. Röllgen and H.D. Beckey, Surface Sci., 23 (1970) 69; 26 (1971) 100.
62 J.H. Block, Proc. 2nd Int. Congr. Solid Surfaces, Kyoto, March 1974, North-Holland, Amsterdam, in press.
63a A.W. Czanderna, J. Phys. Chem., 68 (1964) 2765.
63b P.A. Kilty, N.C. Rol and W.M.H. Sachtler, in J. Hightower (Ed.), Proc. 5th Int. Cong. Catal., Vol. 2, North-Holland, Amsterdam, 1973, p. 929.
63c A.W. Czanderna, Thin Solid Films, 12 (1972) 521.
64 M.M.P. Janssen, J. Moolhuysen and W.M.H. Sachtler, Surface Sci., 33 (1972) 625.
65 A.W. Czanderna, O. Frank and W.A. Schmidt, Surface Sci., 38 (1973) 129.
66 A.W. Czanderna, S.C. Chen and J.R. Biegen, J. Catal., 33 (1974) 163.
67 K. Bergmann, M. Eigen and L. DeMaeyer, Ber. Bunsenges. Phys. Chem., 67 (1963) 819.
68 J.H. Block, Z. Naturforsch. A, 18 (1963) 952.
69 J.H. Block and P.L. Moentack, Z. Naturforsch. A, 22 (1967) 811.
70 W.A. Schmidt, O. Frank and J.H. Block, Abstr. 21st Field Emission Symp., Marseilles, July, 1974.
71 E.W. Müller, Phys. Rev., 103 (1956) 618; E.W. Müller and T.T. Tsong, in S. Davison (Ed.), Progr. Surface Sci., Vol. 3, Pergamon, Oxford, 1973, pp. 1—173.
72 D.G. Brandon, Surface Sci., 3 (1965) 1.
73 T.T. Tsong, J. Chem. Phys., 54 (1951) 4205.
74 D. McKinstry, Surface Sci., 29 (1972) 37.
75 I. Langmuir and K.H. Kingdon, Phys. Rev., 34 (1929) 129.
76 M.D. Scheer, J. Res. Nat. Bur. Stand. A, 74 (1970) 37.
77 N.I. Ionov, Sov. Phys. Techn. Phys., 14 (1969) 542 (and related work).
78 J.H. Block and M.S. Zei, Surface Sci., 27 (1971) 419.
79 P.E. Pickert, J.A. Rabo, E. Dempsey and V. Schomaker, Proc. 3rd Int. Congr. Catal., Vol. 1, 1964, p. 714.
80 P.J. Venuto, L.A. Hamilton and P.S. Landis, J. Catal., 5 (1966) 489.
81 J. Türkevich, F. Nozaki and D.N. Stamires, Proc. 3rd Int. Congr. Catal., Vol. 1, 1964, p. 586.
82 H.G. Karge, Surface Sci., 40 (1973) 157.
83 M.S. Zei, Ber. Bunsenges. Phys. Chem., 78 (1974) 443.
84 P.R. Davis, E. Bechtold and J.H. Block, Surface Sci., 45 (1974) 585.
85 J.H. Block, Proc. 5th Int. Congr. Catal., Vol. 1, 1973, p. E91.
86 I.V. Goldenfeld, V.A. Nazarenko and I. Pokrovsky, Dokl. Akad. Nauk SSSR, 161 (1965) 276.
87 A.R. Anway, J. Chem. Phys., 50 (1969) 2012.
88 F.W. Röllgen and H.D.Beckey, Surface Sci., 27 (1971) 321.
89 L. Onsager, J. Chem. Phys., 2 (1934) 599.
90 J. Berkowitz and W.A. Chupka, J. Chem. Phys., 45 (1966) 4289.
91 D. Detry, J. Drowart, P. Goldfinger, H. Keller and H. Rickert, Ber. Bunsenges. Phys. Chem., 72 (1968) 1054.

92 M. Volmer, in Th. Steinkopff (Ed.), Kinetik der Phasenbildung, Dresden, Leipzig, 1939.
93 J. Stuke, Phys. Status Solidi, 6 (1964) 441.
94 I.N. Stranski and G. Wolff, Z. Elektrochem., 53 (1949) 1; K.A. Becker, K. Plieth and I.N. Stranski, Progr. Inorg. Chem., 4 (1962) 1.
95 K.A. Becker, H.J. Forth and I.N. Stranski, Z. Elektrochem., 64 (1960) 373.
96 K.A. Becker, J. Block and H. Saure, Ber. Bunsenges. Phys. Chem., 75 (1971) 406.
97 W.A. Schmidt, Angew. Chem., 80 (1968) 151.
98 A. Ozaki, H. Taylor and M. Boudart, Proc. Roy. Soc. London Ser. A, 258 (1960) 47.
99 S.R. Logan and J. Philip, J. Catal., 11 (1968) 1.
100 R. Brill, Ber. Bunsenges. Phys. Chem., 75 (1971) 455.
101 J. Block, Z. Phys. Chem. (N.F.), 82 (1972) 1.
102 J. Block and G. Schulz-Ekloff, J. Catal., 30 (1973) 327.
103 G. Ertl and J. Tornau, Z. Phys. Chem., 93 (1974) 109.

Chapter 10

INFRARED REFLECTION—ABSORPTION SPECTROSCOPY

H.G. TOMPKINS

I. Introduction

Infrared spectroscopy (IR) has been used for many years to obtain information about molecular structure. Its primary usage has been with gases, and with liquids or solids which are transparent to most of the wavelengths in the region of interest. Metals, on the other hand, are opaque to infrared radiation simply because they are metals. The free electrons which give the metallic characteristics follow the oscillating electric field and do not allow the field to penetrate the metal to any significant extent. Interest in molecules adsorbed on metal surfaces and in thin films on metals have stimulated many attempts to obtain a method of taking IR spectra of species on metal surfaces.

From the point of view of surface science, infrared radiation is an excellent probe of a surface since it is desirable to perturb the surface as gently as possible during a measurement. Most surface probes, i.e. electrons, ions, or UV photons cause desorption, rearrangement, or total destruction of the surface. Infrared radiation, on the other hand, only causes the molecules to vibrate with a slightly higher energy.

There have been several different methods devised to obtain spectra of molecules on metal surfaces. The classic work of Eischens et al. [1,2] describes a method whereby very small particles of metal are supported in an inert matrix of silica spheres. Gas molecules are then adsorbed on the metal particles. Because the silica is transparent to the radiation and because the metal particles are much smaller then the wavelength, the standard transmission method of IR spectroscopy can be used to obtain spectra of the adsorbed molecules.

Other methods have been devised to obtain small particles such as evaporating the metal into an oil matrix [3] and evaporating the metal in the presence of an inert gas [4] to give a high surface area soot-like deposit. Some work has been done on very thin films where the metal is thin enough to allow sufficient transmission of the radiation to obtain a spectrum [5].

All of these methods are deficient because the surface of the metal is not characterized. Cleanliness is questionable and single crystal work is impossible. It is desirable, then, to be able to use a bulk sample where the area is approximately the geometrical area (as opposed to a powder where porosity gives a much larger surface area) and where the surface is accessible for cleaning and characterizing by other surface measuring techniques.

A logical way in which to pursue this goal is to try a reflection experiment. Efforts have been made in this direction and the state of the science is such that IR spectroscopy is now being used to investigate surfaces. It is the purpose of this chapter to describe a method for obtaining an infrared spectrum of molecules or thin films on a metal surface. This method has been called reflection-absorption (RA) spectroscopy by Greenler et al. [6] because it is an absorption spectrum which is obtained by using a reflection. This is to distinguish this method from reflection spectroscopy [7] which is used to obtain optical constants of metals and is entirely different from the technique described here.

In addition to simply describing the technique, we shall discuss the theory, describe some experimental apparatus used, give some examples, and discuss the applicability of the method, i.e. when it should be used and when it should not. The intensity of the IR absorption bands is low because of the small number of molecules involved, and the methods used to enhance these small bands will be discussed.

It is intended that the reader should be able to decide, after reading the chapter, whether this technique can be applied fruitfully to his specific problem.

II. Theory

A. HISTORY

Several early attempts were made to obtain IR spectra of very thin films (monolayers of gas) on metal surfaces [8—10]. Generally, angles of incidence used were in the intermediate range (45—70°) and, in one case [8], near normal incidence was used (with up to 30 reflections). These early works suffered from small band size (low signal-to-noise ratio) and generally were not very satisfactory. The development of adequate theory about the technique demonstrated rather well why the early attempts were not particularly successful.

B. SINGLE REFLECTION

Theories of the technique have been developed primarily by Francis and Ellison [11], and by Greenler [12,13]. The problem was considered as a boundary value problem by both works and the model is shown in Fig. 1. The nomenclature used here will be that of Greenler [12] and the nomen-

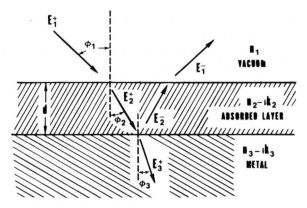

Fig. 1. Boundary value problem arrangement for a thin layer on a metal surface.

clature which follows in discussing Francis and Ellison's work has been changed accordingly. In Fig. 1, the arrow labeled $E_2{}^+$, for example, represents an electromagnetic wave which is the mathematical sum of all multireflected waves in that layer traveling in the direction shown. The problem must be considered separately for radiation polarized parallel to the plane of incidence and for radiation polarized perpendicular to the plane of incidence.

The measurable quantities shown in Fig. 1 are, of course, $E_1{}^+$ and $E_1{}^-$ or, as is usually the case, the reflectivity R which is defined as

$$R = \left[\frac{E_1^-}{E_1^+}\right]^2 \tag{1}$$

which is essentially the ratio of reflected intensity to the incident intensity. The thickness of the adsorbed layer is d, ϕ is the angle of incidence and n and k are the optical constants.

1. Intuitive picture

Before we consider how R is calculated, let us first consider heuristically what we might expect to obtain. Let us ignore the film for the moment. The electrical field intensity E near the surface is determined by a combination of the incoming and the outgoing waves. At normal incidence, there is a $180°$ phase shift of the wave with the result that the combined waves form a standing wave with a node very near the metal surface. The effect of this is that, with a node near the surface, the value of field strength E is extremely small throughout the film (or sample) region, hence there is essentially no interaction between the beam and the film, and correspondingly no IR spectra can be obtained. The situation is not as hopeless as it might seem, however. The phase shift upon reflection is dependent on the angle of incidence and one might expect that for some angles of incidence, the field strength E might have a value large enough for significant interaction with an adsorbed layer or thin film. The fact that this is true is shown in the following section.

2. Calculations

The reflectivity R as defined in eqn. (1) is determined by the optical parameters of the film and metal (which in turn are functions of the wavelength) and by the angle of incidence. In general, however, we are more interested in the difference in R when a film is present and when it is not. Therefore, the calculation is usually made on a difference in values of R for the presence or absence of a film.

We shall deal exclusively with the case of radiation polarized parallel to the plane of incidence since Greenler [12] has shown that the light polarized perpendicular to the plane of incidence does not interact to any extent with an adsorbed layer. Francis and Ellison [11] define a term ΔR, which we shall call ΔR_{FE} to distinguish it from another quantity called ΔR in Greenler's work. They define ΔR_{FE} as

$$\Delta R_{FE} = \frac{R_1 - R}{R_1} \tag{2}$$

where R_1 is the value of reflectivity if the film were absent (i.e. $n_2 = n_1$, $k_2 = 0$).

They use a treatment by Fry [14] and obtain the expression

$$\Delta R_{FE} = \frac{16\pi \, d \, \cos\phi}{\lambda} \left(\frac{k_2 \, \sin^2\phi}{n_2^3 \, \cos^2\phi} - \frac{n_3 \, F(n_2,\phi)}{k_3^3 \, \cos^4\phi} \right) \tag{3}$$

where $F(n_2,\phi)$ is a function of n_2 and ϕ having a value of the order of unity. This expression is valid based on the following assumptions

$$0 \leqslant \phi \leqslant 80°$$

$$k_3^2 \gg n_3^2$$

$$k_3^2 \cos^2\phi \gg 1$$

These assumptions are reasonably valid for good reflectors such as copper and gold. They are not valid for many of the other metals, and they do not deal with angles of incidence greater than 80°. (We shall show later that this eliminates a significant area of interest.)

Keeping in mind the limited validity for this expression, let us pursue eqn. (3) further. For the good reflectors, the second term in the brackets is less than 10% of the first term for most angles of incidence and therefore neglecting it in eqn. (3) gives

$$\Delta R_{FE} = \frac{4 \sin^2 \phi}{n_2^3 \cos \phi} \frac{4\pi k_2}{\lambda} d \tag{4}$$

If $\alpha = 4\pi k_2/\lambda$, as it is often defined, is substituted into eqn. (4), one obtains

$$\Delta R_{FE} = \frac{4 \sin^2 \phi}{n_2^3 \cos \phi} \alpha d \quad \text{or} \quad \frac{R_1 - R}{R_1} = \left[\frac{4 \sin^2 \phi}{n_2^3 \cos \phi}\right] \alpha d \tag{5}$$

To understand the significance of this, let us consider Beer's Law for transmission $I = I_0 e^{-\alpha d}$, which for very weak absorbers ($\alpha d \ll 1$) becomes

$$I/I_0 = e^{-\alpha d} \approx 1 - \alpha d \quad \text{or} \quad (I_0 - I)/I_0 \approx \alpha d \tag{6}$$

so that ΔR_{FE} in reflection-absorption corresponds to $(I_0 - I)/I_0$ in Beer's Law and $[(4\sin^2\phi)/(n_2^3 \cos\phi)]\alpha d$ in reflection-absorption corresponds to αd in Beer's Law. The term in front of αd, then, is a magnification factor obtained by a reflection experiment over a transmission experiment.

The limitations of eqns. (4) or (5) are rather stringent. They do give a closed-form expression which can be used when accuracy is not critical. This expression has the advantage of being simple and it gives some insight into the physical phenomenon.

Greenler [12] has done the complete calculation without the limiting assumptions. He defines a factor A such that

$$A = \frac{R^0 - R}{R^0} \tag{7}$$

where R^0 is the value of R if $k_2 = 0$.

This is similar to Francis and Ellison's ΔR_{FE} except for the fact that R_1 in eqn. (2) is defined as R when $k_2 = 0$ and $n_2 = n_1$ (i.e. no film present). Defin-

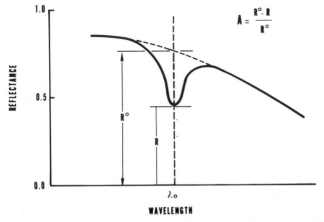

Fig. 2. Definition of absorption factor, A. (From R.G. Greenler [12] with permission.)

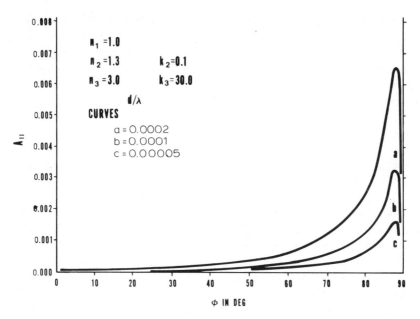

Fig. 3. Absorption factor as a function of angle of incidence for parallel polarization with three different values of d/λ. (From R.G. Greenler [12] with permission.)

ing R^0 as Greenler has, the value of A can be measured for a single reflection from a chart recording as shown in Fig. 2.

If the complete calculation is carried out without approximation, then no simple closed-form expression is available to illustrate the physical picture. The problem can be done easily with a computer, however, and is valid for all angles of incidence and for both good and poor reflectors.

Greenler [12] illustrates the effect of the parameters ϕ, d, n_2, k_2, n_3, and k_3 on the value of A. All of these except n_3 have a significant effect. The reader is referred to the original paper [12] for further details. The effect of ϕ is particularly significant in the design of an experiment and will be illustrated further. The value of A is plotted as a function of ϕ in Fig. 3. The absorption is greatest for an angle of incidence of 88°. It is clear from this figure why some of the early experiments carried out at 45° and even 60° were not particularly successful. This figure shows that if a single angle of incidence is used, it should be 88°.

C. ABSORPTION BAND MAGNIFICATION

If the absorption band obtained on standard available IR equipment for a single reflection were large enough to allow easy measurement, the problem discussed in this section would be minimal. Unfortunately, sometimes, due to the small amount of absorbing species involved, this is not the case and some means of enhancement must be sought. For a given sample, there are basically two methods of enhancing the output of a spectrometer in order to make the absorption bands large enough for measurement. They are: expanding the scale or increasing the number of reflections.

1. Scale expansion

Scale expansion is reasonably straightforward, although some subtleties are often not understood by casual workers in the field. We shall consider the standard, optical-null, double-beam spectrometer available commercially. Scale expansion is simply taking a small part of the vertical scale and displaying it on the entire vertical scale. Most instruments have a scale expansion feature whereby expansion up to 20 × can be obtained by turning a couple of knobs. Another scale expansion often used (sometimes without realizing it) is to use something to reduce the amount of radiation in the reference beam of the standard double-beam spectrometer. The primary consideration when using scale expansion is that although the band size is expanded, the noise level is also expanded by the same amount, hence scale expansion must usually be accompanied by some means of noise reduction. Signal-to-noise enhancement by spectrum averaging with a computer and by lock-in amplifiers are separate subjects and will not be dealt with here. These require equipment construction beyond the standard spectrometer and in the case of very weak absorbers may be needed. The standard spectrometer itself has several means of decreasing the noise level [15]. The pen noise n of a double-beam null balance spectrometer is given by [16]

$$n = \frac{c}{w} \sqrt{\frac{g}{\tau T_r}} \tag{8}$$

where w is the slit width of the monochrometer, g is the amplifier gain, τ is the amplifier period, T_r is the transmittance of the reference beam (defined as unity for an unobstructed beam in regions free of atmosphere absorption), and c is the constant of proportionality. Hence, increasing w and T_r and decreasing g will reduce the noise level. One additional parameter which improves the noise level is to increase the amount of energy in the beam. (It should be pointed out that increasing the slit width, w, essentially increases the amount of energy used.) The beam energy can be increased by operating the IR source as its brightest level. Although this shortens the life of the source, experience indicates that this is money well spent.

Although not usually considered in great detail, scale expansion with noise suppression offers one of the viable ways of obtaining a measurable absorption band.

References p. 472

2. Multiple reflections

It would seem that if one reflection is good, two reflections should be better. In fact, that is the case up to a point. Using multiple reflections is another method of enhancing the absorption bands. However, there are several subtleties involved in band enhancement by multiple reflections. It would, at first, appear desirable to get as many reflections as possible. This is not the case, however, and we shall illustrate this in the following discussion.

For clarity, let us denote the terms for a single reflection as R, R^0, etc., and for multiple reflection as T, T^0, etc. Then, we have $T = R^N$, $T^0 = (R^0)^N$, etc. where N is the number of reflections. The values of T and T^0 can be read off the chart paper whereas R and R^0 must be deduced (except for $N = 1$). The factor corresponding to A in eqn. (7) is $[T-T^0]/T^0 = [(R^0)^N - R^N]/(R^0)^N$. This term is essentially the band depth divided by the background level and would be read off the spectrum directly. If this were the quantity to be enhanced, more reflections would give better spectra. However, this quantity is not the quantity to be enhanced, but instead the signal-to-noise level should be optimized [13]. The signal, in this case, is $T - T^0$, or ΔT, rather than R or T, so the quantity to be optimized is $[(R^0)^N - R^N]$/noise.

In considering this quantity, there is obviously an optimum number of reflections above which the signal-to-noise level decreases. The appearance of a hypothetical absorption band after different numbers of reflections is shown in Fig. 4. The noise level will be the same on all the hypothetical spectra shown. If the reflections were lossless, as in internal reflection spectroscopy, the band would continue to get larger (i.e. $T - T_0$ would continue to increase). For a metal, however, the background decreases (even for no film) and eventually T^0 decreases faster than T hence the quantities

$$\frac{T - T^0}{\text{noise}} = \frac{(R^0)^N - R^N}{\text{noise}}$$

will start to decrease. Greenler [13] has shown that this occurs when $T^0 = 1/e$. The value of N which is optimum, N_{opt}, depends on the optical properties of the metal, but not, in general, on the properties of the film.

Tompkins and Greenler [17], and Kottke et al. [18] have shown experimentally that an optimum number of reflections exists and, in fact, the value of N determined experimentally agrees well with the theoretical predictions. Recently, the value of the optimum number of reflections was tabulated as a function of the angle of incidence for several metals [19]. An example is shown in Fig. 5 for copper. Let us denote $(R^0)^N - R^N$ as ΔR (not to be confused with ΔR_{FE} defined by eqn. (2), i.e. $\Delta R = (R^0)^N - R^N$. The value of ΔR is also shown in Fig. 5 for a single reflection and the value of ΔR is plotted for the optimum number of reflections (i.e. $N = N_{opt}$, which varies with ϕ).

As Fig. 5 shows, if the optimum number of reflections is obtained, the value of ΔR is not as sensitive to the angle of incidence as is the case for a

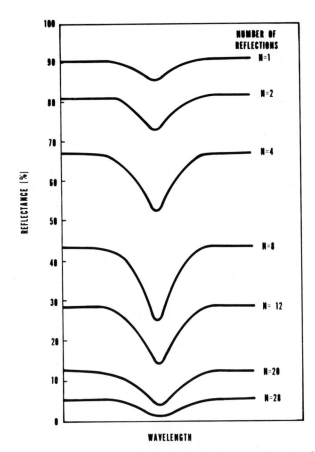

Fig. 4. Appearance of a hypothetical absorption band after different numbers of reflections. (From R.G. Greenler [13] with permission.)

single reflection. It should be pointed out that this is true from 90° to about 70°. Below 70°, the value of $\Delta R(\mathrm{opt})$ decreases considerably [13]. A significant point is that at an angle of 88° the optimum number of reflections for copper is 4. This is to say that the best signal-to-noise ratio is obtained with 4 reflections and that any additional reflections at 88° actually decrease the signal-to-noise ratio. The optimum number of reflections (at 88°) for several metals is listed in Table 1.

Depositing a fresh surface of the metal or cleaning an existing surface prior to adsorbing gas molecules is rather involved when multiple reflections are used whereas in a single reflection experiment, it is quite straightforward. The samples in multiple reflection apparatus can be separated in situ and then repositioned for the IR measurement, but the equipment is cumbersome to use. Unless the band is extremely difficult to detect, the experimenter may decide that the convenience of measuring a single reflection is

456

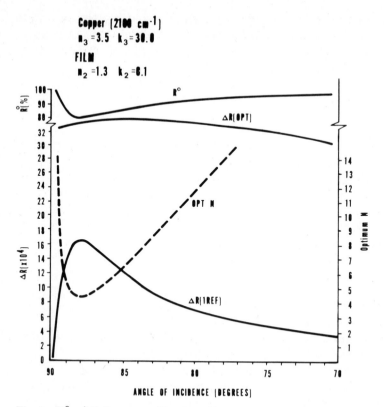

Fig. 5. R^0, ΔR for one reflection, the optimum number of reflections, and ΔR for the optimum number of reflections as a function of angle of incidence. (Provided by R.G. Greenler.)

TABLE 1

Optimum number of reflections for several metals, $\phi = 88°$, $\lambda^{-1} = 2100$ cm^{-1} [19]

Metal	Optimum number of reflections
Silver	6
Copper	4
Gold	3
Nickel	2
Platinum	2
Titanium	1
Palladium	1

more important than the increased signal obtained by using the optimum number of reflections. The band size obtained using the optimum number of reflections is less than a factor of 5 greater than the band size obtained from a properly designed, high angle, single reflection.

III. Applicability

Whether or not an investigator will use reflection-absorption spectroscopy depends on the difficulty in obtaining a measurable absorption band. If standard equipment can be used to obtain the needed information, the technique will be readily used. If, on the other hand, additional noise reducing techniques or other non-standard equipment are required, the investigator will seriously consider whether the desired information is worth the additional effort needed to obtain that information. The subject of this section, then, is to assess the level of difficulty for obtaining a measurable absorption band and, in fact, whether the technique is applicable to whatever problem is at hand.

The size of an absorption band in an observed spectrum depends on several parameters. The primary considerations, however, are

(1) the strength of the interaction between the film material and the electric field, which is determined primarily by the film material;

(2) the magnitude of the electric field at the location of the film, which is determined primarily by the metal (substrate) and the angle of incidence; and

(3) instrumental considerations such as scale expansion and noise level. Let us consider the first two of these in detail.

A. FILM EFFECTS

There are several terms used in the literature to express the interaction of the material with the electric field. Unfortunately, there is no uniform acceptance of terms. Hence, we shall define the terms used here. Beer's Law is sometimes written as

$$I = I_0 e^{-\alpha d}$$

and sometimes as

$$I/I_0 = 10^{-\epsilon C d}$$

where α is the absorption coefficient, ϵ the molar extinction coefficient and C the concentration. These are related to the complex index of refraction $n = n - ik$ by the relationships $\alpha = 4\pi k/\lambda$ and $\epsilon = (\alpha/c)\log_{10} e$. Clearly, α and k are bulk properties whereas ϵ is a molecular property. Because the actual value of ϵ depends to some extent on instrumental conditions (e.g. slit width), handbook values for different materials are generally not used for

References p. 472

458

TABLE 2

Representative apparent extinction coefficients
(Values calculated roughly from Sadtler Standard Spectra [20, 21])

Group	λ^{-1} (cm^{-1})	ϵ (l mole^{-1} cm^{-1})
CH$_3$	2960	20
Cu$_2$O	640	30
Benzene	1480	67
C=O	1710	95
O—H	3300	110
CCl$_3$	720	114
COO$^-$	1445	170
CF$_3$	1460	225
Benzene	671	500

quantitative analysis and values obtained on one instrument are not accurate for use on another instrument. In addition, the chemical environment of the bond also affects the value of ϵ. Our purpose here, though, is not to deal with exact quantitative measurements, but to indicate roughly the band size expected. Molar extinction coefficients for several different species are listed

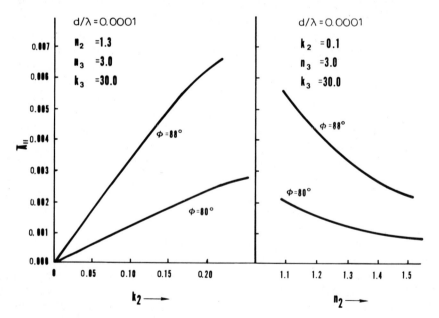

Fig. 6. The variation of the absorption factor with the optical constants of a thin absorbing film on a metal surface. (From R.G. Greenler [12] with permission.)

in Table 2. These values are not exact, but are reasonably representative and should give a rough idea of the relative magnitudes of the interaction of the species with the electric field.

It seems reasonable that higher values of ϵ will give larger absorption bands and this, in fact, is true as shown in Fig. 6, where the band size is plotted versus k_2, which is simply a multiple of ϵ, λ, and C.

From Table 2 and Fig. 6, we can see that it will be considerably easier to obtain a spectrum of a carboxylate species or a fluorocarbon species than a hydrocarbon species. It should be pointed out in passing that in a reflection experiment, the band size does not increase indefinitely as k_2 (or ϵ) increases. As k_2 takes on values comparable with k_3, the surface appears metallic to the beam, giving very small (if any) absorption bands. This is not a limitation, however, since this case seldom occurs.

One additional factor which must be considered in RA spectroscopy is the effect of the index of refraction of the film, n_2. This effect is also shown in Fig. 6, where a decrease in intensity by a factor of 2 might result for n_2 varying from 1.1 to 1.5.

B. SUBSTRATE EFFECTS

It is relatively easy, by comparison with transmission spectroscopy, to understand how the film affects the size of the absorption band. The way in which the metal affects the band size is more complex. The magnitude of the electric field resulting from the combination of the incoming and outgoing wave depends primarily on the optical constants of the metal. A larger value of electric field results in a larger interaction with the film material and yields correspondingly larger absorption bands. The optical constants of several metals are listed in Table 3. Greenler [12] has calculated the effect of the optical constants n_3 and k_3 of the metals on a small absorption band. He finds that the size of the band is rather insensitive to the value of n_3. The effect of k_3 is shown in Fig. 7. For an angle of incidence of 80°, the size of the band is insensitive to k_3 above $k_3 = 10$, whereas for higher angles of incidence, 88°, higher values of k_3 give larger absorption bands for the same amount of material.

C. COMBINING SUBSTRATE AND FILM EFFECTS

Let us now consider some combinations of substrates and films to predict the sizes of some actual bands. The values for the parameters used for the calculation are shown in Table 4 along with the values of the band size. For convenience, we have assumed a density of 1 g/cm³, a thickness of 5 Å for the adsorbed layer, and a value of $n_2 = 1.5$. These are very rough estimates and do not take into account any change in molar extinction coefficient upon adsorption. They also do not consider orientation effects or whether the

TABLE 3

Optical constants for several metals [19]
(λ^{-1} = 2100 cm^{-1}, $n_3 = n_3 - ik_3$)

Metal	n_3	k_3
Silver	2.9	33.7
Aluminum	6.8	32.0
Copper	3.5	30.0
Gold	4.2	27.6
Platinum	5.0	20.0
Nickel	5.4	18.6
Molybdenum	4.2	18.2
Vanadium	5.1	18.2
Niobium	5.1	18.0
Tungsten	4.7	16.7
Cobalt	3.8	15.7
Iron	5.3	15.5
Palladium	6.3	15.3
Chromium	3.8	14.7
Tantalum	6.0	12.7
Titanium	5.5	10.5
Hafnium	3.6	10.4
Zirconium	4.3	10.2
Manganese	6.2	6.7

group will adsorb on the surface. Nevertheless, from the model it would appear that a CH_3 group on titanium would require an enhancement of about 100 for reasonable detection whereas we might expect to see a CF_3 group with only a ten-fold magnification. Using multiple reflections up to

TABLE 4

Rough band size calculations*

Group	Concentration mole l^{-1}	ϵ	λ μm	k_2	Metal	Band size $\Delta R/R^\circ$ (1 reflection)
CH_3	67	20	3.3	0.08	Titanium	0.00049
CH_3	67	20	3.3	0.08	Copper	0.00259
C=O	36	95	5.8	0.36	Platinum	0.0034
CF_3	15	225	6.6	0.42	Copper	0.0058

*Calculations made with the computer program of refs. 12 and 13. Parameters assumed: density 1.0, n_2 = 1.5 thickness = 5 Å, and ϕ = 88°.

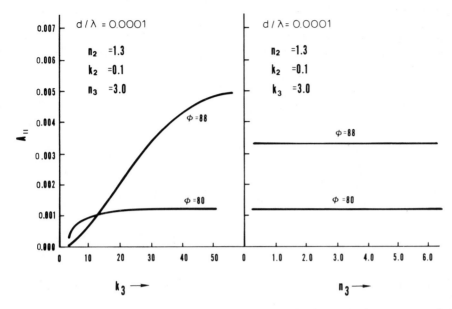

Fig. 7. The variation of the absorption factor with the optical constants of the metal, when covered by a thin absorbing film. (From R.G. Greenler [12] with permission.)

the optimum number can enhance the band sizes by a factor of four for copper, but no enhancement can be obtained for titanium since the optimum number of reflections is 1.

This discussion has primarily considered monolayer films. For multilayer films, the magnification needed is much less, and often the work can, in fact, be done at 1 ×.

We should consider the effect of orientation of the molecule. The coupling of the electromagnetic wave and the molecular vibration occurs through the changing dipole moment of the molecule. The electromagnetic field at the surface is nearly perpendicular to the surface, hence any molecule whose changing dipole moment lies parallel to the surface will not interact with the field and consequently no absorption band will be observed. Another way of considering this is that the extinction coefficient ϵ (for random orientation) has changed to zero upon adsorption (highly oriented parallel to the surface). Examples of this idea would be the symmetric and asymmetric stretching of CF_3. The symmetric stretching mode has a dipole which oscillates parallel to the axis of symmetry while the asymmetric stretching has a dipole which oscillates perpendicular to the axis of symmetry. If the group adsorbed with the axis of symmetry perpendicular to the surface, we would expect to see the symmetric stretching mode and not the asymmetric mode. Microscopic roughness of the surface will modify this to the same degree, however.

462

IV. Experimental arrangements

A. CONCEPTS

The basic requirements for RA spectroscopy are an infrared radiation source, a monochromator, a sample, and a chamber to control the sample environment. Because of the nature of RA spectroscopy, the infrared beam must be diverted from its usual path to obtain reflections. Thus, even standard infrared spectrometers will require some modification. The amount of modification needed ranges from very little to the case where the IR source and the monochromator are separate components and not even part of the same piece of equipment. This latter case is sometimes required when the sample is installed in a large vacuum chamber, which also contains several other methods for investigation of the surface, e.g. AES, LEED, etc. In this case, the optics must be designed to be physically compatible with the other equipment. A laser which is tuneable through the wavelengths of interest might also be used.

In the remainder of this section, we will discuss experimental arrangements which use the standard double-beam, optical-null spectrometers such as the Beckman IR9 or the Perkin—Elmer 621. These spectrometers cover a wide range of wavelengths and are readily available commercially. Existing laboratory spectrometers can be readily modified for this application while retaining the ability to operate in the normal mode.

When using a standard spectrometer, the requirements are that the IR beam be directed to the sample and then back into the monochromator. The beam must return in such a way as to be going in the same direction and focus at the same spot as it would have if it had not been diverted.

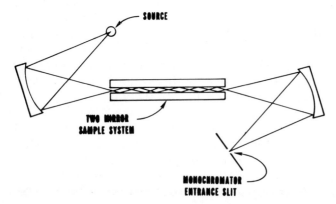

Fig. 8. Schematic representation of a sample system giving more reflections for lower angles of incidence and fewer reflections for higher angles. (From R.G. Greenler [13] with permission.)

The next consideration is whether a single reflection or multiple reflection sample will be used. There are advantages in both cases.

A single reflection sample is much more convenient, since it can be easily cleaned (or deposited). A single crystal sample can be handled much easier. Additional techniques such as AES, LEED or flash desorption can be used simultaneously with the IR techniques.

A multireflection arrangement has added sensitivity. This depends on the metal, but a factor of 4—6 can be obtained for copper [19]. The optimum number of reflections indicated in Table 1 is roughly the added sensitivity factor obtained.

In choosing, one must weigh the need for sensitivity against the inconvenience of multiple reflections. In the past, multiple reflections have been primarily used but this may be due to the fact that the concept of the optimum number of reflections was not well understood.

B. TYPICAL ARRANGEMENTS

Several different arrangements have been used in the early work. We shall, however, concentrate on those used in later work where the concepts of high angle of incidence and optimum number of reflections are better understood.

1. Multiple reflections

When using a thermal source of radiation, we can concentrate a reasonable fraction of the emitted radiation in a small image only at the expense of working with a highly convergent beam. This is to say that a range of angles of incidence will be used rather than a single angle. Our problem is to construct a sample for which rays at all angles of incidence make the optimum number of reflections. Part of Fig. 5 is a plot of the optimum number of reflections versus angle of incidence. This indicates that below 88° it is necessary that rays having lower angles of incidence make more reflections than those having higher angles of incidence. This condition is realized when the sample consists of two parallel plates facing each other and separated by a small distance. As shown schematically in Fig. 8, the angles of incidence would range from 90° to about 84° using a beam 12° wide.

This geometry has been the one used in the most recent investigations. Pritchard [22] used a sample which was hinged to the bottom. A clean film would be deposited by evaporation and the samples then brought parallel via a magnetic linkage. Tompkins and Greenler [17] used a similar geometry with mechanical linkages. Pritchard's cell is shown in Fig. 9. The optics needed to divert the beam to the sample and then back into the spectrometer for the work of Tompkins and Greenler [17] and Kottke et al. [18] are shown in Fig. 10.

It should be pointed out that although these two examples use high angles

464

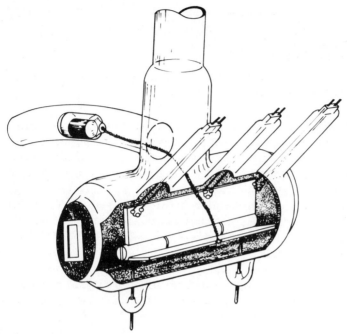

Fig. 9. Diagram of Pritchard's sample cell. (From J. Pritchard and M.L. Sims [22] with permission.)

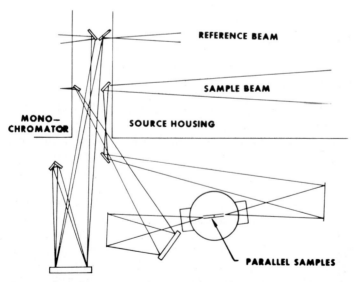

Fig. 10. Diagram of Tompkins and Greenler's optical setup used to divert the beam to the sample and return it to the spectrometer. (From H. Tompkins and R.G. Greenler [17] with permission.)

Fig. 11. Schematic representation of a sample giving the optimum number of reflections for the entire beam using lower angles of incidence than Fig. 8.

of incidence, the optimum condition can be obtained at lower angles (down to about 70°) by designing the sample slightly differently. The sample design shown schematically in Fig. 11 can give the optimum number of reflections for all angles of the beam (beam width 12°). One of the samples must be longer than the other to avoid losing half of the beam.

Another arrangement for multiple reflection is shown in Fig. 12. With this sample, the beam passes through the region twice, being reversed by the retromirror. The advantage of this arrangement is simplicity. The optics needed are available commercially as a simple accessory to a standard spectrometer and only one window in the vacuum cell is needed. Disadvantages are that it is difficult to clean and/or deposit a metal on the sample and only half of the beam is transmitted to the monochromator (the other half returns to the source).

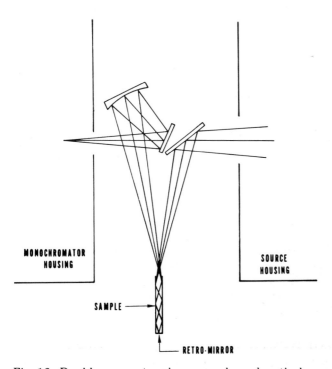

Fig. 12. Double pass retromirror sample and optical arrangement shown schematically.

References p. 472

2. Single reflection sample design

One obvious modification of the sample system shown in Fig. 8 would be to remove one of the samples, reposition the samples, and adjust the beam so that one reflection is made. The angles of incidence would, of course, range from 90° to 78° for a 12° beam.

Another possible way of obtaining a single reflection arrangement is shown in Fig. 13. It would appear that two reflections are involved, but the second reflection is nearly at normal incidence, hence its only function is to return the beam to the spectrometer. If the sample were mounted on a rotatable holder, it could be turned to face other measuring devices, an ion source for sputter cleaning, or an evaporation source for depositing a new metallic surface. The optics for this geometry are also available commercially.

3. General considerations

One of the prime limitations of any arrangement such as those discussed above is physical space. The geometries shown in Figs. 12 and 13 are designed to fit quite closely to the spectrometer, hence, the usage of other techniques along with RA spectroscopy is somewhat limited. They are somewhat simpler and the optics are available as an accessory. The design shown in Fig. 10 has considerably more room for additional measuring techniques. Experience indicates that the entire IR beam must be flushed with dry air to

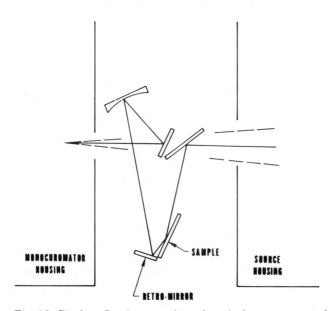

Fig. 13. Single reflection sample and optical arrangement shown schematically.

minimize absorption from water vapor and CO_2 while using a reference beam to compensate for the remaining absorption. Attempts to compensate exactly the reference beam have been less than satisfactory. Purging appears to be the most satisfactory solution to this problem.

The problem of putting an infrared transmitting window in an ultrahigh vacuum system seems to have as many solutions as there are workers in the area. O-Ring seals are satisfactory when ultrahigh vacuum conditions are not needed. When UHV is required, seals are sometimes made with silver chloride [23] or a silicone resin [24].

V. Applications of RA spectroscopy

In this section, we shall show several examples where reflection-absorption has been used to study structure or size. Most of the preceding discussion has dealt with chemisorbed species because this is the limiting case. The technique can also be used for studying film growth such as multilayer formation. In this case, there is more material available and hence the band sizes will be larger than for monolayers of the same material.

The primary example of monolayer coverage (or less) has been carbon monoxide adsorbed on metals, including copper [17,22], gold [18], tungsten [25], nickel [26], and rhodium [27]. The work on tungsten was carried out using a single reflection and the investigators, Yates et al., report detectability of 4% of the total carbon monoxide monolayer. Work on copper has been conducted, by Pritchard and co-workers [28] on single crystal substrates with a single reflection and computer averaging techniques. Spectra obtained from polycrystalline gold by Kottke et al. [18] are shown in Fig. 14. The carbon monoxide is believed to be adsorbed in a linear mode, i.e.

$$
\begin{array}{c}
O \\
||| \\
C \\
| \\
Au - Au - Au \\
| \quad\ | \quad\ | \\
Au - Au - Au
\end{array}
$$

and the vibration observed is the C—O stretching vibration. In Fig. 14, spectra as obtained from the spectrometer chart with no additional signal processing also show the signal-to-noise ratio. Additional noise suppressing techniques would, of course, provide even more sensitivity, if this were deemed necessary.

Multilayer formation (or thin films) has been studied for several chemical systems. The formation of Cu_2O has been studied by both Poling [29] and Greenler et al. [6]. Spectra obtained by Poling are shown in Fig. 15. The smallest band shown is for a film with a thickness of approximately 200 Å.

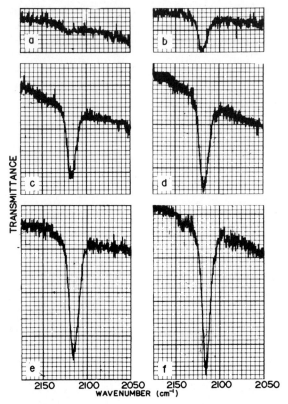

Fig. 14. Spectra showing the variation with pressure of the 2115 cm^{-1} band of carbon monoxide adsorbed on gold. (a) 0.01 torr, (b) 0.11 torr, (c) 0.5 torr, (d) 1.0 torr, (e) 3.0 torr, (f) 30.5 torr. (From M.L. Kottke et al. [18] with permission.)

It should be mentioned that no scale expansion was used to obtain this spectrum.

An example of the RA spectrum of a polymer is shown in Fig. 16 [30]. A thin film of poly(methyl methacrylate) (PMMA) was deposited simultaneously on NaCl discs, a KRS-5 internal reflection element [31] and two copper plates, by spraying a solution of PMMA in acetone with an artist's air brush. By weighing the amount of solution sprayed, the film was estimated to be 570 Å. Spectra were taken by transmission spectroscopy, internal reflection spectroscopy [31], and reflection-absorption spectroscopy on the NaCl disc, KRS-5 sample, and copper plates, respectively. The scale for the spectrum obtained from transmission is expanded by a factor of 10, whereas the scales for the internal reflection spectrum and the RA spectrum are not expanded.

Poling [32] has studied the oxidation of oil on copper oxide and obtains

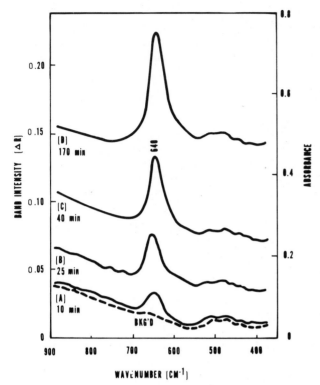

Fig. 15. Spectra from copper oxidized in air at 140°C. (From G.W. Poling [29] with permission.)

the spectrum shown in Fig. 17 which he interprets as a 220 Å film of copper oxalate (CuC_2O_4). This spectrum was obtained with seven reflections at 73°.

Chan and Allara [33] combined RA spectroscopy and internal reflection spectroscopy to show that the degradation of polyethylene in the presence of oxidized copper involves the formation of a carboxylate species at the polymer—metal interface. In conjunction with this, Tompkins and Allara [21] have studied the interaction of acetic acid with a copper oxide surface. An acetate species is obtained and the spectra are shown in Fig. 18. The thickness of the larger band is about 250 Å whereas the smaller band is interpreted to be monolayer coverage of the acetate species.

VI. Summary

In this chapter, we have discussed infrared reflection-absorption spectroscopy. The theory was reviewed for a single reflection and the methods of magnifying the absorption band discussed. These methods include, primarily, multiple reflections and scale expansion with noise reduction. A section on the applicability of the technique deals with the relative difficulty of obtain-

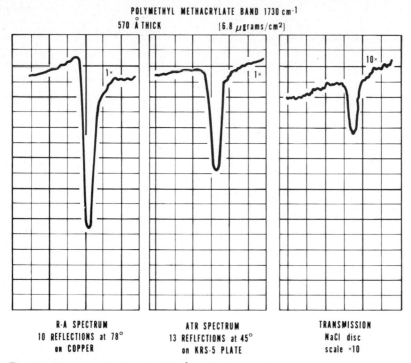

POLYMETHYL METHACRYLATE BAND 1730 cm⁻¹
570 Å THICK   (6.8 μgrams/cm²)

10×

1×

1×

R-A SPECTRUM	ATR SPECTRUM	TRANSMISSION
10 REFLECTIONS at 78°	13 REFLFCTIONS at 45°	NaCl disc
on COPPER	on KRS-5 PLATE	scale ×10

Fig. 16. Spectra of a film (~570 Å) of poly(methyl methacrylate) obtained with transmission, internal reflection spectroscopy and by RA spectroscopy.

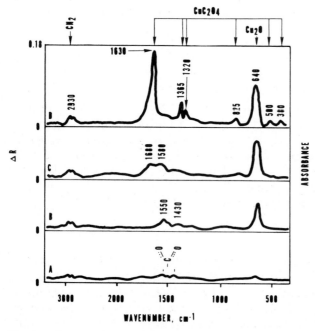

Fig. 17. Spectra from copper exposed to mineral oil at 90°C for various times up to 2543 h. (From G.W. Poling [32] with permission.)

471

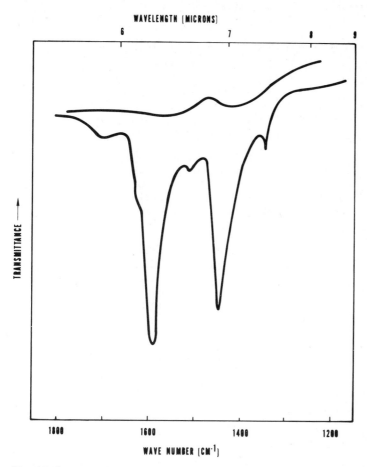

Fig. 18. Spectra of acetate species obtained by exposing copper oxide to acetic acid vapor.

ing absorption bands. It is clear that films down to fractional monolayer thickness can be detected for some species. Several experimental arrangements used in the past are illustrated and examples of actual spectra are shown.

This technique has developed considerably in recent years, particularly with the availability of adequate theoretical considerations and undoubtedly will continue to grow as its potential is realized.

Acknowledgments

The author would like to thank Dr. R.G. Greenler and Mrs. M.G. Chan for critically reading the manuscript.

References p. 472

472

References

1 R.P. Eischens, W.A. Pliskin and S.A. Francis, J. Chem. Phys., 22 (1954) 1786.
2 R.P. Eischens and W.A. Pliskin, Advan. Catal., X (1958) 1.
3 G. Blyholder, J. Chem. Phys., 36 (1962) 2036.
4 C.W. Garland, R.C. Lord and P.F. Troiano, J. Phys. Chem., 69 (1965) 1188.
5 J.F. Harrod, R.W. Roberts and E.F. Rissmann, J. Phys. Chem., 71 (1967) 343.
6 R.G. Greenler, R.R. Rahn and J.P. Schwartz, J. Catal., 23 (1971) 42.
7 J.E. Stewart, Infrared Spectroscopy. Experimental Methods and Techniques, Marcel Dekker, New York, 1970, Chap. 16.
8 H.L. Pickering and H.C. Eckstrom, J. Phys. Chem., 63 (1959) 512.
9 A.A. Babushkin, Russ. J. Phys. Chem., 38 (1964) 1004.
10 M.J.D. Low and J.C. McManus, Chem. Commun., (1967) 1166.
11 S.A. Francis and A.H. Ellison, J. Opt. Soc. Amer., 49 (1958) 131.
12 R.G. Greenler, J. Chem. Phys., 44 (1966) 310.
13 R.G. Greenler, J. Chem. Phys., 50 (1969) 1963.
14 T.C. Fry, J. Opt. Soc. Amer., 22 (1932) 307.
15 See for example, W.J. Potts, Jr. and A. Lee Smith, Appl. Opt., 6 (1967) 257.
16 J.E. Stewart, Infrared Spectroscopy. Experimental Methods and Techniques, Marcel Dekker, New York, 1970, pp. 440—441.
17 H.G. Tompkins and R.G. Greenler, Surface Sci., 28 (1971) 194.
18 M.L. Kottke, R.G. Greenler and H.G. Tompkins, Surface Sci., 32 (1972) 231.
19 R.G. Greenler, private communication.
20 R.N. Jones and C. Sandorfy, in W. West (Ed.), Chemical Applications of Spectroscopy, Interscience, New York, 1956, Chap. IV.
21 H.G. Tompkins and D.L. Allara, J. Colloid Interface Sci., 49 (1974) 410.
22 J. Pritchard and M.L. Sims, Trans. Faraday Soc., 66 (1970) 427.
23 R.W. Roberts, J.R. Harrod and H.A. Poran, Rev. Sci. Instrum., 38 (1967) 1105.
24 M. Kottke and R.G. Greenler, Rev. Sci. Instrum., 42 (1971) 1235.
25 J.T. Yates, Jr., R.G. Greenler, I. Ratajczykowa and D.A. King, Surface Sci., 36 (1973) 739; J.T. Yates, Jr. and D.A. King, Surface Sci., 36 (1973) 739.
26 E.F. McCoy and R.St.C. Smart, Surface Sci., 31 (1973) 109.
27 H.L. Pickering and H.C. Eckstrom, J. Phys. Chem., 63 (1959) 512.
28 M.A. Chesters, J. Pritchard and M.L. Sims, Chem. Commun., (1970) 1454.
29 G.W. Poling, J. Electrochem. Soc., 116 (1969) 958.
30 Spectra taken by the author.
31 Internal reflection spectroscopy is treated at length in N.J. Harrick, Internal Reflection Spectroscopy, Interscience, New York, 1967.
32 G.W. Poling, J. Electrochem. Soc., 117 (1970) 520.
33 M.G. Chan and D.L. Allara, J. Colloid Interface Sci., 47 (1974) 697.

INDEX